Les Liaisons Dangereuses

'Even today *Les Liaisons* remains the one French novel that gives us an impression of danger: it seems to require a label on its cover reserving it for external use only.'

Jean Giraudoux

Routledge Classics contains the very best of Routledge publishing over the past century or so, books that have, by popular consent, become established as classics in their field. Drawing on a fantastic heritage of innovative writing published by Routledge and its associated imprints, this series makes available in attractive, affordable form some of the most important works of modern times.

For a complete list of titles visit
www.routledge.com/classics

Pierre Ambroise François

Choderlos de Laclos

Les Liaisons Dangereuses

Translated by Richard Aldington

London and New York

Les Liaisons Dangereuses originally published in 1782

This translation first published under the title *Dangerous Acquaintances* in 1924
by George Routledge & Sons Ltd

First published in the Routledge Classics 2011
by Routledge
2 Park Square, Milton Park, Abingdon, Oxon, OX14 4RN

Simultaneously published in the USA and Canada
by Routledge
270 Madison Avenue, New York, NY 10016

Routledge is an imprint of the Taylor & Francis Group, an informa business

Typeset in Joanna by
RefineCatch Limited, Bungay, Suffolk

British Library Cataloguing in Publication Data
A catalogue record for this book is available from the British Library

Library of Congress Cataloging in Publication Data
A catalog record for this book has been requested

ISBN 13: 978–0–415–57753–3 (pbk)

Contents

INTRODUCTION

LIFE OF CHODERLOS DE LACLOS, 1741–1789

Until 1905, when two biographies of Laclos and his family correspondence were published, his life was only known in outline; the natural tendency to identify him with the Vicomte de Valmont caused misapprehensions of his real character. Unfortunately for us, the bulk of the information collected by Mm. Dard, Caussy and de Chauvigny relates to the last fifteen years of Laclos's life, so that we know more than enough about Laclos the Revolutionary conspirator and general of artillery and comparatively little about Laclos the author of the "Liaisons Dangereuses." His character is something of an enigma, made rather more mysterious by his voluble biographers.[1] His correspondence is almost valueless as a key to this enigma, because it was written at the end of his life (when he had become precociously senile), was addressed to his children or to his wife (who was not an intelligent woman), was composed either under the constraint of a thirteen months imprisonment or in the bustle of active service; finally this correspondence is infected with 18th century self-consciousness and the pose of "virtuous sensibility." For these and other reasons the man's letters, generally so useful in determining character, are not of much assistance. How trust the correspondence of a man of such deep reserve and dissimulation as Laclos? And even in prison might he not have played the epistolary part of a "good papa"? Who knows? The writer of the *Liaisons Dangereuses*, the calculating Orléanist conspirator, might have feigned anything.

Our knowledge of Laclos's family and early life is slight. He belonged to the *petite noblesse*, the minor "gentry" of France, who were exploited by the *ancien régime* because they were poor, proud and patriotic and who received their reward under the Revolution too often in the shape of imprisonment and the guillotine. The family of Laclos is supposed to have been Spanish in origin; the earliest notice of the name occurs in 1683. The father of the author, Jean Ambroise Choderlos de Laclos, was "sécretaire de l'Intendence de Picardie et Artois"; he married Marie Catherine Gallois. Their son, Pierre Ambroise François Choderlos de Laclos, was baptised at Amiens, 19th of October, 1741. The first definite knowledge we have of him is that in 1759 he began to work at mathematics with a view to entering the Artillery. He was received in January 1760; and made *sous-lieutenant* in March 1761; *lieutenant en deuxieme* in 1762. He was not made captain until 1771. He saw no active service until after the Revolution and, after spending all his life in the army, received his baptism of fire at sixty. He was in garrison at Toul, Strasbourg, Grenoble and other places. The period at Grenoble, 1769–1775, is important. Grenoble was then "a charming town, sparkling with intelligence, where the women did not allow themselves to be forgotten," says Stendhal, who many years later met Laclos at Milan. Laclos went out into the best society and admitted to Count Tilly in London that he had "quelques adventures assez piquantes"; though the reports of his senior officers praise his devotion to duty and military knowledge. At Grenoble he is supposed to have known the originals of the principal characters in the "Liaisons Dangereuses." Stendhal boasted that as a child he had known the original of Madame de Merteuil, a mysterious Mme. L.T.D.P.M., interpreted as Mme. de la Tour de Pin Monmort. Cécile Volanges is supposed to be derived from a Demoiselle de Blacons. But these identifications are shadowy and remote; all that we can say definitely is that Laclos found some of the material for his book in the society of Grenoble.

He is described at this time as tall, thin, narrow-shouldered, with fine, pale features and blue eyes; "under a cold exterior he hid an ardent mind." At twenty he wept over Rousseau; indeed the influence of Rousseau upon him was capital. It is obvious in his work. Possibly, his efforts to force his temperament into a Rousseau-made mould accounted for some of the incoherence and mystery of his personality. He also admired *Clarissa* and was struck with the character of Lovelace, who certainly has a distant affinity with Valmont. At Grenoble he wrote his first poems, some of which were published in the *Almanac des Muses*. In 1777 he wrote the libretto of an unsuccessful opera *Ernestine*, which was laughed off the stage. In 1779 he was detached from his regiment and went to assist the Marquis de Montalembert in the building of a fort on the island of Aix, near La Rochelle, a task upon which he was engaged for some years.

In a few sentences all our essential knowledge of Laclos up to his fortieth year has been related. The reader will see how very meagre are the *data* upon which we may form a conception of the man who wrote *Les Liaisons Dangereuses*, for it was at Aix in the months between July 1780 and September 1781 that Laclos wrote his famous novel. What sort of a man was Laclos when he wrote *Les Liaisons Dangereuses*? What were his motives and his object? If his character had any consistency, if he did not change completely after his marriage and with the Revolution, he cannot have been the mere voluptuary he is often held to be. He cannot have been the original of Valmont. An unscrupulous voluptuary of the Valmont kind would not have been an irreproachable artillery officer and the trusted subordinate of Montalembert. When we come to the period of Laclos's life about which we have ample information we do not find him a dissolute or "immoral" man at all. On the contrary, we find a clever political agent and a sentimental family man. We may perhaps grant that he followed the current of the times which coincided with his own development; that he was gay and "licentious" under Louis XV, ardently reforming under Louis XVI, sentimental and republican during the Revolution; but the key to his character and life, as well as to a correct understanding of *Les Liaisons Dangereuses* lies in the fact that he was a disciple of Rousseau and a man of disappointed ambition. The sentimental Rousseauite in him was the expansive family man and perhaps explains the ardent passages in *Les Liaisons Dangereuses* which procured him the reputation of "immorality"; the disappointed ambition gave him his cold exterior, made him a political intriguer, landed him in a revolutionary gaol, and supplied the motive force of his novel. Laclos intended that work as an attack on the upper classes of the *ancien régime*; it was one of the innumerable straws which showed which way the gathering storm of revolution was blowing.

Tilly, in the interesting account of his conversation with Laclos in London after the outbreak of the Revolution, confirms this reading of Laclos. He says Laclos told him he was tired of rhyming without success, "of studying a profession which could lead him neither to great advancement nor to great consideration," that he resolved "to write a book which should be out of the common way, which would make a stir and be heard of in the world after he had left it." Whether Laclos did or did not make this confession cannot be known certainly; but it is the most plausible explanation of the genesis of *Les Liaisons Dangereuses*. That book was the revenge of a disappointed man of genius, fretting against a system which condemned him to obscurity and monotonous routine in a subaltern position.

Let us try to imagine the life and prospects of a man like Laclos under the *ancien régime*. They were not so brilliant as might be supposed. We hear so much of the luxury and license of 18th century France and forget that, as in all times, these were the privileges of a small minority. Because the Maréchal Duc de Richelieu was an unconscionable libertine and was imitated by the gilded youth of the time, it does not follow that every officer was a *petit maitre*. Laclos never had the means of being a *petit maître*. He had not enough money to marry until he was over forty. Until his father died in 1784, he had only his pay. Even in 1784 his income was only 2,700 *livres* pay and 1,800 *livres* private income. A *petit maître* usually had an income of about 100,000 *livres* and died several millions in debt. There is an absurd story that Laclos, in order to seduce the girl he afterwards married, hired the next door house and had a sliding panel cut through into her boudoir;[2] it has been proved that the unfurnished rental alone of the house was two-thirds of his whole income! Laclos-Valmont is a myth.

This does not mean that Laclos was any more austere than young officers and young Frenchmen usually are. No doubt he had adventures with women and possibly prided himself upon possessing an uncommon "strategy of seduction." Even Montaigne had this weakness.[3] But no doubt these affairs were commonplace enough; to indulge in adultery on the scale vaunted by erotic novelists would demand a lavish expenditure far beyond the means of a humble captain of artillery. Those writers then, who, mistaking the purpose of *Les Liaisons Dangereuses*, have seen merely a libertine in its author, are guilty of an error. The canker in Laclos's character is disappointed ambition. Under the *ancien régime* there was little hope for a poor officer without "protectors." Exceptional merit on active service might indeed lead a brilliant soldier to the rank of general or even marshal of France; but Laclos never had an opportunity for active service. His promotions came with discouraging slowness, the mere result of seniority; moreover, the artillery was a despised branch of the service and in the decade before the Revolution the French War Office reserved the higher commands of the army entirely for noblemen of rank. Laclos saw that a complete reform of government was the only chance of his gratifying his ambition; as a disciple of Rousseau he was angered by the behaviour of the ruling classes; the *Liaisons Dangereuses* was his declaration of war on the *ancien régime*. The book had a prodigious success. Grimm begins his literary letter for April 1782 with a long review which opens with the words: "For several years no novel has appeared whose success has been so brilliant as that of the 'Liaisons Dangereuses.' " Grimm, who did not understand the book, also remarks that Rétif de la Bretonne had been called the Rousseau of the gutter; "M. Choderios de Laclos is the Rétif de la Bretonne of good company." Mme. Riccoboni who engaged in a semi-moral, semi-literary and wholly polite argument with Laclos about the novel, wrote him:

> All Paris is eager to read you, all Paris is talking about you. If there is any happiness in occupying the attention of the people in that immense metropolis, enjoy it. No one has ever been in a better position to enjoy it than you.

Tilly, then page to Marie Antoinette, wrote that it "made a prodigious sensation among the public." After the Revolution a copy of one of the earliest editions was found among the Queen's books; it was richly bound, but, as a sacrifice to modesty, bore no title on the outside. The first edition (April, 1782) of 2,000 copies was followed a month later by another of the same number; the book was frequently reprinted; indeed there were no less than fifty pirated editions.[4] Had the copyright laws of to-day then existed Laclos would have made a considerable income from these sales; as it was he received 1,600 *livres* for the first edition and authorised a second on the same terms. But the pirates must have robbed him of most of his possible gains.

This brilliant success was, however, largely a *succès de scandale*. All the women were angry with the creator of Mme. de Merteuil; that character made the pseudo-virtuous Mme. Riccoboni feel "outraged in her dignity as a woman and a French-woman." Mme. de Genlis (mistress of the dissipated Duc d' Orléans) "nearly died of despair" at the thought that someone had mistaken for hers "that infamous work of M. de Laclos." The Marquise de Coigny (the mistress of Lauzun) closed her doors to the author of *Les Liaisons Dangereuses*. Strangely enough the objections mostly came from such ladies; no doubt they had reason to fear the sharp eyes of this officer-novelist, who observed character as coolly and accurately as he worked out his mathematics and tactics. On the other hand the old Bishop of Pavia (apparently a saintly man and obviously an intelligent one) told Laclos in 1801 that he considered the book a moral one, "very fit to be put into the hands of young ladies." The book was never officially condemned by Louis XVI, any of the Revolutionary governments or Napoleon. It was forbidden in 1825 by the reactionary government of Charles X; at the same time were also forbidden the works of Montesquieu, Voltaire, Rousseau, Beaumarchais, La Rochefoucauld, Vauvenargues, and the *Télémaque* of Fénelon; so the *Liaisons Dangereuses* was not in bad company. The real cause of the scandal in 1782 was the freedom with which "keys" to the novel were circulated in conversation, involving many contradictions, of course, but providing admirable opportunities for malicious gossip.

Laclos did not wholly escape persecution for his book. The sale was not interfered with and no direct censure could be made on him; in 1782 officers had the right to publish works not dealing with military affairs, without submitting them for the approval of the Minister of War. But there are many indirect ways of persecuting an officer and Laclos soon began to feel them; moreover, the Ministry were sharp enough to see the political motives behind the book. Laclos had now no chance of rapid promotion (he was even made to wait a year longer than other officers of his seniority for the Croix de Saint Louis) and when later he published a pamphlet on a military topic, without obtaining formal permission, retribution was swift.

Laclos had received leave of absence from Montalembert to superintend the publication of his novel, and, with his superior's connivance, had outstayed his leave six weeks, presumably to enjoy the pleasures of his sudden fame. The Minister of War could not take notice of this—Montalembert alone was responsible—but he sent a peremptory order for Laclos to rejoin his regiment. The object of this was plain; while Laclos was on detachment he would have few or no regimental duties and hence plenty of leisure for writing, but once back with the regiment he would soon find that duties rained on him. But Montalembert was his friend and Laclos was an old soldier. He rejoined his regiment indeed but immediately "went sick" and remained sick for some months, and even had the skill to exact a gratuity of 800 *livres* for expenses. Meanwhile Montalembert used his influence and when Laclos ceased to be "sick" he returned to La Rochelle. He had won the first round with the Ministry.

Probably, his chief concern was to follow up his success effectually. In 1785 the Academy of Châlons-sur-Marne offered a prize for an essay on "the best means of perfecting the education of women." Laclos began to write an essay and afterwards turned it into a discourse *De L' Education des Femmes*, a sort of pendant to *Émile* and the famous *Discourse on the Inequality of Men*. It was not published until 1903 and adds nothing to Laclos's literary reputation. One sentence in it deserves quotation, because it shows where his thoughts were tending; after he had argued that no improvement was possible in the education of women because society made them "slaves" he adds: "Know that people can only escape from slavery through a *great revolution.*" Perhaps it was lucky for Laclos that his treatise—which by the way is an example of thoroughly false reasoning from the "noble savage" fallacy—was not published in 1785.

Meanwhile Laclos had been putting into practice some of these educational theories. At La Rochelle there lived a family named Duperré, where there was a pretty daughter named Marie-Soulange. Laclos and she fell in love and on the 1st of May, 1784, the lady gave birth to a male child, baptised Etienne. Naturally, in a small Huguenot town like La Rochelle the scandal was considerable. It does not appear to have harmed Laclos, however; he was still received in the best society, made rhymed compliments for Madame de Montalembert and met "le Président Alquier" who was subsequently very useful to him in Revolutionary days. He was also received into the literary society of La Rochelle, which had had the honour to rank Voltaire among its members.

This incorrect proceeding on the part of Laclos was happily repaired by a marriage in May, 1786. The lady was so delighted that she commemorated the event by a lapidary inscription upon a public building. The marriage was indeed a happy one—how could it fail to be after such auspicious preliminaries? One explanation for Laclos's action lies in his poverty. His wife had a dot of 5,000 *livres* a year. In 1784 he had only his pay. A seduction was perhaps his only means of outwitting wealthier rivals and deciding the family. At any rate when he married, the death of his father had made him possessor of a small private income, which with his pay amounted to 4,500 *livres*.

A day or two after the marriage, Laclos obtained a month's leave and went to Paris with his wife. But the honeymoon was destined to interruption. There had been a war of military pamphlets over Vauban, in which certain topics of fortification were debated. Laclos published his views in a pamphlet disparaging Vauban's defensive works; this was highly displeasing to the Minister of War who was a fanatical disciple of Vauban. Laclos also committed the imprudence of a breach of discipline by not submitting his pamphlet for official approval before publication. The Minister at once wrote a severely worded order to Laclos to rejoin his regiment at Metz, although he was on leave. Laclos got wind of this and lay *perdu* for nearly a fortnight with Mme. de Laclos in the Hotel des Milords where the emissary of the War Office finally unearthed him and delivered the order. Laclos had no choice but to obey. Even the charms of Mme. de Laclos, directed upon the Minister of War, failed to melt that functionary's resentment and he refused to countermand the order, only too enchanted to have an opportunity of legally quelling this turbulent and far too clever subordinate.

Laclos's effort to call public attention to himself in his military capacity by this Vauban pamphlet was then a miserable failure. From a military point of view he was no doubt right—at least Napoleon (himself an artillery officer) thought well enough of Laclos to make him a general; but woe to the officer who attempts to pierce through the pedantry of army councils with no weapons but sensible ideas, close reasoning and light sarcasm! The officers who answered Laclos had two irrefutable arguments; first of all, Vauban was a great Frenchman and had discovered everything there was to know about the science of fortification; it was therefore impious to attack his principles and futile to attempt any other means of fortification. The other (of course) was that Laclos had written *Les Liaisons Dangereuses*. Fortunately, the merits of the dispute do not concern us. All we need note is that Laclos gained nothing by his pamphlet and lost a great deal; he received a severe reprimand from the War Office, was sent back from leave to his regiment at Metz, and some pretext was made to delay his Croix de Saint Louis for a year.

This misadventure was decisive in Laclos's life. He realised that the Government and the Ministry were prejudiced against him, and with reason. He saw that he could hope for nothing from the Army but slow promotion by seniority; and this was insufficient for his ambition. He determined to leave the service and to try his fortune elsewhere. Since the publication of his book he had made friends with several distinguished people who had not only admired the author of the "Liaisons Dangereuses" but felt the charm of his personality. Among these were the Vicomte de Noailles, the Duc de Lévis and the Vicomte de Ségur; the last named, son of his persecutor, the Minister for War. Through their influence he tried to obtain a military post with the Turkish army in the summer of 1787, but this, as usual, was blocked by the Ministry who were determined to allow Laclos no opportunity of distinguishing himself on active service. Laclos consoled himself by writing a project for renumbering the streets and houses of Paris, on the soulless and mechanical system of modern American towns.

It was now 1788. The meanest intelligences could perceive that some great change or at least reform was about to take place in the Government of the nation. Voltaire had laughed and reasoned the French into contempt for their political institutions and revolt against Catholicism; Rousseau had filled their heads with "noble savage" chimaeras, sophistries of equality and vague expansive sentimentality; the Encyclopedists and the example of the English had given them an itch for commerce and the useful arts so little honoured by the monarchy. The monarchy itself was in a condition of desolating decrepitude. The wave of sentimental loyalty to Louis XVI at his accession soon spent itself and rolled back into contempt and hatred when it was found that an apathetic *roué* had given place to an apathetic imbecile, the nadir of whose intelligence may be judged by the following entry in his private diary: "14 *Juillet*, 1789, *Rien*." The only ability in the royal family was that of Marie Antoinette; hence the disgusting calumnies of the "patriots". The best of the nobility were in favour of a change and deplored the abuses of the out-worn machine of government; the interest and insolence of the others were but too plain. Every Frenchman had become a social reformer, a friend of man and an inheritor of the mantle of Rousseau. Pamphlets and projects fell like hail, especially after the taking of the Bastille; the most insane "reforms" were propounded in the language of sages, the most impossible ideals were evoked in the name of reason, sensible ameliorations were demanded in the tones of raving sentimentality. It was clear to all that the millennium was at hand.

These troubled waters were exactly the element for a cool fish like Laclos. In 1788 he had taken his observations and had made up his mind. He easily persuaded the Vicomte de Ségur to introduce him to the Duc d' Orléans. The duke was soon fascinated and dominated by Laclos's superior intelligence. It was easy for the King's cousin to obtain a kind of permanent leave of absence for Laclos, who, without resigning his commission, entered the duke's household about the end of 1788 with the post of supernumerary "secrétaire des commandements" a salary of 6,000 *livres* and a lodging on the second floor of 27, Cour des Fontaines.

Life of Choderlos de Laclos, 1789–1803

Jules Janin, writing with all the accuracy of a literary journalist, says; "The life of Choderlos de Laclos ends with the *Liaisons Dangereuses*." On the contrary, it only then began. Laclos's later life as Orléanist conspirator, Jacobin, prisoner in Revolutionary gaols, general under the Consulate, husband and father, is full of interest and quaint ironies. But is not the intention of the present writer to follow Laclos's former biographers who in their zeal have re-written the secret political history of the Revolution, with many discussions upon military affairs, including a description of Kellerman's dispositions at Valmy and a map of the battle.

The part played by Laclos in the Revolution was tortuous and obscure, but not unimportant; the evidence is often indecisive, sometimes untrustworthy—remarks in memoirs, anonymous pamphlets, accusations of enemies. True to his character he himself reveals little or nothing. We catch glimpses of him through the smoke and din of Revolution, we perceive that his name was connected with important events, we observe that he obtained the reward of great abilities—many enemies. His part in the Revolution was not dramatic, like that of his friends Mirabeau and Danton. Like other politicians in other circumstances, he backed the wrong horse in the Duc d' Orléans; when he, rightly, staked all he had left on Napoleon, he was too old and incapable to derive any real benefit for himself or his children.

The reader may be surprised to find an ardent disciple of Rousseau and a "friend of man," at the centre of Orléanist intrigues. Was Laclos duped by Egalité's pose? Did he believe this fat, debauched cousin of Louis XVI to be a true friend of liberty and the people? The answer is decisively, No. Laclos joined the Orléanists from motives of self-interest, after a careful estimation of political possibilities.

In 1789 hardly anyone in France wanted a Republic, no one believed the monarchy could be overthrown. Sincere believers in reform wanted at the outside a constitutional monarchy founded upon that of 18th century England. Laclos was playing a deeper and more far-sighted game, but like so many of the *ci-devants* he entirely underestimated the forces of the Revolution. Apart from that enormous factor, which completely wrecked his calculations, his reading of the situation was subtle and accurate, so accurate that the result he played for was that actually attained in the Revolution of 1830; a Bourbon dynasty gave way to an Orléanist King.

The Dukes of Orléans were always a nuisance to the French government. As the King's cousin, the Duc d' Orléans was naturally an enemy of the Court and at the head of any Fronde or important conspiracy or opposition to government which came along. It was therefore the tradition, rigidly observed, so to order the education and training of the Duc d' Orléans that he became contemptible, debauched in habits, irresolute in character, incapable of public business. Still it was sometimes possible to galvanise an Orléans into activity. Retz succeeded fairly well with Gaston. Laclos determined to play Retz to Egalité and to avoid Retz's mistakes. Egalité hated Marie Antoinette, was ready to do anything to embarrass her, was "not without ambition" but had a strong feeling of "I dare not." Until Laclos came along, Mme. de Genlis had been his virtuous Lady Macbeth.

The plot was a simple one. Orléanist agents were to create as much disturbance and political agitation as possible; Orléans, by lavish charities and bribes (nearly all the early Revolutionaries except Lafayette were in his pay) and by posing as a friend of the people, an advocate of extreme reforms, was to gain popularity. Then, at the decisive moment, some pretext was to be taken to secure the abdication or forcible exile of Louis XVI, and Orléans was to be proclaimed a constitutional King, or, failing that, Regent to the Dauphin, who would be proclaimed Louis XVII. Laclos would of course have obtained a high, perhaps a dominant, post in the new government. In any case his fortune would have been made. To further this end Laclos, (it is said), drew up the democratic "Instructions données aux baillages par le Duc d' Orléans"; wrote pamphlets and got himself elected *Commissaire* on the "assemblée electorale des citoyens nobles de Paris." He is also accused of fomenting riots.

After the fall of the Bastille, Orléanist intrigues increased. Versailles was alarmed by the progress of the conspiracy. Unfortunately for him, Egalité himself became alarmed by the turn of events and allowed Versailles to parry his schemes by sending him on a trumped-up special mission to London, in October, 1789. Laclos went with him and continued his laborious direction of the conspiracy. Meanwhile the Duc d' Orléans distinguished himself in London by offending King George, disgusting the *émigrés* and shocking society by his open affair with Mme. de Boufflers. The Duke, with Laclos, returned to Paris in July, 1790.

Once back in Paris, Laclos plunged directly into political action. He was a prominent member of the early Jacobin club and edited their paper, the "Journal des Amis de la Constitution." Although the Duc d' Orléans had by this time exhausted nearly all his vast wealth, Laclos remained faithful to him, led the debates of the Jacobins in the direction favourable to the Duke, and tenaciously held to his scheme. Danton throughout was the friend of Laclos and was deeply involved in the Orléanist plot. In general, the ideas defended by Laclos were democratic; he attacked the *émigrés* violently, advocated the plebiscite, liberty of the press, eligibility of all citizens to the position of deputy, election of Ministers by universal suffrage, and the sharing of inherited property equally among a man's children. On the recovery of Louis XVI from a short illness, he protested against money being used to pay for special *Te Deums* "as in the days of despotism" and urged that the money should be given as dowries to "destitute daughters of the conquerors of the Bastille." On the other hand, he naturally always advocated the continuance of the monarchy, with strong hints on the advisability of a change of dynasty. Less comprehensibly, he opposed the freeing of negro slaves in the French colonies. But his position among the early Jacobins must have been important, if we may judge by two facts. The first is that he was picked out for violent attacks by other clubs opposed to the Jacobins; the second that he persuaded the Jacobins to enroll Egalité among their members.

The Orléanist plot finally came to a head in June and July, 1791, and failed, chiefly through the incapacity and timidity of the Duc d' Orléans himself. On the morning of the 21st of June,

1791, Paris was in a turmoil at the news of the flight of the King; his arrest at Varennes and ignominious return to Paris completed the confusion. By all except the ardent Royalists this was considered to be an act of war upon the nation and many held that it was equivalent to abdication. Republicanism made great advances. But Republicanism was not what Laclos wanted then. He argued that the King had virtually abdicated, that the throne was vacant and proposed that the Dauphin should be proclaimed as Louis XVII, with the Duc d' Orléans as Regent. This proposal was diligently circulated by the Orléanists, but either from fear or policy, the Duke publically announced that he would not accept the Regency if it were offered him. But Laclos still persisted.

On the 25th of June, the Assembly passed a decree suspending the King from his royal functions until the new constitution was presented. In July, they passed another decree proclaiming the "inviolability" of the King, which was really equivalent to pardoning the flight to Varennes and asserting his position as monarch in the new constitution. Laclos now played his last card. To understand his action, it must be premised that any citizen might present a petition which other citizens would sign if they were in agreement, but the drawing up of petitions by groups, like political clubs, was not allowed. Laclos was addressing a very stormy meeting of the Jacobins on the subject of a petition against this decree, when a mob (probably organised by Laclos's agents), burst into the club and demanded that a petition should be drawn up there and then and carried the next day to the Champ de Mars for signature by the public. Laclos feebly opposed the project as illegal but appeared to yield to force; in fact the document was drawn up by another hand, though Laclos subsequently added an important sentence to it. This led to serious results, including the riot, known as the Massacre of the Champ de Mars. The National Assembly parried Laclos's petition by prolonging the decree of the 25th of June. The Jacobins split up; all but six of the members who were also deputies of the

National Assembly withdrew and went over to the Feuillants. From this moment the Jacobins became the violent Republican force we know them as, and were implacably hostile to Laclos. He had to leave the club; his last visit was on the 10th of August, 1791. He lost all favour with the Duc d' Orléans, though he continued to draw a reduced salary until October, 1792.

It is easy to see Laclos's error in the Revolution. Though he was theoretically and perhaps sincerely a Revolutionary, he remained at heart a man of the *ancien régime*. Like so many theoretically wise men he lacked a certain suppleness; he had observed so closely, calculated so logically, planned with such foresight, that he could not believe any unsuspected event, any unforeseen force, could overthrow his calculations. As Lord Chesterfield, in an earlier generation, persisted in trying to govern George II through his mistresses, according to the long-established rules of Court life, and failed; so Laclos attempted to thrust into power an impotent prince he despised, not realising that the French nation was thoroughly weary of such intrigues, and would have no more incompetent despots.

In other respects Laclos was supple enough. Immediately after the affair of the Champ de Mars, he abandoned the Orléanists and proclaimed himself a Republican. He procured his election as *Commissaire* on the *municipalité provisoire* of the *Section de la Butte des Moulins*; the next day he was appointed vice-president of the permanent Assembly of the section. The Jacobins remained suspicious and resentful; they refused to admit him into the Commune, and thus narrowly the author of *Les Liaisons Dangereuses* avoided complicity in the crimes of Robespierre and Marat, and the inevitable guillotine. Nevertheless the year 1792 was the summit of his career. His friend Danton included Laclos in the list of thirty commissioners of the executive power, (29th of August, 1792). On the same day, another friend, Servan, then Minister of War, made Laclos Commissioner to the army of Luckner, with the rank and pay of *maréchale de camp*.

Thereafter events followed rapidly. Laclos left Paris a day or two after the September massacres; prepared a plan for the defence of Paris, in the very probable event of a defeat of the Republican armies; worked day and night to send troops and munitions to Kellerman and Dumouriez and was to that extent responsible for the wholly unexpected victory of Valmy. On the 22nd of September, 1792, the "Republic, one and indivisible" was proclaimed. In October, Laclos was appointed an infantry general in the non-existent Army of the Pyrenees. In December he returned to Paris from the south. He had been appointed Governor-General of the French *établissements* in India, and he prepared a plan for driving out the English with 15,000 men and 15 ships of the line. The governments changed and events occurred too rapidly once more for the satisfaction of Laclos's ambitions; he was never Governor-General of India outside the walls of Paris. His enemies were now in power and profoundly mistrustful of him as a dangerous, cool conspirator and the heart of the old Orléanist conspiracy. Laclos vegetated for some months in Paris. On the 31st of March, 1793, Laclos was arrested, together with his old opponent, Mme. de Genlis, and the children of the Duc d' Orléans. On the 4th of April Egalité was himself arrested.

Laclos was released under surveillance on the 10th of May; in September he resigned his rank of general and his Governor-Generalship. But nothing could placate his enemies and on the 4th of November he was re-arrested (when engaged in some experiments with "hollow cannon balls" on behalf of the Ministry of War) taken to La Force and subsequently transferred to Picpus. Egalité was guillotined on the 7th of November.

Laclos was in great danger, especially just after the execution of the Duc d' Orléans, until after the fall of Robespierre. Why he was not guillotined remains a mystery. There is nothing intrinsically impossible in the story that while in prison he wrote some of Robespierre's speeches for him, except the one fact that

Robespierre was far too vain of his oratory to think of asking outside assistance.

Laclos was kept in prison for thirteen months. Many of the letters he wrote to his wife are in existence but they are not of much interest. They breathe the monotony of prison life, are severely limited in scope by prison censorship; they show a certain stoicism and a warm family affection. The following two letters to his wife will serve as an example of all the others and at the same time will give a sketch of Laclos's life in prison.

7 prairial An II de la Rep. une et indiv.

This letter will indeed be short, my dear, for I have absolutely nothing to tell you. It is pouring with rain and I am well—that is all the news I can give you. I was wise enough to exercise this morning in spite of the horrible black cold; but I made a little fire for my dinner, I am thoroughly warm at the moment of writing to you (4 o'clock in the afternoon) and I shall spend my evening in killing time with a few companions in misfortune. I shall not recognise any duration of time until I can employ it for you and near you. That indeed will be a happy period since it will be a time of both justice and happiness. God grant, first that it comes and then that it comes soon. Then, however much it rains, all our days will be fine; until then I have no day at noon and I shiver in the sun. It is simply to relieve you of anxiety that I write this completely empty letter; but you will see from it, as from the fullest, that I am well, that I think of you, and that I love you with all my heart; is not that the main thing?

15 prairial An II de la Rep. une et indiv.

You are concerned that I have nothing to read! I can no longer read for pleasure. I read the papers as much as I can, because I have not ceased to be interested in our victories. For the rest, I only read to learn; I have made reading a labour, not a recreation. I have gone through Bezoult's arithmetic again, to be

> sure that I can instruct my daughter in it; I have learned the new method of calculating, according to the new weights and measures, and I am now busy with book-keeping by double-entry, for eventual necessities. I am also looking for notes on rural economy. I think that is enough; perhaps there are superfluous things in it; but that time alone can show.

Gradually Laclos's situation grew better. His brother was released from the Luxembourg on the 30th of September, 1794; Laclos came out on the 3rd of December, 1794. Soon after his release he obtained 10,000 *livres* from the government, being his pay as general and compensation for his imprisonment. But the Directoire kept him in disgrace. For four years and upwards he was engaged in no public affairs, but only held the undistinguished post of trustee of mortgages.

Thus, all Laclos's ambitions were disappointed and his dreams of wealth and power shrank into the dull reality of an unimportant office appointment. For five years he played a daring often a shrewd, always a self-interested game; time after time he seemed about to reach his objective and time after time he was disappointed or received nothing but an empty title of power without either its functions or its control of wealth. The irresolution of the Duc d' Orléans wrecked his earlier efforts; the enmity of the Jacobins and the Directoire kept him from important employment under the Republic. Except for his brief and laborious part in Valmy, his functions as general under the Republic were purely decorative. The truth is, he was regarded almost universally with suspicion, and not a small part of that suspicion was aroused by his reputation as author of *Les Liaisons Dangereuses*. The book's moral and veiled political purpose were overlooked or ignored and Laclos himself was regarded as a kind of political Valmont. When he tried to find out the charges upon which he was imprisoned, he could only discover that he was accused of being a dangerous person, a man of genius, author of *Les Liaisons*

Dangereuses. No doubt genius is a sort of crime in an egalitarian state, but it was the reputation of his novel which caused Laclos to be regarded as a *dangerous* genius. His prison experience was disastrous to him; the interminable months of apprehension and privation were too much for a man of his age. He left the revolutionary prison with all his fire and energy gone, prematurely senile, wrapped up entirely in his family and in a state of mind when he seriously thought of writing a sequel to *Les Liaisons Dangereuses* to prove that "there is no happiness except in a family."

The 18th Brumaire and the rise of Napoleon to power rekindled in Laclos some sparks of energy, and all his old ambitions. It is thought that he served the rising fortunes of General Buonaparte by anonymous pamphlets. At all events Buonaparte became his hero, "a prodigy in the art of ruling men," an "immortal general" alone "worth 30,000 men in the field"; Laclos was "full of confidence in the fortune and genius of Buonaparte." In January, 1800, Buonaparte made him a general of artillery and next month sent him to the army of the Rhine. Still Laclos's evil genius attended him; he did nothing except pursue the advancing army in a post-chaise and was intrigued into the army of Italy. In May he was at Grenoble; in September at Turin and Milan where he met Stendhal and made a considerable impression on that writer's mind. Later in the year he was in action in Italy—after serving all his life in the army, he received his baptism of fire at sixty!—but the great seige train he had prepared and the chance of using his "hollow cannon-balls" in real action were rendered abortive by the unexpected victory of Marengo. Once again fate thwarted him.

During 1802 he was in Paris engaged in some artillery staff work. In May, 1803, he received orders to join the army of Italy and to take command of the artillery at Taranto, a greatly dreaded place owing to its unhealthiness. After a long and wearisome journey in great heat the old man reached his post, only to fall ill of dysentery. He died on the 5th of September, 1803. His last letters were to his wife, General Marmont and Buonaparte. That to Marmont is pathetic and expresses keenly the misery, the anxiety and bitterness of his end:

General,

I am so near to absolute despair and my situation is indeed so terrible that I think I have a sort of right to importune you, not as First Inspector, but as General Marmont, kind and sensitive, a friend to me and mine.

You know already that the command you procured me, and in which you may remember my one desire was active service, you know, I say, that it has dwindled—without any hope of glory or other advantage—to keeping garrison 550 leagues from my family and my friends, in the most unhealthy climate of Italy only to be considered afterwards as not being active service in the present war.

Among the numerous victims to the unhealthiness of the district, I am one, with an illness such that the doctors only hope to bring me to a state where I can go and breathe the air of France, without which they assert I cannot be entirely cured, and in which I foresee I shall need long and costly attention.

The doctors have already reported to General Saint-Cyr and he is ready to authorise my departure when I have the strength, but this is my position:

After having spent in this journey 6000 *livres* of my own money, in posting charges and in the buying and selling of horses, household establishment, etc, I find myself here with thirty *louis* of my running salary, with the expenses of a sickness which will come to three or four thousand francs at the least; moreover I have to travel more than 250 posts, in which the doctors say I cannot go more than ten or twelve leagues a day, which makes an expense of sixty or seventy days at inns, so that I must die at Taranto if I do not receive the assistance of at least 12,000 *livres*.

> Who but the First Consul can do me this service? Who but you can ask him for it on my behalf? I have left my wife in Paris, encumbered with three children and with no pecuniary resources; if she sold all her furniture she could not raise enough to get me away from here . . . when I left Paris I did not think the issue of my journey would be to come to Taranto to ask for charity.
>
> I am very miserable; tears overcome me. Farewell. General, I rely on your friendship for me.

The Character of Laclos

Errors in appraising Laclos's character have two sources; there is a persistent effort to describe him in terms of Valmont and a disregard of his peculiarly 18th century characteristics. To make the dominant trait in his character a sort of Machiavellian *volupté* is merely substituting a literary fancy for such real evidence as we have. Doubtless Laclos pursued *volupté* in his youth—otherwise he could hardly have written *Les Liaisons Dangereuses*—and there is a trace of Machiavellianism in his make-up. But that is far from explaining what appears enigmatic and is complex in the man.

That ambition was his dominant passion through the latter part of his life can hardly be doubted by an impartial observer of his actions. Frustrated ambition was the force behind his writings, his conspiracies, his political theorisings: as, on a larger scale, the frustrated ambition of the minor gentry and *bourgeoisie* created part of the initial force of the Revolution. Love of money was not stronger in Laclos than in other men, but love of power, of *considération*, was very powerful. But merely to say he was ambitious does not explain him.

The singular union of voluptuousness, false sentiment and arid utilitarianism was not uncommon among the makers of the Revolution. The French Revolution was not a puritan movement, like the English revolutions of the 17th century; on the contrary it was avid of pleasure. "La Revolution a été fait par des voluptueux"—the Revolution was made by pleasure-lovers—says Baudelaire somewhere; and it is true. That is one side of Laclos. But the pursuit of pleasure for its own sake takes the sweetness and the freshness out of life, brings *ennui* and distaste and a regret for lost innocence. These 18th century *voluptueux*—so given to reasoning—reasoned themselves into an admiration for sentiment, for wholesome family affection, for benevolence and justice; an admiration, stimulated by Rousseau. This admiration was a matter of abstract conviction rather than spontaneous feeling; consequently the sentiment of these people nearly always appears false, affected and theatrical. It is like Marie Antoinette playing shepherdess because she is weary of the Court; it may be charming, but it is not genuine. When Laclos writes to his wife urging her to "*arroser*"—to sprinkle—their daughter with "her expansive sensibility" he strikes the false note I am trying to define. It is like the flood of tears M. Diderot felt it necessary to shed at first seeing the aged Fontenelle, which (quite rightly) made the old gentleman very annoyed and sarcastic indeed. That gives us another side of Laclos.

What I have called "utilitarianism" was partly a result of Laclos's training, partly a result of the cynicism of the age, partly the creation of Encyclopedist propaganda. Under the *ancien régime*, money and the arts of making money were not the sole and absolute test of merit as they are with us. Military capacity and success, learning and holiness, the dignity of the law, were all at least theoretically recognised as superior to mere wealth in the aristocracies of *noblesse d' épée, noblesse d' église* and *noblesse de robe*. Even literature, the arts and science were recognised as superior to mere wealth, a fact which is due, be it said, to the enlightened despotism of Cardinal Richelieu and Louis XIV. But the result was that France was not a rich country like England and Holland and the useful arts were despised, often foolishly enough. The

Encyclopedists (and Voltaire) argued that a rich country is a happy country and that sound political economy is the mother of the Muses. Across this again came the idealism of Rousseau, vaunting the charms of the small-holder's life. Thus, we have the grotesque spectacle of an elderly artillery officer attempting to learn the science of agriculture by reading in prison a book on rural economy written by an *abbé*, of all people! Even their most ardent advocates will admit that there is no soaring on the viewless wings of poesy with five acres and a cow.

To pursue the matter further; the social life of the 18th century, in which Laclos was readily accepted, with all its charm and ease, was cynical. That statement requires no proof. But cynicism is very apt to consider "commodity" as the one thing needful. Laclos did not escape the contagion. In his case this utilitarianism, this aridity of spirit, were fostered by the military life, by the study of mathematics, and a passion for rigidly logical argument upon false premises accepted as axioms. It is Stendhal's "*Lo-gique*," the Frenchman's delight to "*avoir raison*"; a wordy and arid virtue. To this, in Laclos's case (as in Stendhals') must be added a profound dissimulation, a desire to cover up tracks, to leave no trace of activities, a love of subtle back-stairs manoeuvering. Laclos did not indulge in the childish concealments of Stendhal (pseudonymous signatures, letters in bad English, etc.) but he very effectually succeeded in hiding a large portion of his life. He had the egotist's joy in wearing a mask.

Laclos's life was certainly a failure and the reasons for it are plain enough. He did none of the great things he wanted to do, made no mark on history, achieved no worldly success. But for his novel he would be completely forgotten. His methodical, calculating, tortuous, yet sentimental mind lacked the "something" that gave men like Mirabeau, Danton and Napoleon their influence over others.[5] He was more like Robespierre, fish-like and pedantic, without the reputation of being "incorruptible"—a reputation Laclos certainly did not deserve. He had none of those sudden flashes of intuition, those lightening appeals to men's hearts, that profound knowledge of crowd psychology, which the great Revolutionaries had. That is why he would never have been a great general. No doubt he could solve military problems, as he worked out political problems, theoretically and at leisure. But he had not the gift of action. Perhaps he had worked out too many formulae of velocity and resistance, solved too many tactical problems on paper. At all events, he had a certain mental rigidity which made him a failure as a man of action. Only once did he combine his gifts successfully, his sentiment, his laborious observations, his tactical knowledge, his mathematical precision of construction, his sense of intrigue; and that of course was in *Les Liaisons Dangereuses*.

Laclos's religious beliefs were vague and unobstrusive. Like most educated Frenchmen of his generation, he found the pleasantries of Voltaire unanswerable objections to Christianity. But since he could not possibly be acquainted with the far more effective criticism of modern times, his attitude was perhaps the most intelligent which could be adopted. It would be absurd to expect a man with so little imagination and sense of mystery, a man so rigid and pedantic between the parallel lines of his logic, to perceive even, far less enjoy, the poetry of religion. He had no real sense of beauty, no ardent passion for *τό κὰλον*, not a spark of Platonism; elegance, *convénance*, the formal or delicately *polissons* beauties of 18th century art sufficed him.

He appears to have taken his paternal duties solemnly. He was a doting husband in his old age, which is perhaps an argument for his having been a *roué* in his youth—when he had no longer the energy and presence to capture other men's wives, his own became doubly precious. There is some reason to believe that he was a *mari complaisant*, that the creator of Valmont was himself Valmontised. Mme. de Laclos's third child was born (5th June, 1795) only six months after Laclos left prison (3rd December, 1794); a letter from Laclos in prison to his wife, dated 4 sans-culottides An II (i.e. 20th September, 1794) contains this curious and suggestive passage:

> No, assurely, my dear, your silence of five days (not three days) created in me no doubt as to your feelings, but keen anxiety as to your health. It would be easier for me to think you dead than to think you had done wrong; but as both would forever annihilate all hope of happiness for me, the expressions might have been the same.

But this is only a conjecture; and perhaps it is unkindly in a biographer to scrutinize such matters too closely.

The final judgment on Laclos's character is, I think, that he followed the spirit of his age almost slavishly; he did not lead it and he does not personally incarnate it in a complete and striking manner. He was a great novelist but not a great man.

Laclos's Minor Works

Laclos's minor writings deserve but the briefest notice. It is sufficient to state that he wrote a pamphlet on Vauban, an unfinished tract on the education of women, and diverse political pamphlets. His speeches, his journalism and his letters have only a biographical interest. The "Education des Femmes" has been published by M. Edouard Champion; M. Caussy printed the Pamphlet, *La Guerre et La Paix* as an appendix to his biography.

The poems of Laclos perhaps call for more notice; a collection of them was published in 1908 by Mr. Arthur Symons and M. Louis Thomas. These poems range from the four-line epigram to the *conte* in the manner of the innumerable 18th century imitations of La Fontaine; they contain songs, epistles and a rondeau. They are no better and no worse than the occasional pieces of scores of French writers of the age. Voltaire and Gentil-Bernard appear to have been his models. One Conte, *Le Bon Choix* has a biographical interest, since it relates an anecdote of an amorous undertaking in which a "Bel Esprit", who talks charmingly and goes for his books to lend the lady is worsted by a "*sot*" who merely makes love to her. The lady informs the "Bel Esprit" (who surprises her with the "*sot*") that she is not afraid of his writing a satire:

> Car entre nous, ce que vous pourrez dire
> Ne vaudra pas ce que Cléon a fait.[6]

And Laclos reflects in a characteristic vein:

> Soit à la Ville, à la Cour, à l' Armée,
> Les gens d'esprit n'ont jamais les bons lots:
> Les sots ont tout, même la renommée.[7]

The other *contes* are stories of light love told in nimble, humourous verse, very much like that of Prior in English and Voltaire in French, though of course without the easy mastery of those great poets. There is a mediocre poem on Death which avoids none of the commonplaces and strikes out no new thought. There is a would-be crushing epigram on Madame de Genlis which has not much point and one of those carefully composed "impromptu" quatrains habitual with occasional poets—the explanatory title contains exactly as many words as the poem itself. The songs have been praised but are mediocre. The Rondeau is perhaps as good as any and will suffice to show the quality of Laclos's verse:

> *Elle est à moi*; cette aimable Rosine,
> Son coeur sensible a couronné mes feux;
> Elle est à moi . . . puis-je etre plus heureux?
> O volupté! J'éprouve ton délire,
> Et mon bonheur a surpassé mes voeux.

Sein palpitant, voluptueux sourire,
Lèvres de rose et regards langoureux,
Rosine a tout: cette aimable Rosine
Elle est à moi.

En vers naïfs dans mes chants amoureux,
Ainsi, d'Amour je célébrois l'empire.
Las! Peignez-vous, mes transports douloureux,
Lysis arrive: il accorde sa lyre,
Puis chante aussi: 'cette aimable Rosine,
Elle est à moi'.[8]

It is neat, charming, conventional verse, but other French poets of the 18th century—Dorat, Lebrun, Desmahis, Bernis, Gentil-Bernard, have done the same kind of thing much better. Those who have directed attention to these minor works of Laclos have not increased his literary reputation, may even have harmed it. He is a man of one book.

Les Liaisons Dangereuses

The 18th century in France produced at least four great novels, all love stories. They are Marivaux's *Marianne*, Rousseau's *La Nouvelle Héloïse*, Prevost's *Manon Lescaut*, Choderlos de Laclos's *Les Liaisons Dangereuses*. A timid and prudish criticism has long witheld from Laclos the praise he deserves and, in spite of enthusiasts like Mr. Arthur Symons, M. Van Bever, M. Suarès (not to mention Stendhal), the novel is not yet correctly esteemed by the general reader. Behind *Les Liaisons Dangereuses* there is a keen if limited intelligence and wonderful observation. The novel is constructed with almost faultless precision, the precision of a mathematician; the intrigue and situation are developed with the forethought and care of a man skilled in tactics, a man accustomed to neglect no possibility. The characters are observed so nicely and rendered so vividly that they become real persons to us—indeed they are "drawn from the life." The story itself is so ingeniously unfolded, the interest so well kept up, that once a reader is caught by it, he is compelled to read on almost feverishly to see what will happen next. Finally, the style, artificially varied for each character, is subtle, expressive and exactly adapted to its purpose; its very negligencies, which are intentional, reveal some trait of character, some psychological detail. Its faults are slight. A little too complaisant a dwelling upon erotic incidents; an occasional lengthiness and over-development of detail, the one inseparable from a novel in letter-form, the other inevitable in psychological novels. It is the most "modern" of 18th century novels, less remote from us than *La Nouvelle Héloïse* or even than *Manon*. There is a possibility that the reader who picks up this book will have some feeling of prejudice against it; I beg him not to be misled by the timidities and prejudices of "official" criticism—so often made without direct contact with the book criticised[9]—but to read it and to judge it for himself with an open mind. I am confident Laclos's book can endure triumphantly any reasonable test.

Les Liaisons Dangereuses is an "erotic" novel; but that is not saying much, seeing that nine-tenths of the world's novels circle round the same perennial topic. It is also a "psychological" novel, if by that rather loose term we mean a novel in which the main interest is directed towards the analysis of character, rather than the narrative of events. This type of psychological love-novel, so common in this century, was very largely the creation of 18th century France; perhaps it goes back as far as Mme. de La Fayette's *Princesse de Cléves*. The Greek novel, the mediaeval romance, the Renaissance pastoral all treated of love indeed, but in a conventional, superficial, more or less romantic way. The hero and heroine were in love, got into all sorts of difficulties and adventures in consequence, were eventually united and left to be happy ever after. There was very little "psychology", or at least only of a very primitive sort; nothing resembling the complicated, not to say tedious, subtlety of the love-poetry of Dante, Guido Cavalcanti and their numerous satellites. The 18th century novelists made a complete change of venue; instead of taking the "love interest" as a necessary but conventional theme or embroidery, they concentrated upon the actual relations of living men and women with ever increasing precision of observation and minuteness of analysis. This was a most important departure. It not only attracted a more mature and intelligent set of readers; it made novel-writers a more considerable part of the world of literature; it attracted writers who might otherwise have been moral essayists or dramatists. And it had the further consequence of a greater openness, a more frank discovery of the ordinary (or extraordinary) habits of lovers; the female busts of the older type of novelist became full-length statue.

There was and is no rational objection to this, so long as the writer is really an artist. No one has ever yet succeeded in drawing an exact line between what is permissible and what is not permissible in these matters; certainly if we excised from literature everything which appears improper to the rabid puritan we should castrate more than half the masterpieces of literature. But in dealing with 18th century French novels we have to use cur judgment and to discriminate between those works which have literary importance and those which were written merely to tickle the appetite of "amateurs of rare and curious *facetiae.*" I think that I should certainly rank among the last named the *Imirce* of the Abbé Dulaurens, the *Fecilité ou Mes Fredaines* of Andrea de Nerciat and the *Dialogues ou Tableaux des Moeurs du Temps* attributed (but most doubtfully in my opinion) to Crébillon fils.[10] The hasty judgment which consigned so many 18th century novels to this undesirable category has been revised time and again. Voltaire and Marivaux scarcely needed justi fication; but the works of Casanova, Louvet de Couvray, Rétif de la Bretonne, Crébillon, Diderot, Prévost, and Florian even, have been one by one rescued from the "Enfer" of improper books, examined impartially, and found to contain talents and sometimes even genuis. *Les Liaisons Dangereuses* emerged as triumphantly as any. No doubt, the novels discussed here are not to be handed to the Podsnappian Young Person along with her morning tract, but they will be found interesting and serious enough by intelligent men and women.

The sea of 18th century novels is large enough, but it is now moderately well charted. The writers hitherto mentioned are about the most important (one might add Duclos and Marmontel) and are those principally relevant to a discussion of the "Liaisons Dangereuses." Three writers may be picked out as the principal influences on Laclos: Rousseau, Marivaux and Crébillon fils. I do not say there are not others, but either they are not obvious influences or they are the same type of writer as one of these three.

Marivaux's *Marianne* is not only a delightful book by a delightful author, but had a decisive influence on many later 18th century novels; Richardson was deeply indebted to it, and one cannot but regret that *Marianne* is not now accessible to English readers and that so little justice is done to it. The book is tremendously long and by its adventure side is related to the older type of novel; by its form of fictitious memoirs it decidedly influenced successors—Duclos, Louvet, even Casanova who perpetually hovers between fiction and memoirs. But its real departure was in bringing the novel away from princely and noble characters and giving the main part to a simple "Marianne," an *enfante trouvée*,[11] and above all in its treatment of psychology. Marivaux was so far ahead of his contemporaries in delicacy of observation and character-study that they simply did not understand him. Almost the only other 18th century novelist who is so shrewd and profound is Laclos himself. Marivaux is one of the few authors Laclos quotes[12] and though what he says is not wholly approving one sees the influence of Marivaux in almost every letter of *Les Liaisons Dangereuses*. Here is a page taken almost at hazard from "Marianne" which will show how deeply Laclos had meditated that novel:

> Valville[13] doubtless knew where I was living; yet I heard nothing said about him and my heart could not understand it. Even if Valville had found a way to send me news of him, he would have gained nothing by it; I had renounced him but I did not want him to renounce me; what fantasticality of sentiment!
>
> One day when I was thinking of this and in spite of my doing so (it was in the afternoon) I was told that a lackey wished to speak to me; I thought he came from my benefactress and I went down to the parlour. I scarcely looked at this pretended servant, who only showed himself sideways and handed me a letter with a trembling hand. 'From whom?' said I. 'Look, Mademoiselle,' he replied in an agitated voice, which my heart recognised before I did, since I became agitated myself.
>
> I looked at him when I took the letter and found his eyes upon me; what eyes, Madame! Mine were fixed on him; we remained a short time without speaking; only our hearts had spoken when the convent door-keeper came in and told me that my benefactress was coming up, that her carriage had just entered the courtyard. Notice that she did not mention her name; 'It is your good Mamma,' she said, and then left.
>
> 'Ah! Monsieur, leave me,' I cried in distress to Valville (for you see it is he), who only replied by a sigh as he went out.
>
> I hid my letter as I waited for my benefactress, who appeared a moment afterwards . . .

Might not that be a quotation from *Les Liaisons Dangereuses*, from one of Cécile's or Madam de Tourvel's letters? Its tone is much more the tone of the *Liaisons* than that of *La Nouvelle Héloïse*. I am tempted to think that it was Marivaux who taught Laclos to avoid all those extraneous digressions (about English travellers, Alpine scenery, Italian music and what not) which so charmed the first readers of *La Nouvelle Héloïse* and which we find an impediment to our pleasure in the book. Nevertheless Rousseau's mind had probably more action upon Laclos than that of any other writer; but we see this in the general Rousseauism of his outlook and philosophy, not so much in any definite quality of his novel. The device of a novel in letter-form was very common in 1780; Laclos may have adopted it from fondness for *La Nouvelle Héloïse*, but there were dozons of other models at hand. *Les Liaisons Dangereuses* is an attack upon civilised woman and the social customs which control her life. The characters all become unfortunate and unhappy. Why? Because all more or less have erred. And how have they erred? By a greater or less departure from the Rousseauesque "laws of nature."

Laclos as a disciple of Rousseau would have us consider his characters somewhat in this way. Valmont makes seduction his aim in life because the sentiment of nature has been perverted by society to a cold and unnatural vanity which prefers to triumph rather than to enjoy. The malice of Madame de Merteuil is equally the result of a perversion of natural sentiment through a wrong education and the *mariage de convenance*. These two types of "civilised" wickedness prey upon the others—Cécile, Danceny, Madame de Tourvel—who, though naturally inclined to "virtue", are rendered vulnerable to their attacks through defects in the social system. Danceny has been condemned to an unnatural celibacy; Cécile comes ignorant from a convent and is to be compelled to marry a man she has never seen; Madame de Tourvel is left alone by her husband and is childless; hence for all her "virtue" she falls a victim to Valmont. And this after all has common sense in it; and it was not feigned; these characters and these incidents existed in real life. Laclos, then, was justified in the contention which emerges from his book—that the social customs of the upper classes in the 18th century were such that they condemned many women almost inevitably to one or another form of unhappiness and often to "vice."

It is from Crébillon fils that we can learn something of these conditions. He is a much maligned man who so far from being a monster of Don Juanism, loved his ugly little English wife dearly and was faithful to her all her life. His novels are a long pleading against cynicism in love and on behalf of sentiment. They are unrestrained in certain details, it is true, but far less so than is generally supposed and they are really so moral they are a little on the dull side. So far from its being true that "there is nothing moral about 'The Sofa, A Moral Tale' except its title," the whole book is "moral" in both senses of the word; moral because it is filled with observations and reflections on human life; and moral because the author's intention is to urge his readers to choose the better part. But before touching any further on Crébillon it will be advantageous to give a brief sketch of the 18th century idea and practice of love and marriage.

The outward, established and lawful form in all classes was, of course, the indissoluable Catholic marriage. With the aristocracy it had certain peculiarities; a girl was educated in a convent, away from men and any knowledge of the world; at fifteen or sixteen she was suddenly informed by her mother that "a marriage had been arranged" and a month later, with her complete ignorance of life, of herself, of marriage, of men, she would find herself wedded for life to a man she had not seen three times. The following account of an 18th century wedding gives the situation perfectly:

> M. de Rinville had proposed to M. de Bellegarde a husband for his daughter Mimi in the person of one of his half-cousins who was said to be a very worthy young man. As M. de Bellegarde was an excellent father and above all things desired that the young man should "please his daughter"—that was the usual phrase—a day was fixed; and, Mimi having been well instructed

> because she had a habit of never paying attention to anyone, they went to dine with Mme. de Rinville where they found all the Rinvilles and all the d'Houdetots in the world. First of all the Marquise d'Houdetot kissed the whole Bellegarde family. They sat down to table, Mimi beside young d'Houdetot; M. de Rinville and the Marquise d' Houdetot took possession of M. de Bellegarde; at desert they talked openly about marriage. After the coffice was served and the servants had gone: "Come!" said old M. de Rinville, "Here we are in the family circle, don't let us be so mysterious. It's a matter of Yes or No. Does my son suit you? Yes or No; and your daughter the same, Yes or No; that's the main point. Our young *Comte* is already in love; your daughter has only to see whether he does not displease her, let her say so . . . Make your decision, my goddaughter." Thereupon Mimi blushed. And Mme. d'Esclavelles trying to halt matters, asking for time to breathe, M. de Rinville went on; 'Yes, we had better deal with the settlement first, and meanwhile the young people can talk to each other.' 'Well said, well said.' And at that, they went to a corner of the drawing-room. And here was M. de Rinville announcing that the Marquis d'Houdetot gave his son an income of 18,000 *livres* in Normandy, and the company of cavalry he bought him the year before; here was the Marquise d'Houdetot giving her diamonds, which are fine ones, and all of them. M. de Bellegarde retaliated by promising 300,000 *livres* as a dowry and a share in the inheritance. And they rose saying: 'Agreed,' let us sign the contract this evening. We will publish the banns on Sunday; we will get the rest dispensed with and the marriage shall be on Monday'."[14]

Marriages so obviously made in Heaven could not be very successful on earth. The young people so brusquely bundled into matrimony were at the mercy of the first *amourette* which came along. The *mariage de convenance* has produced some singular freaks of morals. In Italy it brought with it the custom known as "*cicisbeism*," which was carried so far that the name of the lover was actually inserted in the marriage-contract.[15] In mediaeval Provence the elaborate amatory code of "*l'amour courtois*" was created, an essential article of which was the assertion that the lovers must not be married to each other, though the lady might be and generally was, married to someone else; one Court of Love distinctly laid it down that a husband could not be a lover. It is a perfectly understandable result of these arbitrary matings. A delicate-minded woman could not possibly satisfy all her natural coquetry and sentiment with a mere "lord and master," who looked on her as a part of his property, a means of continuing his family. The ingenuity of women invented this kind of "Platonic" love to gratify their finer sensibilities, while remaining faithful to their husbands in the flesh; the lover might sigh and serenade, protest his devotion, wear his lady's colours, defend her in word and deed, sing her praises, converse with her and be smiled upon, but he might not aspire to more substantial favours.

It was plainly a false position and a perilous one for matrimonial honour. In Italy, the problem of the erring wife was first settled by poison or dagger, later by the more amiable toleration of the "*cicisbeo*" after a year of marriage. In France, where "*galanterie*" has always existed, matters were rather different and even more illogical. Husbands might do as they pleased, but wives must be virtuous; a detected wife might be clapped into a Convent for the rest of her life on the mere complaint of her husband to the King. Plainly, then, the virtue of women consisted in that invaluable quality of not being found out. By the 18th century the delicacy and fine sentiment of "*l'amour courtois*" had disappeared and gave way to a refined but sensual promiscuity, in which the Kings of France from Henri IV to Louis XV had led the van.[16] But that imperious and sometimes absurd vanity of the French stepped in and made matters even worse. They were not content with a real and lasting passion (which, however illegal or reprehensible, had a real excuse and cannot be refused human sympathy); they abandoned all true sentiment, though they continued to use its language; and their "affairs" degenerated from sentiment to sensuality, from sensuality to the gratification of a vanity which had in it elements of base selfishness, cruelty and perfidy. Don Juan became the Duc de Richelieu, became Valmont, became the insufferable *petit maitre*. The object was to win as many women as possible, simply to boast of it; to "gain a reputation" as "irresistible"; to attack "our most fashionable women" of the moment; and, when once the attack was successful, the triumph was to break away as quickly as possible, (especially if the woman were really in love) in as public and humiliating a manner as the amiable lover could contrive.

That is what Stendhal was denouncing when he declared again and again that only the Italians understood love and that the French ruined everything by their vanity, as the English did by their "phlegm." And, indeed, could there be a more paltry and inexcusable perversion of the most delicious of human sentiments, a worse crime against that "charme de la vie"—the charm of life—so praised by 18th century poets? Perhaps no more startling example of self-centred vanity ever existed than the 18th century *petit maître*. But as all things in life, even malicious vanity, have some compensation, the *petits maîtres* were at least a considerable cause of the remarkable intelligence noticed in 18th century Frenchwomen. It is true they were often "cruel, courteous, smooth, inhuman"—Mme. de Merteuils, in fact—but the necessity for a constant defence against enterprising Don Juans, possessed of a whole strategy of seduction, sharpened their wits. All the odds were against them in this unequal battle. Sentiment and senses, frustrated by forced matches, warred against them; if they yielded, and were found out, an inconsistant and malicious "world" congratulated the man on a "triumph" and treated the unhappy woman as "dishonoured"; if they yielded and were not found out, they had to suffer the arrogance, infidelities and exactions of lovers who held their reputation, freedom and happiness in their hands; if they did not yield they were looked upon as hypocrites, or "*fausses prudes*," or were credited with lovers they had not, or were openly claimed by men they had refused. The 18th century woman needed all her charm and intelligence, all her perfidy and dissimulation, all the arsenal of feminine weapons to hold her own; and in a society infested with Valmonts, a Mme de Merteuil is a necessary consequence, almost a deserved revenge.

Of course all the upper classes were not Valmonts and Merteuils.[17] There were plenty of "*honnêtes gens*" among the aristocracy; and the *bourgeoisie* were then almost inexorably strict in morals. But the tone of society was both cynical and libertine; wonderfully refined indeed and glazed over with all the charm of the arts, but essentially false and worldly. Nothing was so ridiculous as for a husband and wife to be in love; it was "*du dernier bourgeois*." The husband went his way, the wife hers; they lived in separate apartments, met only at meals and always on terms of ceremonious politeness.[18] It is hardly a matter of surprise that the women followed the men in libertinage, that they in turn went from lover to lover, amusing themselves by "taking away" a husband from a young wife, a lover from the woman who really loved him. Among this 18th century fast set there was fierce competition for the "man of the hour," whether he were Field-Marshal, or *petit maître*; the gratification was almost wholly one of vanity. The women even had their secret "*petites maisons*"—small houses kept secretly for amorous rendezvous. The man in power was surrounded by women and if one of them could attract him, dominate him and keep him (as sometimes happened) she avenged herself and her sex by a tyranny of caprice; during the 18th century France was governed, and not unskilfully, by women who were the real force behind their powerful lovers.[19]

This was the society and these the situations studied by Crébillon *fils* in his novels, which were carefully read by Laclos. It is true that something must be deducted on account of the imagination of the novelist, it is true that he darkened the shades, but when we find his scenes so abundantly confirmed by memoirs and other evidence, they can be accepted as accurate enough. There is no reason to suppose that Crébillon was especially depraved in choosing these subjects; he was simply pursuing the study of morals rather further and with more freedom than his predecessors. The following page from "The Sofa" will give an idea of his *verve* in satire and at the same time complete this long digression on 18th century morals. The scene is between Nassés and Zulica. Zulica had a rendezvous with a *petit maître* who, with all the insolence of his race, sends a friend, Nassés (whom she had never seen before), to take his place. Zulica is first indignant, then tearful and pathetical, always insincere; Nassés flatters her, condoles with her, makes love to her and is successful. The following conversation takes place later in the same interview:

> 'I am afraid,' said Zulica, 'that you are not very delicate.' 'You wrong me,' he replied tranquilly, 'I am naturally very susceptible to love. But I admit I have possessed more women than I have loved!' 'But that is infamous!' she retorted, 'I cannot conceive how you can boast of that!' 'I do not boast of it,' he answered, 'I simply say that it is so.' 'I think,' said she, 'that you have deceived many women.' 'I have abandoned some but not deceived them,' he answered. 'They did not ask me to be constant, consequently I did not promise to be, and you will see that when there are no conditions, neither side can complain that any have been broken!'
>
> 'I should be very interested to know everything you have done,' said Zulica. 'Must I give you a detailed history?' replied Nassés, 'It would be long and I fear it would weary you. But I can obey you without risk by suppressing details. I have been

ten years in society; I am twenty-five; and you are the thirty-third beauty I have conquered in a regular affair!'

'Thirty-three!' she cried. 'It is true I have had no more than that,' he replied, 'but do not be surprised; I have never been fashionable.'

'Ah! Nassés!' she said, 'how I am to be pitied for loving you, and how hard it will be for me to depend on your constancy!' 'I fail to see why,' he said; 'do you think because I have had thirty-three women I love you any the less?' 'Yes,' she answered, 'the less you had loved the more I could believe that you would have resources left to love again and that your sentiment would not be entirely destroyed.' 'I think,' he replied, 'I have proved to you that my heart is not exhausted; besides, to speak frankly, there are very few affairs in which sentiment enters. Opportunity, convenience, idleness, create almost all of them. People tell each other (without feeling it) that they are amiable; they become intimate, without believing in each other; they see it is vain to expect love and they separate for fear of being bored. Sometimes they are deceived in what they feel; they thought it was passion, it was only inclination; a movement, consequently not durable, which disappears in pleasures, while love seems to be re-born from them . . .'[20]

* * *

The last few pages are merely a sketch to indicate the literary and social background of *Les Liaisons Dangereuses*; they do not pretend to any extensive examination of the novel or of the life of the time. If I have dwelt rather upon the defects and shortcomings of 18th century society, the reason is not that I am blind to its charm, but lies in my theme; if we are to understand Laclos's book we must perpetually keep in mind that it is an attack upon abuses inherent in the 18th century marriage system, code of morals and education of women. It was necessary to indicate these abuses. I know that they are not the whole of 18th century life, but they are the conditions under which his characters live, through which they dominate and torture others or by which they suffer. Again, it would be easy to find other literary influences in Laclos—Voltaire, the comedy writers, Montesquieu—but these are sufficient to show his principal literary affinities and the *genre* of novel which he adopted and developed. If he has less delicacy of feeling and less subtlety of analysis than Marivaux, he has more unity of action and a nearer approach to tragic power. He has not Rousseau's eloquence and passion but he is less verbose and his characters are truer to real life. And though the form Laclos selected shows he had little reliance on his powers as a writer of dialogue, (the hardest part of novel-writing to a novice); and though Crébillon *fils* is a skilful, if verbose, writer of dialogue, and a sharp observer; nevertheless Laclos as a novelist is superior to Crébillon in almost every respect.

One test of a novelist's power is the creation of characters so vivid that they seem to us not inventions but real persons. It is an inherent gift which may indeed be developed by practice but cannot be attained by any amount of labour if it is not already in the man; either he has it or he has not. The characters of the "Liaisons Dangereuses" possess this quality of vividness, this life of their own, which make them as real to us as the personages we meet in the brilliant memoirs of the age. They have a fidelity to what one might call traditional human characteristics which allows us to recognize similar characters under the changed exterior of modern life; but their fidelity to 18th century characters is the source of their striking vitality. Laclos lays bare layer after layer of their souls until we know them as an anatomist knows the structure of a body. There is practically no one who now dares to make the suggestion that the characters of the "Liaisons Dangereuses" are not true to the life of the time. A learned French writer, while compelled by the evidence to admit this, makes the quibble: "They existed, but not in the same circles." How does he know? There were feuds and cliques enough in this society, but any "gentleman" could, if he so desired, be introduced into almost any circle, even the most aristocratic, even the most literary which last was the most jealously guarded.

Valmont is the most striking character in the book, perhaps because he is the most familiar He is one of the innumerable avatars of the protean Don Juan. As someone recently suggested (I think it was Mr. Arnold Bennett) there is still a great deal to be done with Don Juan; he is capable of infinite development, infinite explanation. Sometimes, when reflecting on the character, one is tempted to think of him as one of those unkillable survivals from a primitive and remote civilisation; a survival of a once victorious type from an age of rapine and perpetual war when it was expedient that the nimblest male should capture as many females as possible, a relic of primitive instincts which have long become obsolete, like those of gamblers, or of the gentlemen who shoot tame pheasants in the fields. But Don Juan is never a fool, on the contrary; and for brute force he substitutes an infinite cunning, an abysmal perfidy. The type is strikingly vigorous in France; in England it occurs, but not so openly and with less feline grace. France has but rarely been afflicted with such gentry as the Restoration "cullies" who, at sight of a woman, uttered a sort of amorous "View Halloo! Hark forrard!", like the fox-hunters they were. An excellent sketch of the 17th century Don Juan occurs in the Memoirs of the Comte de Grammont Hamilton represents Saint-Evremond (who was a sage and an Epicurean) rebuking Grammont in these words,

> Is it not true that as soon as a woman pleases you, your first step is to find out if someone else loves her; and the second to exasperate her, for the last thing you trouble about is to make her love you? As a rule, you only enter the lists to disturb someone else's peace. A mistress who had no lovers would have no charms for you, and be without value for herself, though she possessed them.[21]

But the Comte de Grammont was a gentleman and a wit; there was no cruelty in his vanity, no perversity in his infidelities. He was fickle but not cold-blooded. It was reserved for the 18th century to rob "gallantry" of all its charm by importing into it a lawless arrogance, a kind of misdirected ambition, seeking to triumph over and crush women, losing all respect, all poetry, almost all pleasure even, and retaining nothing but a cynical gaiety and a morose sense of superiority. Vauvenargues has intimated a character of the Valmont sort in his "Phalante":

> He knows neither love nor fear, neither faith nor pity. He scorns honour as much as pity; he hates the Gods and the laws. Crime in itself is a pleasure to him . . . Dissimulating and implacable in his hatred, fierce and barbarous in his vengeance, eloquent only to persuade to crime and to pervert innocence, his ferocious and indomitable nature loves to trample under foot humanity, prudence and religion; he lives sullied with infamy; he walks with his head up; he threatens the wise and virtuous with his glances; his insolent temerity triumphs over the laws.[22]

Is not that the mental physiognomy of Valmont? The same defects can be recognised in Valmont's cynicism, in his admission that he is always most roused to action by opposition, in his diabolical perversion of Cécile's ignorance, in his extraordinary temerity, in his delight at the agonies of Mme. de Tourvel, in his perfidy, and his contempt for religion, notably in feigning contrition and piety in order to get in touch with his mistress again. Contemporary opinion, indeed, identified Valmont with several French noblemen. The Goncourts say:

> The only difficulty was that too many models were found. Did not Valmont bring to people's lips the name of a famous man? Did not M. de Choiseul begin his great career by his *rôle*

> of a man of amorous successes, of a pitiless *méchant*, of a consummate *roué*, achieving his aim with an air of carelessness, never making a step or uttering a word without a project against some woman, dominating women by sarcasm, threatening them with his wit, triumphing over them through their fear? But why speak of Choiseul? Had not Laclos the prototype of his creation under his eyes in the terrifying figure of the Marquis de Louvois, in the figure of that Comte de Frize amusing himself by torturing Mme. de Blot?[23]

Some writers on Laclos, strong in their own innocence, have protested that Valmont is an exaggeration, that such a man did not, could not, have existed. Even in the 18th century Grimm asserted that the "motives" of Valmont and the Merteuil woman were inadequate; but Grimm was a German baron and not a Frenchman and he did not know everything about the life of his time. Here is a desolating little tragedy from real life, quoted by the Goncourts, a miserable little tragedy, recalling the "projects" of Valmont on Mme. de Tourvel, where the protagonist and villain is no less a personage than the Duc de Richelieu.

> So lived, so died, the wife of a mirror-maker in the rue Saint Antoine, Mme. Michelin, the eighteen year old fair haired girl seduced by Richelieu. At first it was only the habit of seeing every morning at mass at Saint Paul's a good-looking unknown man. Then, as soon as she had blushed at some ordinary compliment, Richelieu was at her shop, buying mirrors from the husband. And almost immediately, deceived by a letter purporting to come from a Duchess which brought her to Richelieu's *petite maison*, she was face to face with the man she loved, but loved innocently, and, as she said, "without wanting to do wrong." From that day, how many tears were dried only by the vanity of saying "Monsieur le Duc" to the lover who played so cruelly and with such effrontery with her scruples, her tortures,

> her last innocences! The poor little *bourgeoise* began to pine away. Richelieu himself noticed she was changing. She tried to forget herself; but, even in pleasure, this wail escaped her, "Ah! it is all over. I am miserable!" And kissing her lover's hand, she left him forever, left him to die . . . Some time later, Richelieu's carriage over-took a man in deep mourning; it was Michelin, who two days before had buried his wife. Richelieu took him into the carriage to watch him weeping.[24]

Is it necessary to quote more evidence, like the extraordinary account of the lover who bribes servants, conveys letters, even breaks into a woman's bedroom at night, given by Clélie in *Le Hasard du Coin du Feu*?[25] Such perversity, such wantonness of cruelty to the most charming of living things; such a monstrous injustice in repaying the tenderest and most disinterested of affections with contempt, humiliation, death; such tampering with innocence, and such abuse of power (the *grand seigneur* remaining tranquilly aloof in the privileges and immunities of his rank); such a crime makes one shudder. Who can fail to feel indignant at such a story, who can wonder that the streets of Paris were red with blood of "aristos" and the walls echoed with shouts of "A la Lanterne" when such abuses of power occurred? Who will not feel that the very grave-stones of such victims cried for vengeance? And who, after such evidence from real life, can assert that Valmont is a fiction, that he is false to human nature that such cruel arrogance never existed? But many who admit the verity of the "Liaisons Dangereuses" ask us to believe that Laclos was himself a Valmont, that his book is a kind of manual of seduction based upon the practice of these pirates of love. It is an absurd argument, though it is the current interpretation of the book. Lax as sexual customs were, the most eminent and powerful of the race of Valmont would scarcely have dared to issue a written confession of his own infamy in his lifetime; how could a poor artillery captain have ventured to do so? His book was an attack on, not a defence of, the Valmonts. Whatever obtuse people like Grimm and Mme Riccoboni might say, the government of the day was sharp enough to see the satirical and reforming purpose hidden under the book's cold effrontery; had Laclos been really a Valmont he would have been disgraced from the army. Besides, how is it that we find him immediately afterwards beginning a serious treatise on the education of women, denouncing abuses; how is it that a few years later he is an amiable husband and expansively sentimental father; if he were a Valmont? Can one conceive of a Valmont or a Richelieu loving his wife and children? Admit that reformed rakes make good husbands—but has not the cynicism of a Valmont or a Richelieu gone beyond reformation?

And what of Mme. de Merteuil, Valmont's female counterpart, whom necessity has quickened to a yet more subtle cunning, in whom there is an even more diabolical ingenuity in "ruining" her enemies? Is she an invention, offensive to the virtue and French blood of Mme. Riccoboni? By no means. She is the female *riposte* to Valmont's attack on her sex. It is less easy to trace through the century the Merteuils than the Valmonts; complete secrecy was essential to them as the noise of scandal was to the men. Their dominant passion was not so much vanity—though of that they had plenty—as malice (*méchanceté*) that frightfully efficient and unscrupulous female malice which "makes a goblin of the sun." Sensuality they had, indeed, but not the indulgent good nature which so often goes with it; theirs was a hard sensuality, hard and bitter as their roughed mouths, whose gratification could hardly exist except at the expense of a man's bitterness and humiliation or a woman's tears.[26] How cynically Mme. de Merteuil dismisses Belleroche, entraps Prévan, takes Danceny from Cécile! Where she would be almost incredible, were it not for her youth and subtlety, is in her ability to keep her actions secret. But look at this portrait of a Mme. de Merteuil *à l'age de retour*, when self indulgence has marked itself on her face and long immunity has rendered her careless of constraint.

> Madame de Sénanges had been pretty, but her features were worn; her languid and weary-looking eyes had neither fire nor brilliance. The paint which completed the decay of the poor remnants of her beauty, her outrageous clothes, her immodest bearing, only rendered her more insupportable. She was a woman who retained none of her former graces except that immodesty which youth and charm may render pardonable, although it dishonours both; but a woman who, at a more advanced age, offered to the eyes nothing but a picture of corruption, which cannot be looked at without horror.
>
> She had some wit, I mean that sort so commonly found in society; what she said was nothing, but she spared nothing, always was malicious; and, never thinking good, was not afraid to say what she thought. She had the Court turns of phrase, bizarre, negligent, new or renewed; she helped them on by a nonchalant, drawling tone, an affected laziness which is sometimes thought natural and, in my opinion, is merely a method of boring more slowly; in spite of her rare talents for frivolousness, she sometimes abandoned it and discoursed in a self-opinionated way; without judgement and without knowledge, she did not fail to judge; battening, moreover, on sentiment and probity and always amazed at the excessive "*déréglements de son siécle*,"[27] over which she frequently lamented.[28]

The young man who thus describes Mme. de Sénanges is a kind of Danceny, still innocent, and in love with a girl, who is a better and more intelligent Cécile. The harpy decides to "take the young man away" from his Cécile (and from another older woman who is also after him) and the wretched youth is paralysed by her attack, yields weakly (as Danceny does to Merteuil), loses the girl he really loves and is reduced to misery. Those who think Crébillon's novels are not evidence should consult the secret *mémoires* and other documents referred to by the Goncourts. In their rather hysterical style, they sum up the Merteuil type of woman in a couple of pages,[29] from which the following phrases may be taken:

> Woman equalled man, if she did not surpass him, in this libertinage of amorous wickedness . . .[30] In certain rare and abominable women[31] deceit rose to an almost satanic degree. A natural insincerity, an acquired dissimulation, a controlled gaze, regulated features, an effortless lie in the whole being, a profound observation, a penetrating glance, a domination of the senses, a curiosity and a desire for knowledge, which allowed them to see in love nothing but facts to collect and meditate; to such faculties and such dangerous qualities these women owed, from their youth, the talents and policy which might have made the reputation of a Minister.[32]

These two, Mme. de Merteuil and Valmont, are the active characters whose strong malignant force controls to a greater or less extent the other passive characters, delivered into their hands either through their own weaknesses or through the imperfections of social customs. Danceny, Cécile, Madame de Tourvel, Madame de Volanges, and Madame de Rosemonde are, therefore, less striking characters but are none the less sketches from 18th century life. Danceny is rather a conventional character—the poor young man of good family who has little or no experience of life. He is, perhaps, the most literary of Laclos's characters; I suspect he was taken directly from Crébillon. In any case his situation between Cécile and Mme. de Merteuil is practically equivalent to the situation of young Meilcour between Mme. de Sénanges and Mademoiselle de Théville in "Les Egarements," though Mlle de Theville is a much more estimable character than Cécile. Danceny is "still at the age of sentiment," as people then put it; he really loves Cécile, but he is too young to feel the full strength of passion, too inexperienced to resist the wiles of Mme. de Merteuil. As Florian puts it in his *Mémoires d'un Jeune Espagnol*, (which are his own *mémoires*) "I did not much regret Henriette; in paying her attentions it was not she I loved, it was the pleasure of loving a woman I had sought; as soon as my mind was occupied by something else, I ceased to think of love." Florian was sixteen at that time; Danceny is several years older and hence not quite so casual in deceiving Cécile.

As for Cécile herself, she is a masterly study of the "*Caillette*" in process of formation. She is one of those pretty girls so often represented in contemporary engravings, with their delicate oval faces, facile smiles and coquettish brainlessness. The same mixture of childishness and precosity may be seen in some of Greuze's heads. She is a creation of the 18th century treatment of women. As very young children they caught a glimpse of fashionable life in their parents' house; then they were suddenly taken away, shut up for years in a convent, and kept in ignorance of life and particularly of sexual matters, while they acquired the "accomplishments." Meanwhile, the occasional glimpses of the outside world during brief outings, the visits of well-dressed, rouged, perfumed, patched Mammas accompanied by fashionable friends, whisperings of marriage and the delights of the "*monde*," filled their silly little heads with a fever of curiosity and longing for freedom, amusement, *les agréments, les dissipations*. Once free from the convent restrictions they often enough became "*caillettes*." Here is Duclos's description of the *caillette:*

> A woman of this character, or rather of this species, has neither principles, passions nor ideas. She does not think, she believes she feels; her mind and her heart are equally cold and sterile. She is occupied by little things only, talks only in commonplaces which she takes for new expressions. She brings everything back to herself, or to a detail which has struck

> her. She likes to appear well-informed and thinks herself necessary to people. Vain bustle is her element; clothes, decisions on fashions, and ornaments are her occupation. She will cut short the most important conversation to say that this year's taffetas are dreadful and in a taste which shames the nation. She takes a lover as she does a dress, because it is the fashion. She is troublesome in affairs, and boring in her pleasures. The *caillette* of quality is only distinguished from the *bouregoise caillette* by certain words of a better usage and different objects; the first tells you about a visit to Marly and the other bores you with details of a supper at the Marais.[33]

That is what Cécile would have become a year or so after her marriage with Gercourt, had it taken place. As to Mme. de Volanges, she is the French "ma mère," out of touch with her daughter whose sympathy she had never gained, yet sincerely anxious for the child's happiness. She is rather stupid, and her "virtue" is not particularly clear-sighted; the man against whom she warns Mme. de Tourvel so unctuously seduces her own daughter, under her very nose, as it were. Her sudden gush of feeling towards Cécile and her design of marrying her to Danceny are easily thwarted by the astute Mme. de Merteuil, in whose hands she is as wax. Mme. de Merteuil plays on her weaknesses and prejudices with consummate skill. And here we can only stand back and admire the subtlety of Laclos's psychology; in every case where Valmont and Mme. de Merteuil "triumph," it is through their acute perception of the weaknesses in others and their own self-command. The weakness in Cécile is ignorance, sensual curiosity, lack of character; in her mother worldly prudence, conventional virtue and limited intelligence; in Danceny inexperience and credulity; in Mme. de Tourvel a number of very minor faults all arising from a kind of vanity; it is through flattery of a very subtle sort that Valmont makes her love him, and that love once created he has a cardinal weakness to play on. Mme. de Volanges is a very ordinary sort of woman; her ideas move in a narrow circle, she is the slave of convention. She is in no respect remarkable, but merely an 18th century embodiment of that perpetual type, the well-intentioned mother with no particular intelligence.

Mme. de Rosemonde is, as it were, a kind of chorus to the tragedy, aloof in the detachment of her great age, but with that keen mind and sympathetic indulgence for human weakness so noticeable in many old ladies of the time. She is not a Mme. du Deffand or a Mme. de Tencin, but she is of their class and type; she has not aged stupidly, but with wisdom.

Mme. de Tourvel is a little apart from Laclos's other characters because in her he essayed to create an ideal type, the Rousseau-like "virtuous woman," all sentiment, compassion, benevolence,[34] trust. To some extent, then, she is an artificial creation and although her character is worked out with minute care it is not wholly convincing. She does not carry that startling conviction of reality such as we feel with Valmont and Mme. de Merteuil. Yet many of her traits are perfectly in harmony with what we know of 18th century women. Among the Merteuils and Sénanges, the *fausses prudes* and the *coqueties* and the *caillettes* were many women of character and intelligence, some capable of an almost sublime devotion in love.

In that century are to be found some touching examples of women's passion. There is Mlle Aissé with her whole-hearted love of the Chevalier Aydie, of whom she wrote:

> There are many people who never knew the satisfaction of loving with sufficient delicacy to prefer the happiness of what we love to our own.

There is Mlle de Lespinasse with her passion for Guibert and the famous love-letter:

From every moment of my life, 1774.
My dear, I suffer, I love you and I am waiting for you.

There is Mme. de la Popelinière whose love-letter to the faithless Richelieu is an unconscious masterpiece of pathos, with its rush of unpunctuated sentences, its humility, its ardour, its repetition, "Mon amant mon coeur" like a despairing litany of love.[35] There is that Princesse de Condé stooping from her great rank to love passionately, disinterestedly, purely, a young officer. Even the cynicism of relatives is touched by this unselfish, ardent love; Condé himself relents; there is talk of a marriage; but suddenly the Princess is attacked by religious scruples, tears herself from her lover in a heart-rending letter and hides her life and tears for ever in the solitude of the Nuns of the Perpetual Adoration. Mme. de Tourvel is not the only woman of her age who rose above its easy cynicism, its facile pleasures and coldness of heart, to the grandeur of a great passion. For, in spite of the Pharisees, there is something very touching, something noble in these profound passions which hesitate at no sacrifice, suffer all things, forgive all things. The depravity of the object of Mme. de Tourvel's passion only makes it the more tragic; until that cruel letter dictated by the malignant Marquise opens her eyes, she believes Valmont to be worthy. Her death is perhaps a violation of probability; yet people have died of a broken heart.

Such are Laclos's characters, taken either from perpetual human types or from the rarer types created by the life of the 18th century. The catastrophe in which he involves them is tragic enough to justify his freedom of speech. It is objected by some critics that the more or less conventional "judgment" on Valmont and especially on Mme. de Merteuil is an artificial ending and a weakness. It is artificial, so artificial that Laclos must have intended it to be thought so. If we remember that man's character and bitterness, we shall see that this clumsy sounding ending is a last consummate *finesse*. What it says in effect is:

"People of France, these are your rulers, see how they behave, see their cynicism, their depravity, the cold effrontery of their sensuality, their baseness. That is what they are; this is what they do; *and they do it unpunished.*" That is one reason why Tilly said the book was a "wave on the ocean of Revolution."

But the "Liaisons Dangereuses" contains two other inter-dependent moral themes—perhaps two political-social problems would be more accurate—which Laclos leaves his reader to ponder upon, with a shrewd guess as to what the solution will be. Several times I have insisted on the fact that the "Liaisons Dangereuses" is the revenge of a man with disappointed ambitions and the polemic of a disciple of Rousseau. I repeat it, because it is essential if we are to understand the real object of this pitiless, ironic tragedy and not to be misled into thinking it a mere aimless and wanton harrowing of our feelings, an almost devilish display of cynicism.

The first of these problems, and the most important in Laclos's view as a moralist, was the education of women. We can see that his Rousseau "back-to-nature" remedy was mere clap-trap; but his diagnosis was masterly. What he points out in effect is this.

> You educate your girls in cloistered ignorance. You fit them for nothing but to be delicious playthings. You abruptly take them from a convent and throw them in their trembling ignorance and against their will into the arms of a man they have barely seen. You thwart their profoundest sentiments; you outrage their most elementary delicacy; you mock at their "little tears" and "childish sentiments"; you violate a moral law, because marriage without love is a hideous thing; you violate a natural law, for the meanest of females has a right to a choice of the father of her children. What is the result? The ignorance of some makes them an easy prey to an unscrupulous seducer, with an inevitable corruption of character and misery. The

> stronger character of others turns to malignancy; the natural desires of a woman become a shameless sensuality, their natural vanity becomes a preposterous egotism, their natural jealousy of other women a devilish hate, their intelligence creates a whole system of hypocrisy to guard against your anomalous standards, and their cunning becomes an almost superhuman perfidy in their battles with you and you, gentlemen, worthy emulators of that 'très haut et très puissant seigneur,' Monsieur le Vicomte de Valmont. And, supremest wrong of all, the most virtuous and tender of women, capable of a profound and unselfish passion which can never be satisfied by your *mariages de convenance*, is surprised, tortured, victimised, driven to her grave, by this same scoundrel.

Laclos's remedies for this really do not concern us, they have the rigidity and pedantry of the military *philosophe* in them, the illusions of the "friend of man." But his denunciation was not an unjust one. And sometimes, when we are tempted to smile at the sentiment of the early 19th century, to feel irritated at its absurd prudery, we might remember that it was a natural reaction from the rule of the Valmonts and Merteuils. That bath of sentiment had its value. It is possible our own reaction against it has gone too far. There is something revolting in the "*goujaterie*" of a certain type of wealthy young man, which reminds one of a vulgar Valmont.[36]

The other theme has already been touched upon; it is almost entirely political and might be defined as the problem of the *petite noblesse*, to which Laclos belonged. The members of this class were in some respects the most victimised sufferers under the *ancien régime*.[37] Many of them were rigidly virtuous, most were poor; their poverty and principles prevented their taking much share in the excesses of the *grande noblesse* and financiers;[38] they were all profoundly loyal to the Monarchy. This loyalty and principle were quite shamelessly abused by the government. The *petite noblesse* filled most of the minor parts of the services and upon them fell all the hard and unpleasant work and very little of the rewards. The sale of regiments and Court influence made it almost impossible for one of the *petite noblesse* to rise to high rank in the army or navy, unless he was lucky enough to attract the fancy of some Court favourite. Officers who had twenty years service, the Croix de Saint Louis, the scars of half a dozen campagnes, had the profound humiliation of serving under a colonel of sixteen or seventeen, an embryo *petit maître*, whose one merit was (in Voltaire's words) the fact that "he had given himself the trouble of being born."

As we have seen, Laclos was an able artillery officer, especially in fortification; and, as the years passed and he reached the "dangerous" age of forty, his disappointed ambition hardened into bitterness. The "Liaisons Dangereuses," then, in effect, addresses the *petite noblesse* thus:

> We are the effective officers of the Services, the men upon whom the real defence of the country falls; yet we live in poverty, have no hope of high rank, while others no better born than ourselves step over our heads through wealth and influence. While we are with our regiments, these our superiors (worthy descendants of the Crusaders!) are doing what? Indulging in heartless debauchery, seducing our sisters and cousins, perverting the women of France in their *petites maisons*. These are the fine flower of civilisation, these the Paladins of France! What leaders! Godfrey de Bouillon took Palestine from the Saracens—Danceny is left to idleness in the boudoirs of Paris. Du Guesclin, Montluc, Condé, Turenne, captured cities and provinces at the sword's point; M. de Valmont beleaguers a helpless child and triumphs over a modest woman.[39] See, gentlemen, the heroes who command you![40]

* * *

I have one more point to make before bringing this long examination to a close; and that is Laclos's penetration as a "moralist," by which I mean a detached and unbiassed observer of human manners. There is, indeed, a certain coldness and cynicism in his observations which remind one distantly of the "Maximes" of La Rochefoucauld; but even when they are a little warped by prejudice they have enough point and truth at least to set one thinking. Many of these observations are of course brought out indirectly as traits of character, but quite often Laclos allows himself to analyse through the remarks of his letter-writers. Perhaps some of them are the merest *boutades*, but they have a sting. Turn over the pages of the "Liaisons"; you find things like these:

> Have you forgotten that love, like medicine, is simply the art of aiding nature?
>
> *A propos* negligence, you are like the people who send regularly for news of their sick friends, and never ask what the reply was.
>
> But I did not think I ought to lose an opportunity of allowing myself to be given an order; being convinced on the one hand, that she who commands, commits herself, and, on the other hand, that the illusory authority we appear to let women take is one of the snares they avoid with the most difficulty.
>
> Old women must not be angered—they make young women's reputations.
>
> When a woman's heart has been exercised for some time, it needs rest; and I have noticed flattery is the softest pillow one can offer any of them.

Perhaps one of the best is Mme de Merteuil's little essay on the two main classes of women who have outlived their beauty (Letter CXIII). It is rather too long to quote, but the reader should not overlook it. If all the observations of this kind were extracted from *Les Liaisons Dangereuses*, they would make quite a little sheaf of *Maximes et Pensées*—to stand on that side of La Rochefoucauld which is more remote from Pascal.

* * *

Such is my view of one of the greatest of 18th century French novelists and of his single masterpiece. It is not a view hastily adopted and perversely defended. I am aware that it is not the general view of Laclos and *Les Liaisons Dangereuses*, but I trust the reader will consider my arguments and the evidence I bring before rejecting it. This view was gradually formed from a close study of the book and its author, made with all the evidence I could collect and whatever knowledge of French history, manners and literature I happened to possess. I cannot, of course, hope that I have avoided all errors of fact and interpretation in so long a study; but I have been as scrupulous as I knew how. Certainly, had the evidence led me to other conclusions I should have recorded them; and had I not thought the book a good one I should not have translated it. I believe it will surprise many who have not read the novel and know it only by the common defamation. Those who expect to find a light, amorous tale will be disappointed; those of more robust mental calibre and acuter taste will find a tragic story well told and a subtlety of psychological analysis not unworthy a countryman of Stendhal, Flaubert and Balzac.

RICHARD ALDINGTON

There are several other translations of *Les Liaisons Dangereuses;* one made as early as 1784, and one about thirty years ago by that accomplished poet, Ernest Dowson. The present translator has not consulted any of his predecessors.

A list of "Books Consulted" for this Introduction will be found on page lxiv.

All notes *not signed* "C. de L." are the translator's.

List of books consulted

Choderlos de Laclos: *Liaisons Dangereuses*. Notice par Ad. van Bever. (*see above*) 2 vols. Paris, 1920.
Choderlos de Laclos: *Poésies*. Publiées par Arthur Symons et Louis Thomas. Paris, 1908.
Choderlos de Laclos: *Lettres Inédites*. Publiées par M. Louis de Chavigny. Paris, 1904.
Emile Dard: *Le General Choderlos de Laclos*. Paris, 1905.
Fernand Caussy: *Laclos* 1741–1803. Paris, 1905.
Edmond et Jules de Goncourt: *La Femme au Dix-Huitième Siècle*. Paris. (Edition définitive, 1924).
Henri Carré: *La Noblesse de France et l'opinion Publique au XVIIIe Siècle*. Paris, 1920.
Servais Etienne: *Le Genre Romanesque en France depuis l'Apparition de la "Nouvelle Héloïse" jusqu'aux Approches de la Révolution*. Paris, 1922.
Crébillon fils: *Œuvres*. Londres. 7 vols., 1772.
Anthony Hamilton: *Mémoires du Comte de Grammont*. 2 vols. (No place given). 1760.
J. J. Rousseau: *Julie ou la Nouvelle Héloïse*. Paris. 2 vols. (Flammarion).
Marivaux: *Marianne*. Paris (Garnier).
Casanova: *Mémoires*. Paris. 6 vols. (Flammarion).
Florian: *Mémoires d'Un Jeune Espagnol*. Introduction de André Bouis. Paris, 1923.
Louvet de Couvray: *Les Adventures du Chevalier de Faublas*. Paris. (Garnier).
Rétif de la Bretonne: *Les Plus Belles Pages*. Paris, 1906.
Diderot: *Les Bijoux Indiscrets*. Au Monomatapa. (No date, but 18th century).
La Réligieuse. Le Neveu de Rameau. (Flammarion).
Abbé Dulaurens: *Imirce ou la Fille de la Nature*. Paris, 1922.
Andrea de Nerciat: *Félicité*. Paris, 1910.
Duclos: *Confessions du Comte de* ***. Œuvres. Tome VIII, 1806.
(Other books used are mentioned in the notes).

Dangerous Acquaintances

(*Les Liaisons Dangereuses*)

PUBLISHER'S NOTE[1]

The Public is notified that despite the title of this work and what is said by the editor in his preface, we do not guarantee the authenticity of the collection and we have good reason to believe it is only a Novel.

Moreover we think that the author, in seeking verisimilitude, has himself destroyed it in a very clumsy way, by the period in which he has placed the events he makes public. Indeed, several of the characters he describes have such abominable morals that it is impossible to suppose they could have lived in our own century—that century of philosophy in which enlightenment, spreading on all sides, has rendered (as everyone knows) all men so worthy and all women so modest and reserved.

Our opinion therefore is that if the events recorded in this work have a basis of truth they can only have happened in other countries or in other times; and we greatly blame the author who, apparently seduced by the hope of being more interesting by coming nearer his age and country, has dared to display, in our dress and with our customs, morals which are so foreign to us.

To preserve as far as we can the too credulous reader from any mistake in the matter, we support our opinion by a line of reasoning which we put forward with confidence, because it appears to us triumphant and unanswerable;

it is that the same causes could not fail to produce the same effects, and that to-day we never see a young Lady, with an income of sixty thousand livres, become a nun, nor the wife of a Président[2] *die of grief while she is still young and pretty.*

EDITOR'S PREFACE

This work, or rather this collection, which the public may perhaps think too voluminous, only contains a very small number of the letters composing the whole correspondence from which it is taken. When I was requested to put in order this correspondence by those into whose hands it had fallen and whose purpose I knew was to make it public I asked no compensation for my pains beyond the permission to cut out everything which might appear to me unnecessary, and I have tried to preserve only those letters which seemed to me needed for the understanding of events or to reveal character. If to this slight task is added that of rearranging the letters I have preserved in an order which is almost always that of their dates, and a few short notes whose purpose as a rule is merely to point out the origin of quotations or to explain some of the omissions; the whole of my share in this work will be known. My task went no further.[1]

I had proposed more considerable changes, almost all relating to the purity of diction or style, which will be found very faulty. I had also desired authority to cut down certain letters which are too long, several of which treat separately, and with practically no continuity, of matters that have no connection with each other. This labour, which was not accepted, would certainly have been insufficient to make the work valuable but would at least have freed it from some of its defects.

It was pointed out to me that the intention was to publish the letters and not a work founded upon the letters; that it would be as contrary to probability as to truth for the eight or ten people who took part in the correspondence to have written with the same purity. And when I pointed out that far from this being the case, there was none among them guiltless of serious mistakes which could not fail to be censured, I was told that every reasonable reader would naturally expect to find mistakes in a collection of letters by a few private persons since in all those hitherto published by different authors of eminence (and even by certain Academicians), there was not one wholly free from this reproach. These reasons did not convince me and I thought them, as I still think, easier to give than to accept; but I had no authority and was forced to give way. But I reserved the right to protest against the decision and to state that I was of a different opinion; which I now do.

As to any merit which this work may have, perhaps it is not for me to give an opinion, since it ought not and cannot influence anyone's. However, those who like to know what they are to expect before they begin to read a book, may continue to read this preface; others will do better to turn straight to the work itself; they know enough about it.

What I have to say at first is, that although my advice was, as I have admitted, to publish these letters, I am far from hoping they will be successful. But do not take this sincerity on my part for the affected modesty of an author; for I declare with the same frankness that if this collection had not seemed to me worth offering to the public I should not have troubled with it. Let us try to reconcile this apparent contradiction.

The merit of a work consists in its utility or the pleasure it gives, and sometimes in both when that is possible; but success (which is not always a proof of merit) often arises rather from the choice of a subject than from its execution, from the aggregate of objects presented rather than from the manner in which they are treated. Now this collection includes, as its title indicates, the letters of a whole group, and this causes a diversity of interest which weakens that of the reader. Moreover, since almost all the sentiments here expressed are feigned or dissimulated they can only arouse the interest of curiosity which is always far below the interest of sentiment, which excites less indulgence and allows faults of detail to be more easily perceived, because these details are continually opposed to the one desire for which satisfaction is wished.

These defects are perhaps partly atoned for by a quality which lies in the very nature of the work; that is, variety of styles, a merit which an author attains with difficulty but which here arises spontaneously, and at least preserves from the boredom of uniformity. Some persons may count as a merit the considerable number of observations, either new or little known, which are scattered through these letters. That, I think, is all the pleasure that can be hoped for from them, even when they are judged with the greatest indulgence.

The utility of the work will perhaps be more contested but seems to me easier to establish. To me at least, it appears a service rendered to good morals to unmask the methods employed by those whose morals are bad, in corrupting others who are good, and I think that these letters effectively contribute to that end. There will also be found the proof and example of two important truths which might be thought unknown, seeing how little they are regarded; one is that every woman who consents to receive into her society an immoral man ends by becoming his victim; the other is that every mother who allows someone else to possess her daughter's confidence is at least imprudent.

Young people of either sex may also learn here that the friendship which seems to be granted them with such facility by persons of bad morals is never anything but a dangerous snare, as fatal to their happiness as to their virtue. Yet abuse, which is always so near to what is good, seems to me particularly to be dreaded here; and, far from advising this book to youth, I think it very important to deprive the young of all books of this kind. The period when it may cease to be dangerous and become useful appears to me to have been grasped—for persons of her own sex—by a good mother who not only possesses intelligence but a *good* intelligence. "I think," she said to me, after she had read the manuscript of this correspondence, "I should be rendering a real service to my daughter by giving her this book on her wedding-day." If all mothers of families are of a similar opinion I shall always congratulate myself upon the publication of this work.

But even arguing from this favourable supposition it still seems to me that this collection can please but a few. Depraved men and women will find it to their interest to decry a work which may harm them; and since they are not lacking in skill, they may be clever enough to attach the rigourists to their party through their alarm at the picture of evil morals here presented.

Those who pretend to irreligion will take no interest in a female devotee because that very fact will make them look upon her as a foolish little woman, and at the same time the devotees will be angry that virtue succumbs and will complain that religion is represented as having too little power.

On the other hand, persons of a delicate taste will be disgusted by the over-simple and faulty style of several of these letters, while the bulk of readers who are convinced that everything they see printed is the fruit of long labour will think they see in other letters the laboured manner of an author showing himself beneath the personage through whose mouth he speaks.

Finally, the general opinion will perhaps be that each thing is of use only in its place and that if, as a rule, the over-careful style of authors deprives familiar letters of the effects of grace, the negligences in real familiar letters become serious faults and make them unendurable when they are printed.

I frankly admit that all these objections seem to me to be founded; I also think I could reply to them, and that without exceeding the length of a preface. But it will be felt that the necessity of replying to everything would mean that the book responded to nothing; and that, if I had thought this, I should have suppressed both preface and book.

PART I

LETTER I

Cécile Volanges to Sophie Carnay at the Ursuline Convent of . . .

You will see, my dear, that I have kept my word and that bonnets and pom-poms do not take up all my time—there will always be some left over for you. Yet I have seen more clothes in this single day than in the four years we spent together; and I think the haughty Tanville[1] will be more angered by my first visit (when I intend to ask for her), than she thought we were when she came to see us in *fiocchi*.[2] Mamma asks my opinion in everything and treats me much less like a school-girl than she used to do. I have my own maid; I have a room and a study at my disposal and I am writing this to you at a very pretty writing-table whose key was given to me so that I can shut up anything I want in it. Mamma says I am to see her every day when she gets up; that I need not arrange my hair[3] until dinner time because we shall always be alone, and that she will tell me every day when I am to join her in the afternoon. The rest of my time is at my disposal and I have my harp, my drawing and my books just as in the convent, except that Mother Perpetue is not here to scold me and that I need do nothing unless I wish; but since my Sophie is not here to talk and laugh with me, I may just as well occupy myself.

It is not yet five o'clock; I do not see Mamma until seven; plenty of time, if I had anything to say to you! but so far nothing has been said and except for the preparations I see being made and the numbers of sewing-women who all come for me I should think there is no intention of marrying me and that it was one more delusion on the part of the good Josephine.[4] Yet Mamma has told me so often that a young lady should remain at the convent until she is married that Josephine must be right, since Mamma has taken me away.

A carriage has just stopped at the door and Mamma sends me a message to come to her at once. Suppose it were *he*? I am not dressed; my hand trembles and my heart beats. I asked my maid if she knew who was with my mother. "Why," said she, "it is M. C . . ." And she laughed. Oh! I think it *is* he! I shall surely return to tell you what has happened. This, at all events, is his name. I must not keep them waiting. Adieu, for a moment . . .

How you will laugh at your poor Cécile! Oh! I was very much ashamed, but you would have been as helpless as I was. When I entered Mamma's room I saw a gentleman in black standing beside her. I saluted him as well as I could and remained rooted to the spot. You can imagine how I looked at him! "Madame," said he to my mother, as he bowed to me, "This is a charming young lady and I feel more than ever the value of your favour." At this plain remark I began to tremble, to such an extent that I could not stand up; I found an arm-chair and sat down in it, blushing deeply and very disconcerted. I had scarcely sat down when this man was at my knees! Your poor Cécile then lost her head; as Mamma said, I was thoroughly scared. I sprang up with a piercing cry . . . just like the day of the thunder-storm Mamma burst out laughing and said: "Why! What is the matter with you? Sit down and give Monsieur your foot." My dear, "Monsieur" was a shoemaker! I cannot tell you how ashamed I was; fortunately no one was there but Mamma. I think that when I am married I shall never employ this shoemaker.

You must admit we are well informed! Good-bye. It is nearly six o'clock and my maid says I must dress. Good-bye, dear Sophie; I love you as if I were still at the convent.

P.S. I do not know by whom to send this letter, so I shall wait until Josephine comes.

Paris, 3rd August, 17—.

LETTER II

The Marquise de Merteuil to the Vicomte de Valmont at the Château de . . .

Come back, my dear Vicomte, come back; what are you doing, what *can* you be doing with an old aunt whose property is entailed on you? Come at once; I need you. I have a wonderful idea and you must carry it out. These few words should be enough for you and, but too honoured by my choice, you should come eagerly to take my orders on your knees; but you abuse my favours even since you have ceased to make use of them; and in the alternative of an eternal hatred or an excessive indulgence your good luck decides that my kindness should win the day. I shall therefore tell you my plan; but swear to me like a faithful knight that you will engage yourself in no other adventure until you have accomplished this. She is worthy of a hero; you will serve both love and vengeance; and it will be a *rouerie*[1] the more to put in your *mémoires*; yes, in your *mémoires*, for I wish them to be printed one day and I undertake to write them. But let us leave this and return to what concerns us.

Madame de Volanges is marrying her daughter; it is still a secret, but she told me about it yesterday. And whom do you think she has chosen for her son-in-law? The Comte de

Gercourt. Who would have thought that I should become Gercourt's cousin? I am in a rage . . . Have you not guessed why? Dullard! Have *you* forgiven him the adventure with the *Intendante*? And have I not even more reason to complain of him, monster that you are?[2] But I calm myself and the hope of vengeance soothes my mind.

You have been a hundred times annoyed, as I have myself, by the consequence Gercourt attaches to the wife he is to have and by the silly presumption which makes him think he will avoid the inevitable fate. You know his ridiculous prejudices in favour of a cloistered education and his still more ridiculous preconception of the modesty of fair-haired women. Indeed I would wager that, in spite of the income of sixty thousand *livres* which goes with the Volanges girl, he would never have consented to the marriage if she had been dark or if she had not been to a convent. Let us prove to him that he is a mere fool; no doubt he will be one day—that is not what troubles me—but it would be amusing to have him begin by being one. How it will delight us to hear him boasting on the morning after (for he will boast); and then, if you once mould this girl, it would be very unlucky if Gercourt, like anybody else, does not become the talk of Paris.

Moreover, the heroine of this new adventure is worthy of all your attention; she is really pretty, she is only fifteen; a rosebud; ignorant to a degree and entirely unaffected, but you men are not afraid of that; in addition, a certain languid gaze seems to promise a great deal. Add to this that I recommend her to you; you have no more to do than to thank and obey me.

You will receive this letter to-morrow morning. I insist on your being with me to-morrow evening at seven o'clock. I shall receive nobody until eight, not even the reigning Chevalier; his mind is not equal to so important an affair. You see that love does not blind me. At eight o'clock I will grant you your liberty and you will return at ten to sup with the fair creature; for she and her mother are taking supper with me. Good-bye, it is after midday; very soon I shall cease to take an interest in you.

Paris, 4th August, 17—.

LETTER III

Cécile Volanges to Sophie Carnay

I am still kept in ignorance, my dear. Yesterday Mamma invited a number of people to supper. In spite of the fact that it was to my interest to observe them, especially the men, I was very bored. Men and women, everybody, looked closely at me and then whispered in each others' ears. I saw they were talking about me; this made me blush; I could not prevent it. I wish I could have, for I noticed that when people looked at the other women they did not blush; or else the rouge they put on prevents one seeing the colour caused them by embarrassment; for it must be very difficult not to blush when a man looks steadily at you.

What made me most uneasy was that I did not know what they thought about me. I think I heard two or three times the word "Pretty"; but I very distinctly heard "Awkward"; and the latter must be true because the woman who said it is my mother's relative and friend; she even seemed to have a sudden friendship for me. She is the only person who talked to me at all during the evening. To-morrow we go to supper at her house.

After supper I heard a man who, I am sure, was speaking of me, say to another: "We must let her ripen; we shall see, this winter." Perhaps it is he who is to marry me; but then it will not be for four months! I wish I knew what is to be.

Here is Josephine who tells me she is in a hurry. Yet I must tell you another *awkwardness* of mine. Oh! I am afraid the lady was right!

After supper gambling began. I was beside Mamma; I do not know how it happened, but I went to sleep almost at once. I was

awakened by a loud burst of laughter. I do not know if they were laughing at me, but I think they were. Mamma gave me permission to go, which gave me great pleasure. Imagine! It was after eleven o'clock! Good-bye, my dear Sophie; always love your Cécile. I assure you the world is not as amusing as we imagined.
Paris, 4th August, 17—.

LETTER IV

The Vicomte de Valmont to the Marquise de Merteuil at Paris

Your orders are charming; your manner of giving them is still more amiable; you would make despotism attractive. As you know, this is not the first time that I regret I am no longer your slave; and however much of a *monster* you say I am I never think but with pleasure of the time you honoured me with softer names. Quite often I even hope to deserve them again and to end up with you by giving the world an example of constancy. But more important interests must occupy us; to conquer is our fate, and fate must be obeyed; perhaps we shall meet again at the end of our career; for be it said without offence, most fair Marquise, you follow me step by step; and since, after separating for the happiness of the world, we preach the faith separately, it seems to me that you make as many proselytes as I, in this mission of love. I know your zeal, your ardent fervour; and if this God estimated us by our works, you would one day be the patroness of some large town, while your friend at best would be a village saint. This form of expression surprises you, does it not? But for the last eight days I have heard and spoken none other; and in order to grow perfect in it I am forced to disobey you.

Do not be angry; and listen to me. You have shared all the secrets of my heart and I am about to confide to you the greatest project I have ever formed. What was it you proposed to me? To seduce a girl who has seen and knows nothing, who (so to speak) would be handed over to me defenceless, who could not fail to be intoxicated by a first attention and whom curiosity would probably lead more rapidly than love. There are twenty others who could succeed as well as I. But this is not the case with the enterprise which now occupies me; its success assures me as much fame as pleasure. The Love who is preparing my crown himself hesitates between myrtle and laurel, or rather he will unite them to honour my triumph. You yourself, my fair friend, will be seized by a holy respect and you will say enthusiastically: "Here is a man after my own heart."

You know Madame de Tourvel, her religious devotion, her conjugal love, her austere principles. That is what I am attacking; that is the enemy worthy of me; that is the end I mean to reach;

"And if I do not carry off the prize of obtaining her,

At least I shall have the honour of having attempted it."

Bad verses may be quoted when they are by a great poet.[1]

You must know that her husband is in Burgundy on account of some big law-suit (I hope to make him lose a still more important one). His inconsolable spouse is compelled to spend here the whole time of her distressing widowhood. A mass every day, a few visits to the poor in the district, morning and evening prayers, solitary walks, pious conversations with my old aunt, and sometimes a dismal rubber of whist, are her only distractions. I am preparing more effectual ones for her. My good angel led me here for her happiness and for my own. Madman that I was! I regretted the twenty-four hours I sacrificed to the demands of convention. How I should be punished were I forced to return to Paris! Luckily, four people are needed for a hand of whist, and since there is no one here but the local *curé* my eternal aunt pressed me to sacrifice a few days to her. You may guess that I consented. You cannot imagine how much she has flattered me since then and above all how edified she is to see me regularly at prayer and at mass. She does not realise who is the divinity I adore.

For the last four days I have given myself up to this powerful passion. You know I always desire keenly and sweep away obstacles; but what you cannot know is how much solitude adds to the ardour of desire. I have but one idea; I think of it by day and dream of it by night. I must have this woman, to save myself from the ridiculous position of being in love with her—for how far may not one be led by a thwarted desire? O delicious possession! I need you for my happiness and still more for my peace of mind. It is fortunate for us that women are so weak in their own defence! Otherwise we should be nothing but their timid slaves. At this very moment I have a feeling of gratitude for facile women which quite naturally brings me to your feet. I cast myself before them to obtain forgiveness and I conclude this long letter. Good-bye, fair lady—all in good part.

From the Château de . . . , 5th of August 17—.

LETTER V

The Marquise de Merteuil to the Vicomte de Valmont

Do you realise, Vicomte, that your letter is extremely insolent and that I ought to be angry with you? But it shows me you have lost your head and that fact alone saves you from my indignation. I am a generous and compassionate friend and forget my own injury to concern myself with your danger; and, however tiresome it may be to argue, I must yield to your present need of it.

You possess Madame de Tourvel! What a ridiculous caprice! I see it is your usual obstinacy which never wants except what seems impossible to obtain. What sort of a woman is she? Regular features, if you like, but a complete lack of expression; fairly well-made but entirely without grace; always ridiculously dressed with her bunches of neckerchief on her breast and her bodice up to her chin! I warn you as a friend, two women like her will ruin your reputation for you. Think of that day when she made the collection in Saint Roch, when you thanked me so often for having procured you such a spectacle! I can still see her, giving her hand to that hop-pole of a man with long hair, ready to sink down at every step, with her four yards of hoop-skirt continually on someone's head, and blushing at every bow. Who would have said then that you would desire such a woman? Come Vicomte, blush, and recover your senses. I promise you secrecy.

And then, think of the annoyances awaiting you. Who is the rival you must combat? A husband! Do you not feel humiliated by the mere word? What a disgrace if you fail! And how little glory if you succeed! I will go further—you must not expect any pleasure. Can there be any with prudes? I mean with those who are really so. They are reserved in the very midst of pleasure and can offer you nothing more than a half-enjoyment. That complete abandonment of self, that delirium of delight wherein pleasure is purified by its excess, those treasures of love are unknown to them. I warn you—supposing the best, your Madame de Tourvel will think she has done everything for you if she treats you like a husband, and in the tenderest conjugal interview the parties always remain two. Here it is still worse; your prude is religious with the sort of religion which condemns a woman to perpetual childishness. Perhaps you will surmount this obstacle, but do not flatter yourself with the idea that you can destroy it; you may conquer the love of God, but not the fear of the devil; when you hold your mistress in your arms and feel her heart beating, it will be from terror, not from love. Perhaps if you had known this woman sooner you might have made something of her; but she is twenty-two and has been married nearly two years. Believe, me, Vicomte, when a woman is *encrusted* to that extent she must be left to her fate; she will never be anything but a poor creature.

Yet it is for this fair object that you refuse to obey me, that you bury yourself in your aunt's tomb, that you give up a most delicious adventure which would do you the utmost honour! How does it happen that Gercourt always has the advantage over you? Come, I am talking to you good-humouredly; but at the moment I really am tempted to think you do not deserve your reputation, I am tempted to withdraw my confidence from you. I could never grow accustomed to confiding my secrets to Madame de Tourvel's lover.

You are to know that the Volanges girl has already turned one head. Young Danceny is madly in love with her. He has sung with her; and indeed she sings better than a girl just from a convent should. They are to go through a number of duets and I feel sure she will gladly be in unison; but Danceny is a mere child who will lose his time in making love and will never complete anything. The girl herself is quite shy and, in any event, it will be much less amusing than you could have made it; it puts me out of humour and I shall certainly quarrel with the Chevalier when he arrives. I advise him to be gentle; for, at the moment, I could break with him without reluctance. I am sure that if I had the good sense to leave him now he would be in despair; and nothing amuses me like a lover in despair. He would call me false, and the word "false" always gives me pleasure; after "cruel" it is the sweetest in a woman's ear and the least difficult to deserve. Seriously I must think about breaking off this affair. See what you are the cause of! I lay it on your conscience. Good-bye. Recommend me to Madame de Tourvel's prayers.

Paris, the 7th of August, 17—.

LETTER VI

The Vicomte de Valmont to the Marquise de Merteuil

So there is no woman in the world who does not abuse her power! Even you, whom I so often called "an indulgent friend," cease to be so; you do not shrink from attacking me through the object of my affection! With what strokes you dared to paint Madame de Tourvel! A man would have paid for such insolence with his life and any woman but you would have been repaid at least by some revenge. I beg you will not subject me to such harsh tests; I cannot promise to endure them. In the name of friendship, wait until I have had the woman if you want to disparage her. Do you not know that pleasure alone has the right of loosening the bandage from Love's eyes?

But what am I saying? Does Madame de Tourvel need illusion? No, she needs but to be herself and she is adorable. You censure her for dressing badly and I agree with you; all clothes do her injustice, whatever hides her disfigures her. In the unconstraint of her morning-dress she is indeed delightful. Thanks to the extremely hot weather we are having, a morning-dress of simple linen allows me to see her round supple figure. Her breasts are hidden by a single fold of muslin and my furtive but keen glances have already spied out their enchanting shape. Her face, you say, lacks expression. And what should it express at a time when nothing speaks to her heart? No doubt, unlike our coquettes, she has not that delusive gaze which sometimes seduces and always deceives. She cannot cover the emptiness of a phrase by a false smile, and although she has the finest teeth imaginable she only laughs at what amuses her. But you should see her in her playful moments—what an image of frank, natural gaiety! What pure joy and pitying kindness are in her gaze when she hastens to help the unfortunate! Above all you should see her when at the least word of praise or flattery her divine face is coloured by the touching embarrassment of an unfeigned modesty! She is chaste and religious and therefore you think her cold and lifeless? I think very differently. What an amazing sensibility she must possess to be able to shed it even on her husband and to love continually a person who is continually absent! What stronger proof could you desire? Yet I have been able to obtain another.

I arranged a walk so that we came upon a ditch which had to be crossed; and although she is very active, she is even more timid; you may imagine that a prude is afraid to take a leap.[1] She was compelled to accept my help. I have clasped this modest woman in my arms! Our preparations and the crossing of my old aunt sent our gay devotee into peals of laughter; but as soon as I took hold of her, by an intentional awkwardness on my part our arms became mutually entwined. I held her breast against my own and, in that brief moment, I felt her heart beat faster. Her face was suffused by a charming blush and her modest embarrassment showed me *that her heart had beaten with love and not with fear.* And yet my aunt was as mistaken as you and said: "The child was afraid"; but the *child's* charming candour does not permit her a lie and she answered naively: "O no, but . . ." That single word enlightened me. From that moment my cruel uneasiness gave way to a pleasing hope. I shall have this woman; I shall carry her away from the husband who profanes her; I shall even dare to ravish her from the God she adores. What a delicious pleasure to be alternately the cause and the conqueror of her remorse! Far be it from me to wish to destroy the prejudices which torture her! They will add to my happiness and my fame. Let her believe in virtue, but let her sacrifice it to me; let her slips terrify her without restraining her; let her be agitated by a thousand terrors and not be able to forget and to crush them save in my arms. Then I agree, she may say: "I adore you"—and she alone among all women will be worthy to say so. I shall indeed be the God she has preferred.

Let us be frank; in our arrangements, as frigid as they are facile, that which we call happiness is scarcely a pleasure. Shall I confess it to you? I thought my heart withered up and, finding I had nothing left but my senses, I pitied myself for a premature old age. Madame de Tourvel has given me back the charming illusions of youth. Near her I do not need to enjoy her to be happy. The only thing which terrifies me is the time this adventure will take; for I can leave nothing to chance. However much I remind myself of my lucky audacities I cannot resolve to put them into practice. For me to be really happy she must give herself—that is no small matter. I am sure you will admire my prudence. The word "love" I have not yet spoken; but already we have got to "confidence" and "interest." To deceive her as little as possible and especially to forestall any gossip which might reach her I have myself told her, as if accusing myself, some of my best known exploits. You would laugh to see the candour with which she reproves me. She says she wishes to convert me. She does not guess the price she would have to pay for attempting it. She is far from thinking that "by pleading" (as she puts it) "for the unfortunate women I have ruined," that she is pleading beforehand on her own behalf The idea came to me yesterday in the middle of one of her sermons and I could not refuse myself the pleasure of interrupting her, in order to tell her she spoke like a prophet. Good-bye, fairest lady! You see I am not irrecoverably lost.

P.S. By the way, has the poor Chevalier committed suicide in despair? Really, you are a hundred times worse than I am and you would humiliate me, were I conceited.

From the Château of . . . , 9th of August 17—.

LETTER VII

Cécile Volanges to Sophie Carnay [1]

If I have told you nothing about my marriage, the reason is that I know no more about it than I did the first day. I am growing used to not thinking about it and I find this kind of life quite suits me. I spend a good deal of time working at my singing and my harp; I seem to like them more now that I have no master, or rather now that I have a better one. The Chevalier Danceny, the gentleman of whom I spoke, and with whom I sang at Madame de Merteuil's house, is kind enough to come here every day and to sing with me for hours on end. He is extremely agreeable. He sings like an angel and composes most elegant airs to which he writes the words. What a pity he is a Knight of Malta! It seems to me that if he married, his wife would be very happy . . . His gentleness is charming. He never seems to be paying one a compliment and yet everything he says is flattering. He corrects me continually, as much in music as in anything else; but he mingles such interest and such gaiety with his criticism that it is impossible not to feel grateful to him. Only, when he looks at you, he seems to be saying something agreeable. And in addition, he is very unselfish. For example, yesterday, he was invited to an important concert—and he preferred to stay the whole evening here. It gave me a great deal of pleasure; for when he is not here, nobody speaks to me and I grow languid; but when he is here, we sing and converse together. He has always something new to tell me. He and Madame de Merteuil are the only two people I think agreeable. But good-bye, my dear, I have promised to learn for to-day an arietta with a very difficult accompaniment and I do not want to break my word; I shall practise it until he comes.

From . . . , 7th of August, 17—.

LETTER VIII

Madame de Tourvel to Madame de Volanges

No one could be more touched than I am, Madame, by the confidence you show me, nor take more interest than I in the future of Mademoiselle de Volanges. With all my heart I wish her the happiness of which I am sure she is worthy and I rely upon your prudence to obtain it. I do not know the Comte de Gercourt; but, since he is honoured by your choice, I cannot but esteem him highly. I limit myself, Madame, to the wish that this marriage may be as happy as my own, which was likewise your work, and for which I am every day more grateful. May your daughter's happiness be your reward for the happiness you procured me; and may the best of friends be also the happiest of mothers!

I am indeed distressed that I cannot convey to you these sincere wishes in person and make the acquaintance of Mademoiselle de Volanges as soon as I should desire. Since I have received from you marks of kindness that were indeed maternal, I have the right to hope from her the tender friendship of a sister. I beg, Madame, you will be good enough to ask it of her on my behalf until I am in a position to merit it

I have arranged to stay in the country for the whole time of Monsieur de Tourvel's absence. I have made use of this time to enjoy and profit by the society of the respectable Madame de Rosemonde. She is a woman who has remained charming; her great age has robbed her of nothing: she keeps her memory and her gaiety intact. Her body alone is eighty-eight; her spirit is only twenty.

Our solitude is enlivened by her nephew the Vicomte de Valmont, who has been kind enough to spare us a few days I only knew him by hearsay and what I heard made me little desirous to know more of him; but I think him better than rumour pretends. Here, where he is not spoiled by the whirl of society, he talks of Reason with astonishing facility and confesses his faults with rare candour. He speaks to me with great confidence and I lecture him with great severity. You, who know him, will admit that his conversion would be a great achievement; but in spite of his promises I have no doubt that a week of Paris will make him forget all my sermons. His stay here will at least be that much time taken from his ordinary conduct; and I really think after the way he has lived that the best thing he can do is to do nothing at all. He knows I am writing to you and asks me to convey to you his respectful regards. Accept mine also with your usual kindness and never doubt the sincere feelings with which I have the honour to be, etc.

From the Château of . . . , 9th of August, 17—.

LETTER IX

Madame de Volanges to Madame de Tourvel

I have never doubted, my fair young friend, either your friendship for me or your sincere interest in my affairs. It is not for the purpose of clearing up this point—which I hope is forever settled between us—that I reply to your reply; but I cannot avoid discussing with you the subject of the Vicomte de Valmont.

I must admit I never expected to find that name in your letters. What can there be in common between you and him? You do not know this man; where could you have acquired the notion of a libertine's soul? You speak of his *rare candour*; ah yes! Valmont's candour must indeed be very rare. The more false and dangerous in that he is amiable and seductive, he has never from his earliest youth taken one step or said one word without a purpose, and he has never had a purpose but was wicked or criminal. My dear, you know me; you know that among the virtues I have tried to acquire, forbearance is the one I most prize. And so, if Valmont were carried away by impetuous passions; if, like a thousand others, he were seduced by the errors of his age; while I blamed his conduct, I should pity his person and should await in silence the time when some fortunate change would acquire him the esteem of virtuous people. But Valmont is not that kind of man; his conduct is the result of his principles. He calculated all that a man may permit himself in wickedness without compromising himself; and, in order to be cruel and wicked without danger, he chose women for his victims. I will not stop to count those he has seduced; but how many has he not ruined?

In the quiet, retired life which you lead, such scandalous adventures do not reach you. I could tell you some which would make you shudder; but your gaze, as pure as your soul, would be sullied by such pictures; with the certainty that Valmont will never be dangerous to you, you do not need such weapons for your defence. The only thing I have to tell you is that among all the women to whom he has paid attentions, whether successful or no, there is not one but has had reason to complain of him. The Marquise de Merteuil is the single exception to this general rule; she alone was able to resist him and to restrain his wickedness. I must confess that this episode in her life is that which sets her highest in my estimation; and it suffices in the world's eyes as a full justification for certain imprudences she was reproached with when she was first a widow.[1]

At all events, my dear, I am authorised by my age, my experience and above all by my friendship, to point out to you that people in society are beginning to notice Valmont's absence; if it is known that he has spent some time alone with you and his aunt your reputation will be in his hands—the greatest misfortune which can happen to a woman. I advise you to persuade his aunt not to detain him any longer; and if he persists in remaining I think you ought not to hesitate to leave. Why should he remain? What is he doing in that part of the country? If you watch his movements I am sure you will find out that he has simply chosen a convenient place for some evil purpose he is meditating in the neighbourhood. But, since it is impossible to remedy the evil, let us rest satisfied with preserving ourselves from it.

Good-bye, my dear; my daughter's marriage is postponed for a little while. The Comte de Gercourt, whom we were expecting every day, writes me that his regiment is ordered for Corsica; and since the war still drags on, it will be impossible for him to get away before the winter. It is vexatious; but still it lets me hope that we shall have the pleasure of seeing you at the wedding, and I was sorry it should take place without you. Good-bye: without compliment and without reserve I am entirely yours.

P.S. Mention me to Madame de Rosemonde, whom I love as much as she deserves.

From . . . , 11th of August, 17—.

LETTER X

The Marquise de Merteuil to the Vicomte de Valmont

Are you sulking with me, Vicomte? Or are you dead? Or, which would be much the same thing, do you live only for your Madame de Tourvel? That woman, who has given you back *the illusions of youth*, will soon give you back its ridiculous prejudices. You are already timid and slavish; you might as well be in love. You have given up your fortunate audacities. You are acting without principles, leaving everything to chance, or rather to caprice. Have you forgotten that love, like medicine, is *simply* the art of aiding nature? You see I am beating you with your own weapons; but I feel no pride in it; it is beating a fallen man. *She must give herself,* you say; doubtless she must! And she will give herself like the others, with this difference: It will be with a bad grace. But in order that she may end up by giving herself, the best means is to begin by taking her. This ridiculous distinction is the merest raving in love! I say "love" because you are in love. To speak to you otherwise would be deceiving you; it would be hiding your disease from yourself. But tell me, O languishing lover, do you suppose you raped the other women you have had? But, however much a woman wants to give herself, however much of a hurry she may be in, some sort of pretext is necessary; and what could be more convenient for us than a pretext which makes us appear to yield to force? For my own part, I must confess that one of the things which most flatters me is a sharp and well conducted attack, where everything is carried out with order but with rapidity; which never puts us to the painful embarrassment of having ourselves to repair an awkwardness by which we ought to have profited; which preserves an air of violence even in those things we grant, and cunningly flatters our two favourite passions—the glory of defence and the pleasure of defeat. I confess that this talent, which is much rarer than is generally believed, has always given me pleasure even when it has not attracted me; sometimes I have yielded myself simply as a recompense. As in our ancient tournaments, Beauty awarded the prize of valour and skill.

But as to you, you are no longer yourself, and you act as if you were afraid of succeeding. How long is it since you began to travel by short stages and side-tracks? My friend, when you want to get somewhere—post-horses and the mainroad! But let us leave this subject, which gives me the more annoyance since it deprives me of the pleasure of seeing you. At least you might write to me more often than you do and give me news of your progress. Do you know that for more than a fortnight you have been occupied by this ridiculous adventure and have neglected everyone?

A propos negligence; you are like those people who send regularly for news of their sick friends and never ask what the reply was. You ended your last letter by asking me if the Chevalier is dead. I did not reply and you troubled no more about it. Have you forgotten that my lover is your sworn friend? Don't be uneasy, he is not dead; or if he were, it would be from excess of joy. Poor Chevalier, how tender he is! What a lover he is! How keenly he is affected! My head is in a whirl. Seriously, the perfect happiness he enjoys in being loved by me really attaches me to him.

The very day I wrote you that I was preparing to break with him, how happy I made him! I was actually thinking of the means to reduce him to despair when he was announced. Either from caprice or good sense, he never had appeared so well. However, I received him in a bad humour. He hoped to pass two hours with me before my door opened for the rest of the company. I told him that I was going out; he asked where I was going; I refused to tell him. He insisted. "Where you will not be," I replied tartly. Happily for him he was petrified by this reply; for if he had said a word there would have inevitably followed a scene which would have brought about the rupture I had planned. Astonished by his silence, I turned my eyes on him with no other purpose, I swear, than to see what countenance he was keeping. I perceived on his charming face that profound and tender sadness which you yourself admit is so difficult to resist. The same cause produced the same effect; I was conquered a second time. From that moment I only thought of how to avoid his thinking I had been disagreeable. "I am going out on business," said I, with a slightly more gentle air, "and this business partly concerns you; but don't ask me any questions. I shall dine at home; come back and you shall know all about it." He then regained the use of speech; but I did not allow him to use it. "I am in a great hurry," I continued, "Go away until this evening." He kissed my hand and left.

Immediately, to compensate him, perhaps to compensate myself, I decided to show him my little house[1] about which he knew nothing. I called my faithful Victoire, I had my usual headache; the servants were told I was in bed; and, at last when I was alone with "the faithful servant," I dressed myself as a waiting-woman while she disguised herself as a lackey. She then brought a cab to my garden gate and off we went. When we reached the Temple of Love, I chose the most seductive dishabille. It was really delicious; it is my own invention; it lets nothing be seen and yet allows everything to be guessed at. I promise you a pattern for your Madame de Tourvel—when you have rendered her worthy of wearing it.

After these preparations, while Victoire was occupied with the other details I read a chapter of the "Sopha," a letter of "Héloïse" and two tales of La Fontaine, to rehearse the different tones I desired to take. Meanwhile my Chevalier comes to my door with his usual eagerness. My door-keeper stops him and tells him I am ill—the first incident. At the same time he gives him a letter from me, but not in my handwriting, according to my prudent rule. He opens it and finds in Victoire's handwriting: "At nine o'clock precisely, on the Boulevard in front of the cafés." He goes there; and a young lackey whom he does not know (at least whom he thinks he does not know, for it was Victoire) comes and tells him to send away his carriage and follow him. This whole romantic walk over-excites his brain and that never does any harm. He arrives at last; and surprise and love positively enchant him. To give him time to recover we take a turn in the shrubbery; then I bring him back to the house. He sees a table laid for two and a bed made up; we then go into the boudoir, which has all its decorations displayed. There, half out of premeditation, half from sentiment, I threw my arms around him and fell at his knees. "To prepare you the surprise of this moment," I said, "I reproach myself for having troubled you with an appearance of ill-humour, with having veiled for an instant my heart from your gaze. Forgive these faults, I will expiate them by my love." You may imagine the effect of that sentimental discourse. The happy Chevalier raised me and my pardon was sealed on the same ottoman upon which you and I so gaily and in the same way sealed our eternal separation.

Since we had six hours to spend together and I had determined that the whole time should be equally delicious to him, I moderated his transports, and my tenderness was replaced by amiable coquetry. I do not think I ever took so much trouble to please or that I was ever so satisfied with myself. After supper I was successively youthful and rational, playful and emotional, sometimes even wanton, and I pleased myself by considering him as a sultan in the midst of a harem in which I was successively the different favourites. Indeed, although his reiterated regards were always received by the same woman, it was always by a new mistress.

At last at daybreak we had to separate; and in spite of what he said and did even to prove to me the contrary, he needed it as much as he desired it little. At the moment when we left and as a last farewell, I took the key of this happy dwelling and put it into his hands. "I only acquired it for you," said I, "it is but right that you should be its master; the sacrificing priest should have the disposal of the temple." By this manoeuvre I forestalled any reflections which might have occurred to him from my ownership of this little house—which is always a suspicious thing. I know him well enough to be sure that he will only use it for me; and if I take a fancy to go there without him, I still have another key. At all costs he wished to fix another day there; but I still like him too much to want to use him up so quickly. One should only permit excess with those one intends to leave soon. He does not know that; but, for his happiness, I know it for two.

I see that it is three o'clock in the morning and that I have written a volume when I meant only to write a line. Such is the charm of confiding friendship; it is that which makes you still the person I like best, but to tell you the truth the Chevalier gives me more pleasure.

From . . . , 12th of August, 17—.

LETTER XI

Madame de Tourvel to Madame de Volanges

Your severe letter would have terrified me, Madame, if I had not fortunately more grounds for confidence than you find for fear. This dangerous Monsieur de Valmont, who must be the terror of all women, appears to have laid down his murderous arms before entering this *Château*. Far from forming plans here he has not even brought any affections; and the quality of charm which even his enemies allow him has almost disappeared here, leaving him only good nature. Apparently, the country air has caused this miracle. I can assure you that, although he is constantly with me and even appears to take pleasure in it, he has never let slip a word resembling love, not one of those phrases which all men indulge in, without having—as he has—something to justify it. He never forces one into that reserve which every respectable woman nowadays is compelled to adopt towards the men by

whom she is surrounded. He knows how to refrair
using the gaiety he creates. Perhaps he overpraises a l
done with such delicacy that he would accustom n
to panegyric. Indeed, if I had a brother, I should wish him to be such as Monsieur de Valmont shows himself here. Many women perhaps would wish him to pay them more marked attentions; I must confess I am infinitely grateful to him for having realised that I am not to be confounded with them.

Certainly this portrait is very different from the one you made me; and yet both may be likenesses at different periods. He himself admits he has had many faults and that he has been credited with more. But I have met few men who speak of virtuous women with so much respect, I might almost say, enthusiasm. What you tell me shows that at least in this matter he is not deceitful. His conduct with regard to Madame de Merteuil proves it. He often speaks of her, and always with so much praise and the appearance of so genuine an attachment that, until I received your letter, I thought that what he called the friendship between them was really love. I plead guilty to a hasty judgment here, and I was the more to blame since he himself has often taken the trouble to justify her. I confess that I thought this artifice when it was really honest sincerity on his part. I don't know; but it seems to me that a man who is capable of so constant a friendship for so estimable a woman cannot be an abandoned libertine. I cannot say if we owe his good conduct here to certain intrigues in the neighbourhood, as you suppose. There are indeed several charming women round about; but he seldom goes out, except in the morning, and then he says he goes shooting. It is true he rarely brings back any game; but he assures me he is unskilful in that exercise. In any case, what he does out of doors does not concern me much; and if I did wish to know, it would only be to have one more reason for sharing your view or bringing you to mine.

As to your proposal that I should take measures to shorten

Monsieur de Valmont's stay here, it appears to me very difficult to presume to ask his aunt not to have her nephew in her house, the more so since she is very attached to him. Yet I promise you, merely from deference and not from necessity, to take the first opportunity of making this request either to her or to him. As to myself, Monsieur de Tourvel knows I have arranged to stay here until he returns and he will rightly enough be surprised at my inconsistancy if I change my plans.

This is a very long explanation, Madame, but I thought I owed it to truth to give a favourable account of Monsieur de Valmont, which he appears to stand in great need of with you. I am none the less sensible of the friendship which dictated your advice. I owe to it also the obliging things you say to me about the postponement of your daughter's marriage. I thank you sincerely; but, however great my pleasure would be in passing that time with you, I would gladly sacrifice it to my desire to see Mademoiselle de Volanges sooner happy, if she can ever be happier than at the side of a mother so worthy of all her tenderness and respect. I share with her these two sentiments which attach me to you and I beg you to receive kindly this assurance of them.

I have the honour to be, etc.

From . . . , 13th of August, 17—.

LETTER XII

Cécile Volanges to the Marquise de Merteuil

Mamma is indisposed, Madame; she will not go out and I must keep her company; so I shall not have the honour of going to the Opera with you. I assure you that I far more regret not being with you than missing the performance. I beg you will believe me. I like you so much! Would you be so kind as to tell Monsieur le Chevalier Danceny that I do not possess the album of which he spoke and that it would give me great pleasure if he would bring it to me to-morrow? If he comes to-day he will be told we are not at home; but that is because Mamma does not wish to receive anybody. I hope she will be better to-morrow.

I have the honour to be, etc.

From . . . , 13th of August, 17—.

LETTER XIII

The Marquise de Merteuil to Cécile Volanges

I am very sorry, my dear, to be deprived of the pleasure of seeing you, and for the cause of this deprivation. I hope the opportunity will occur again. I will deliver your message to the Chevalier Danceny, who will certainly be very sorry to know that your Mamma is unwell. If she will receive me to-morrow, I will come and keep her company. She and I will attack the Chevalier de Belleroche[1] at piquet; and while we win his money we shall have as an extra pleasure that of hearing you sing with your charming master, to whom I shall propose it. If this suits your Mamma and you, I will answer for myself and my two Chevaliers. Good-bye, my dear; my regards to dear Madame de Volanges. I kiss you tenderly.

From . . . , 13th of August, 17—.

LETTER XIV

Cécile Volanges to Sophie Carnay

I did not write to you yesterday, my dear Sophie, but it was not on account of my pleasures; I assure you. Mamma was ill and I did not leave her all day. At night when I left her I had no heart for anything; and I went to bed as quickly as I could, to be certain the day was over; I have never spent such a long one. It is not that I don't love Mamma—I don't know what it was. I was to go to the Opera with Madame de Merteuil; the Chevalier Danceny was to have been there. You know they are the two persons I like best. When the time came that I should have been there too, my heart shrank in spite of myself. I felt disgusted with everything and I cried and cried without being able to stop myself. Fortunately, Mamma was in bed and could not see me. I am sure the Chevalier Danceny was sorry too; but he had the distraction of the performance and everybody who was there—it was a very different thing.

Happily Mamma is better to-day, and Madame de Merteuil is coming with someone else and the Chevalier Danceny; but Madame de Merteuil always comes very late; and it is very tiresome to be all alone so long. It is only eleven o'clock. It is true that I must play on my harp; and then it will take me a little time to dress, for I want to have my hair well done to-day. I think Mère Perpetue was right and that we become coquetish as soon as we are in the world. I have never wanted so much to be pretty as in the last few days and it seems to me I am not so pretty as I thought; and then one is at a disadvantage beside women who are rouged. Madame de Merteuil, for example—I can see that all the men think she is prettier than I am; that does not annoy me very much because she likes me and then she assures me that the Chevalier Danceny thinks I am prettier than she is. It was very kind of her to have told me! She even seemed to be pleased by it. I can't understand how she could be. It is because she really loves me! And he! . . . Oh! How it pleased me! It seems to me that just to look at him would be enough to make one handsome. I could look at him forever if I were not afraid of meeting his eyes; for every time that happens it puts me out of countenance and hurts me, as it were; but it doesn't matter.

Good-bye, my dear; I must begin to dress. I love you as ever.

Paris, the 14th of August, 17—.

LETTER XV

The Vicomte de Valmont to the Marquise de Merteuil

It is very kind of you not to abandon me to my sad fate. The life I lead here is truly fatiguing, with the excess of its repose and its insipid uniformity. When I read you letter and the details of your charming day, I was twenty times tempted to invent an excuse, fly to your feet and ask of you in my favour an infidelity to your Chevalier, who after all does not deserve his happiness. Do you know you have made me jealous of him? What do you mean by speaking of our "eternal separation?" I deny the oath, spoken in delirium; we should not have been worthy to make it if we were to keep it. Ah! May I one day avenge myself in your arms for the involuntary vexation caused me by the Chevalier's happiness! I confess I am indignant when I think that this man, without thought, without giving himself the least trouble, by stupidly following the instinct of his heart, should find a felicity to which I cannot attain. Oh! I shall disturb it . . . Promise me I shall disturb it. Are you not humiliated yourself? You take the trouble to deceive him and he is happier than you are. You think he is in your chains! It is you who are in his. He sleeps calmly while you are awake for his pleasures. What more could his slave do?

Come, my fair friend, as long as you share yourself between several, I am not in the least jealous; I simply see your lovers as the successors of Alexander, incapable of holding among them all that Empire where I reigned alone. But that you should give yourself entirely to one of them! That there should exist another man as happy as I! I will not endure it; do not think that I will endure it. Either take me back or at least take someone else as well; and do not let an exclusive caprice betray the inviolable friendship we have sworn each other.

No doubt it is quite enough that I should have to complain of love. You see that I agree with your ideas and admit my faults.

Indeed, if to be in love is not to be able to live without possessing that person one desires, to sacrifice to her one's time, one's pleasures, one's life; then I am really in love. I am no nearer success; I should not have anything at all to tell you about the matter, but for an occurrence which makes me reflect a good deal and from which I do not know yet whether I should fear or hope.

You know my man, a treasure for intrigue and a real comedy-valet; as you may suppose, my instructions included his falling in love with the waiting-woman and making the serving men drunk. The rascal is luckier than I am; he has succeeded already. He has just discovered that Madame de Tourvel has ordered one of her serving-men to obtain information about my conduct and even to follow my morning walks as much as he can without being perceived. What does the woman mean? And so the most modest of them all yet dares to risk things we should scarcely dare allow ourselves! I swear . . . But, before I think of avenging myself for this feminine ruse, let me think how to turn it to my own advantage. Hitherto these suspected walks have had no purpose; I must give them one. It deserves all my attention, and I must leave you and reflect on it. Good-bye, my dear friend

Still from the Château de . . . , 15th of August, 17—.

LETTER XVI

Cécile Volanges to Sophie Carnay

Ah! my Sophie, here is news! Perhaps I ought not to tell you; but I must speak to someone; it is stronger than I am. The Chevalier Danceny . . . I am so upset that I cannot write; I don't know where to begin. After I had told you about the pleasant evening[1] I spent here with him and Madame de Merteuil, I did not speak of him to you again; it was because I did not wish to speak of him to anyone; but I kept thinking about him. After that evening he became sad, so sad, so very sad, that it hurt me; and when I asked him why he said he was not sad, but I could see he was. Yesterday he was worse than ever. It did not prevent his having the kindness to sing with me as usual; but every time he looked at me, my heart was wrung. After we had finished singing he went to put my harp in its case; and when he brought me back the key he begged me play on it again this evening as soon as I was alone. I did not suspect anything; I did not even want to play but he begged me so hard that I said yes. He had his reason. When I went to my own room and my maid had left me I went and got my harp. In the strings I found a letter, folded only and not sealed, from him. Ah! if you knew all that he wrote to me! Since I have read his letter I have been so delighted that I could not think of anything else. I read it over again four times in succession and then locked it in my writing-desk. I knew it by heart; and when I was in bed I said it over so often that I did not think of going to sleep. As soon as I closed my eyes, I saw him there saying to me himself everything I had just read. I did not go to sleep until very late and as soon as I woke up (it was still very early) I went and got his letter to read it over again. I took it into bed and kissed it as if . . . Perhaps it is wrong to kiss a letter like that but I could not help it.

And now, my dear, although I am very glad, I am also in a great difficulty; for surely I ought not to reply to this letter. I know I ought not and yet he asks me to; and if I don't answer I am sure he will go on being sad. It is very unlucky for him! What do you advise me to do? But you know no more than I do. I am very much tempted to speak about it to Madame de Merteuil who is very fond of me. I should like to console him but I do not want to do anything wrong. We are always being told to be kind-hearted! And then we are forbidden to carry out what a kind heart inspires in us when it is for a man! It is unjust. Is not a man our neighbour as much as a woman, and even more? For, after all, have we not a father as well as a mother, a brother as well as a sister? And then there is always the husband, in addition. And yet if I did something which was not right perhaps Monsieur Danceny himself would not think well of me! Oh! I would rather he were still sad; and then, after all, I have plenty of time. Because he wrote yesterday, I am not obliged to write to-day; and then I shall see Madame de Merteuil this evening and if I can pluck up courage I shall tell her all about it. If I do what she tells me I shall have nothing to reproach myself with. And then perhaps she will tell me that I may reply to him just a little, so that he will not be so sad! Oh! I am very unhappy.

Good-bye, my dear. Tell me what you think.

From . . . , 19th of August, 17—.

LETTER XVII

The Chevalier Danceny to Cécile Volanges

Mademoiselle, before I yield to the pleasure (shall I say?) or to the necessity of writing to you, I begin by begging you to listen to me. I feel I need your indulgence to dare the declaration of my sentiments; if I wished merely to justify them, indulgence would not be necessary. For, after all, what am I about to do but to show you your own work? And what have I to say to you which has not been said before by my looks, my embarrassment, my behaviour and even my silence? And why should you be angry at a sentiment you alone caused to exist? It arose from you and doubtless is worthy of being offered to you; though it be burning as my soul, it is as pure as yours. Can it be a crime to have appreciated your charming features, your seductive talents, your enchanting graces and that touching candour which adds an inestimable value to qualities of themselves so precious? No, without doubt; but one may be unhappy without being guilty, and unhappiness will be my fate if you refuse to accept my regard. It is the first my heart has offered. But for you I should still be—not happy—but at peace. I saw you; repose fled far from me and my happiness is uncertain. Yet you are surprised by my sadness; you ask me the reason; sometimes I have even thought that it grieved you. Ah! Say but one word and my felicity will be your handiwork! But, before you speak, remember that a word can also overwhelm me with misery. Be the arbitress of my destiny. Because of you I shall be eternally happy or unhappy. To what dearer hands could I confide a greater trust?

I shall end, as I began, by imploring your indulgence. I asked you to listen to me; I will dare more, I will beg you to answer me. To refuse, would be to let me think that you feel yourself offended and my heart reassures me that my respect is equal to my love.

P.S. To answer, you can use the same means by which I conveyed this letter to you; it seems to me both certain and convenient.

From . . . , 19th of August, 17—.

LETTER XVIII

Cécile Volanges to Sophie Carnay

What, Sophie! You condemn beforehand what I am going to do! I had already enough anxieties, and now you add to them. It is clear, you say, that I ought not to reply. It is very easy for you to talk; and besides, you don't understand the whole situation; you are not here to see. I am sure if you were in my place, you would do what I am doing. Of course, as a general rule, one ought not to reply; and you must have seen from my yesterday's letter that I did not want to; but I think nobody was ever in the position I am in now.

And then to be forced to decide all by myself! Madame de Merteuil, whom I counted on seeing yesterday evening, did not come. Everything combines against me; it was she who was the cause of my knowing him. It is almost always with her that I have seen and spoken to him. Not that I bear her any malice for it; but she leaves me at the very moment of difficulty. Oh! I am very much to be pitied!

You must know that he came yesterday as usual. I was so upset that I dared not look at him. He could not speak to me because Mamma was there. I was very much afraid he would be distressed when he saw I had not written to him. I did not know which way to look. A moment afterwards he asked me if I wanted him to get my harp. My heart beat so hard that it was all I could do to answer yes. When he returned, it was much worse. I only glanced at him for a moment. And he did not look at me; but he really appeared as if he were ill. It really hurt me to see him. He began to tune my harp and then, when he brought it to me, he said: "Ah! Mademoiselle! . . ." He only said those two words, but it was in a tone which completely distracted me. I played a prelude on my harp without knowing what I was doing. Mamma asked if we were not going to sing. He made an excuse and said he was a little unwell; and I, who had no excuse, was forced to sing. I wished I had never had a voice. I expressly chose an air I did not know; for I was quite sure I could not sing any one and I should have aroused suspicion. Fortunately, there came a visitor; and, as soon as I heard the carriage arrive, I stopped and asked him to take back my harp. I was very much afraid he would leave then and there; but he came back.

While Mamma and the lady who had just come were talking together, I wanted to glance at him again only for a moment. A second later I saw his tears falling, and he was obliged to turn aside in order not to be seen. This time I could endure it no longer; I felt I was going to cry too. I went out and immediately wrote in pencil on a scrap of paper: "Do not be so sad, I beg you; I promise to reply." Surely, you cannot say there was anything wrong in that; and then it was stronger than I am. I put the piece of paper in my harpstrings, as his letter was, and returned to the drawing-room. I felt calmer. I was very anxious for the lady to go away. Fortunately, she was only making a call and left soon afterwards. As soon as she had gone, I said I wanted my harp again and I asked him to go and get it. I could see, from his expression, that he suspected nothing. But when he came back, oh! how happy he was! As he put my harp in front of me, he placed himself in such a way that Mamma could not see, and he took my hand and pressed it . . . but in such a way! . . . it was only a moment, but I cannot tell you what pleasure it gave me. However, I withdrew it; so I have nothing to reproach myself with.

And so, my dear, you see I cannot help writing to him, since I promised him; and then, I can't bear to make him unhappy again; for I suffer from it more than he does. If it were for something wrong, of course I should not do it. But what wrong can there be in writing, especially if it is to prevent someone from being unhappy? What troubles me is that I cannot turn my letter well; but he will feel that it it not my fault; and then I am sure that the mere fact it is from me will please him.

Good-bye, my dear. If you think I am wrong, tell me; but I do not believe I am. As the moment approaches when I am to write to him, my heart beats inconceivably. And yet I must do it, since I promised. Good-bye.

From . . . , 20th of August, 17—.

LETTER XIX

Cécile Volanges to the Chevalier Danceny

You were so sad yesterday, Monsieur, and I was so grieved by it, that I allowed myself to promise you I would reply to the letter you wrote me. To-day I feel just as much that I ought not to; but, since I promised, I do not want to break my word, and that ought to prove to you the friendship I have for you. Now that you know it, I hope you will not ask me to write to you again.

I also hope you will not tell anyone I have written to you; because I should certainly be blamed for it and it might cause me much distress. Above all, I hope you will not think badly of me for doing it; for that would distress me more than anything. I can assure you that I should not have done this favour for anyone but you. I wish you would do me the favour of not being sad as you have been; it takes away all the pleasure I have in seeing you. You see, Monsieur, I speak to you sincerely. I ask nothing better than that our friendship should last forever; but, I beg you, do not write to me again.

I have the honour to be,

Cécile Volanges.

From . . . , 20th of August, 17—.

LETTER XX

The Marquise de Merteuil to the Vicomte de Valmont

Ah! You rogue, so you are coaxing me for fear I shall mock you? Come, I must pardon you; you write me so many follies that I am forced to forgive you the sobriety in which you are kept by your Madame de Tourvel. I do not think my Chevalier would be as indulgent as I am; he is the kind of man who might not approve the renewal of our lease and who would find nothing amusing in your wild idea. Yet I laughed at it and was really sorry I was forced to laugh at it alone. If you had been there I do not know how far this gaiety might have led me; but I have had time to reflect and I have armed myself with severity. I do not refuse for always; but I delay, and rightly. Perhaps I should bring vanity into it and, once I became interested in the game, there is no knowing where it might stop. I might enchain you again and make you forget your Madame de Tourvel; and suppose I, the unworthy, should disgust you with virtue—what a scandal! To avoid this danger, these are my conditions.

As soon as you have had your fair devotee, and can furnish me with a proof, come, and I am yours. But you know that in important matters only written proofs are accepted. By this arrangement, on the one hand I shall become a recompense instead of being a consolation—an idea which pleases me better—and on the other hand, your success will be the more piquant by becoming itself a means of infidelity. Come, come as soon as you can and bring me the proof of your triumph—like our noble knights of old who laid at their ladies' feet the brilliant fruits of their victory. Seriously, I am curious to know what a prude would write after such a moment and what veil she will throw over her words after having left none upon her person. It is for you to see whether I rate myself at too high a price; but I warn you it will not be reduced. Until then, my dear Vicomte, you must allow me to remain faithful to my Chevalier and to amuse myself by making him happy, in spite of the little distress it causes you.

And yet, if I were less moral, I believe he would have at this moment a dangerous rival—little Cécile Volanges. I am passionately fond of the child; it is a real passion. If I am not deceived she will become one of our most fashionable women. I can see her little heart developing, and it is a delightful sight. She is already madly in love with her Danceny, but does not yet know it herself. He himself, though very much in love, still has the timidity of his age and dares not declare himself too plainly. Both adore me. The little girl especially is very anxious to tell me her secret; particularly in the last few days I have noticed she is really oppressed by it and I should have done her a great service by helping her a little; but I do not forget that she is only a child and I do not want to compromise myself. Danceny has spoken to me a little more plainly; but I have made up my mind not to listen to him. As for the girl, I am often tempted to make her my pupil; it is a service I should like to render Gercourt. He leaves me plenty of time since he is in Corsica until October. It occurs to me that I might make use of this time and that we will give him a fully formed woman instead of his innocent convent-girl. What insolent confidence on the part of this man, to dare to sleep tranquilly while a woman who has reason to complain of him has not yet avenged herself! Ah! if the girl were here now I do not know what I might not say to her.

Good-bye, Vicomte; good-night and good luck; but, for Heaven's sake, make some progress. Remember, if you do not have this woman, that others will blush at having had you.

From . . . , 21st of August, 17—. 4 o'clock in the morning.

LETTER XXI

The Vicomte de Valmont to the Marquise de Merteuil

At last, my fair friend, I have made a step forward, a large step, which, although it has not led me to the goal, has allowed me to see that I am on the road and has banished my fear that I had missed the way. I have at last declared my love; and although she kept the most obstinate silence I obtained perhaps the least equivocal and most flattering reply possible; let us not anticipate, but go back to the beginning.

You remember that my walks were being watched. Well! I wanted to turn this scandalous circumstance to public edification, and this is what I did. I ordered my man to find in the neighbourhood some unfortunate person in need of help. It was not a difficult task to accomplish. Yesterday afternoon he informed me that this morning there would be a destraint on the furniture of a whole family which could not pay its taxes. I made certain that the family contained no girl or woman whose age or features might render my action suspicious; and, when I was certain of this, I announced at dinner my intention of going shooting the next day. Here I must render justice to my Madame de Tourvel; no doubt she regretted the order she had given and, not having the strength to conquer her curiosity, she had at least enough to oppose my desire. It would be excessively hot; I ran the risk of getting ill; I should kill nothing and tire myself in vain; and during this dialogue, her eyes, which perhaps spoke more than she wished, showed me that she desired I should take her bad reasons as good ones. I was careful not to yield to them, as you may suppose, and in the same way I resisted a little diatribe against shooting and sportsmen and also a small cloud of ill-humour which darkened that heavenly face the whole evening. For a moment I feared she might countermand her orders and that her delicacy might harm me. I reckoned without the curiosity of a woman—and so I was wrong. My man reassured me the same evening and I went to bed satisfied.

At dawn I rose and set out. I was scarcely fifty yards from the Château when I saw my spy following me. I set out across the field as though I were shooting, in the direction of the village I wished to reach; my only pleasure on the way was to set the pace to the fellow who was following me and since he dared not leave the roads he often had to cover at top speed triple the distance I went. I exercised him so much that I became very hot myself and sat down at the foot of a tree. And did he not have the insolence to slip behind a bush about twenty yards from me and sit down too? For a moment I was tempted to take a shot at him, for although my gun was only loaded with pellets it would have given him enough of a lesson on the dangers of curiosity;[1] luckily for him I recollected that he was useful to me and positively necessary to my plans and this reflection saved him.

When I reached the village I saw there was a disturbance; I went forward; I asked questions; I was told what was happening; I called the Tax-Collector; and, yielding to my generous compassion, I nobly paid fifty-six *livres*, for which five persons were to be reduced to straw and despair. After so simple an action you cannot imagine what a chorus of benedictions echoed round me from the spectators! What tears of gratitude flowed from the eyes of the aged head of the family and embellished this patriarchal face which a moment before was rendered truly hideous by the wild imprint of despair! I was watching this spectacle when another younger peasant, leading a woman and two children by the hand, rushed towards me, saying to them: "Let us all fall at the feet of this Image of God"; and at the same moment I was surrounded by the family prostrate at my knees. I must admit my weakness; my eyes filled with tears and I felt an involuntary but delicious emotion. I was astonished at the pleasure there is in doing good; and I should be tempted to believe that those whom we call virtuous people are not so virtuous as they are pleased to tell us. However that may be, I thought it only just to pay these poor people for the pleasure they had just given me. I had brought ten *louis* with me; which I gave to them. Their thanks recommenced but without the same degree of pathos; necessity had produced the great, the true effect; the rest was only a simple expression of gratitude and astonishment for superfluous gifts.

However, in the midst of the wordy benedictions of this family I was not unlike the hero of a play in the last scene. You are to know that my faithful spy was in the forefront of the crowd. My object was fulfilled; I got free from them and returned to the Château. Taking it all round, I congratulate myself on my invention. This woman is certainly worth all the trouble I have taken; one day it will form my claim upon her; and having, as it were, thus paid for her beforehand I shall have the right to dispose of her as I fancy, without reproaching myself.

I forgot to tell you that, to turn everything to my profit, I asked these good people to pray to God for the success of my plans. You will see that their prayers have already been partly recompensed . . . But I am told that dinner is served and it will be too late for this letter to go if I finish it after dinner. And so—the rest by the next post. I am sorry, for the rest is the finest part. Good-bye, my fair friend. You steal from me a moment of the pleasure of seeing her.

From . . . , 20th of August, 17—.

LETTER XXII

Madame de Tourvel to Madame de Volanges

No doubt you will be glad, Madame, to hear of an act of Monsieur de Valmont which, it seems to me, is in strong contrast to all the traits by which he has been represented to you. It is so painful to think disadvantageously of anyone, so grieving to find nothing but vices in those who should have all the qualities needed to make virtue beloved! And then you are so glad to be indulgent, that it is doing you a favour to furnish you with motives for reversing too harsh a judgment. It seems to me that Monsieur de Valmont has a right to hope for this favour. I might almost say this justice; and this is why I think so.

This morning he went on one of those walks which might lead one to suspect some intrigue on his part in the neighbourhood, as you thought; a thought which I blame myself for having accepted too quickly. Fortunately for him, and above all fortunately for us, since it preserves us from being unjust, one of my men-servants had to go the same way as he did[1]; and in this way my reprehensible but fortunate curiosity was satisfied. He related to us that Monsieur de Valmont found in the village of . . . an unhappy family whose goods were being sold because they could not pay their taxes; Monsieur de Valmont not only hastened to discharge the debt of these poor people but even gave them a considerable sum of money. My servant was a witness of this virtuous action; and moreover he tells me that the peasants, talking among themselves and with him, said that a servant, whom they described, and whom my servant thinks is Monsieur de Valmont's, yesterday made inquiries about such inhabitants of the village as might be in need of help. If this is true, it is not merely a passing compassion provoked by accident; it is a pre-conceived plan of doing good; it is the solicitude of charity; it is the fairest virtue of the fairest souls; but whether it was chance or method, it is still a good and praiseworthy action, the mere relation of which moved me to tears. In addition I must say, still from a sense of justice, that when I spoke to him about this action, of which he did not say a word, he began by denying it, and, when he did admit it, seemed to set so little value upon it that his modesty doubled its merit.

And now, my respectable friend, tell me if Monsieur de Valmont is indeed an abandoned libertine? If this is all he is and if he acts in this way, what is there left for the virtuous? What! Should the wicked share with the good the sacred pleasure of charity?[2] Would God permit a virtuous family to receive from a wicked man's hand that aid for which it returned thanks to His divine providence? And could He be pleased to hear pure mouths shed their benedictions upon a reprobate? No. I prefer to believe that his faults, though long continued, are not eternal; and I cannot think that he who does good is the enemy of virtue. Monsieur de Valmont is perhaps only one more example of the danger of acquaintances. I dwell on this idea, which pleases me. If, on the one hand, it may serve to justify him in your mind, on the other, it renders more and more precious to me that tender friendship which unites me to you for life.

I have the honour to be, etc.

P.S. Madame de Rosemonde and I are just going to see this honest and unfortunate family and to add our tardy succour to that of Monsieur de Valmont. We are taking him with us. At least we shall give these good people the pleasure of seeing their benefactor again; I am afraid it is all he has left us to do.

From . . . , 21st of August, 17—.

LETTER XXIII

The Vicomte de Valmont to the Marquise de Merteuil

We broke off at my return to the *Château*; I continue my story.

I had only time for a brief toilet; I went into the drawing-room where my fair one was embroidering, while the local Curé was reading the Gazette to my old aunt. I went and sat down beside the frame. Glances, softer than usual, and almost caressing, soon allowed me to guess that the servant had already given an account of his mission. Indeed my amiable Inquisitive was unable to keep for long the secret she had stolen from me; and she was not afraid to interrupt a venerable pastor whose utterance closely resembled that of a sermon, by saying: "And I too, have news to tell you"; and she immediately related my adventure with a precision which did honour to the intelligence of her historian. You may imagine how I displayed all my modesty; but who could stop a woman when she is praising what she loves, although she does not yet know it? I decided to let her proceed. It was as if she were preaching the panegyric of a saint. During this time I observed, not without hope, all that was promised to love by her animated looks, her freer gestures, and above all by the tone of her voice which by its perceptible alteration betrayed the emotion of her soul. She had scarcely finished speaking when Madame de Rosemonde said: "Come, nephew come, let me embrace you." I felt immediately that the pretty preacher could not prevent herself from being embraced in turn. However, she tried to escape; but she was soon in my arms; and far from having the strength to resist she had scarcely enough to stand upright. The more I observe this woman the more desirable she seems to me. She hastened back to her frame and seemed to the others to recommence her embroidery; but I could easily perceive that her trembling hand would not permit her to continue her work.

After dinner the ladies desired to visit the unfortunate persons I had so piously succoured. I accompanied them. I spare you the boredom of this second scene of gratitude and praise. My heart, urged by a delicious memory, hastens the moment of the return to the *Château*. On the way, my fair lady, more preoccupied than usual, did not say a word. I was busy trying to find means to profit from the effect produced by the event of the day and kept equally silent. Madame de Rosemonde alone spoke and received only short and rare answers from us. We must have bored her; that was my plan and it succeeded. So, when we got out of the carriage, she went to her room and left my fair lady and me alone together in a dimly lighted drawing-room; soft obscurity, which emboldens timid love!

I did not have the trouble of turning the conversation in the direction I wished it to go. The amiable lady's fervour served me better than my skill could have done. "When a person is so worthy of doing good," said she, fixing her gentle gaze on me, "how can he pass his life in doing ill?" "I do not deserve either this praise or this censure," I replied, "and I cannot understand that with your intelligence you have not understood me. Even though my confidence should do me harm with you, you are too worthy of it for me to be able to refuse it to you. You will find the key to my conduct in a character which is unfortunately too compliant. Surrounded by the immoral, I have imitated their vices; I may even have gratified my vanity by surpassing them. Seduced in the same way here by the example of virtues, without hoping to attain to your level, I have at least tried to follow you. Ah! Perhaps the action for which you praise me to-day would lose all its value in your eyes, if you knew its real motive!" (You see, my fair friend, how near the truth I was). "It is not to me," I continued, "that these unfortunates owe my succour. Where you think you see praiseworthy action, I only sought a means of pleasing. I was only, (since I must say it), the weak agent of the divinity I adore." (Here she tried to interrupt me; but I did not give her time). "At this very moment," I added, "my secret only escaped me through my weakness. I had promised myself to withhold it from you; I held it a happiness to render to your virtues as to your charms a pure regard of which you would forever remain ignorant; but, incapable of deceiving, when I have the example of candour beneath my eyes,

I shall not have to reproach myself with culpable dissimulation towards you. Do not suppose that I insult you by a criminal hope. I shall be unhappy, I know it; but my sufferings will be dear to me; they will prove to me the excess of my love; it is at your feet, and in your bosom that I shall lay my troubles. There I shall gather strength to suffer anew; there I shall find compassionate kindness, and I shall think myself consoled because you have pitied me. O you whom I adore, hear me, pity me, help me!" I was now at her knees and clasped her hands in mine; but she suddenly snatched them away and pressed them over her eyes with an expression of despair. "Ah! unhappy woman!" she exclaimed and then burst into tears. Fortunately I had worked myself up to such an extent that I was weeping too; and taking her hands again, I bathed them in tears. This was a very necessary precaution; for she was so occupied by her own grief that she would not have perceived mine, had I not thought of this method of advertising her. In addition I had the advantage of gazing at leisure upon that charming face, still more embellished by the powerful attraction of tears. I grew warm, and was so little master of myself that I was tempted to profit by this moment.

How great is our weakness; how great is the Power of circumstances, if I myself, forgetting my plans, risked losing by a premature triumph the charm of long struggles and the details of a painful defeat; if, seduced by desires worthy only of a young man, I almost exposed the conqueror of Madame de Tourvel to the fate of gathering as the fruit of his labours nothing but the insipid advantage of having had one woman more! Ah! Let her yield herself, but let her struggle! Let her have the strength to resist without having enough to conquer; let her fully taste the feeling of her weakness and be forced to admit her defeat. Let the obscure poacher kill the deer he has surprised from a hiding place; the real sportsman must hunt it down. A sublime plan, is it not? But perhaps I should now be regretting I had not followed it, if chance had not come to the aid of my prudence.

We heard a noise. Somebody was coming into the drawing-room. Madame de Tourvel rose precipitately in terror, seized one of the candlesticks and left the room. I had to let her go. It was only a servant. As soon as I was certain of this, I followed her. Scarcely had I taken a few steps when either because she recognised me or from a vague feeling of fear, I heard her increase her pace and throw herself rather than enter into her apartment whose door she closed behind her. I went up to it; but the key was in the inside. I was careful not to knock; for that would have given her the opportunity for too facile a resistance. I had the happy and simple idea of attempting to look through the key-hole and I saw this adorable woman on her knees, bathed in tears and praying fervently. What God did she dare to invoke? Is there any sufficiently powerful against love? In vain does she now seek outside aid; it is I who control her fate.

Thinking I had done enough for one day, I also retired into my apartment and began to write to you. I hoped to see her again at supper; but she sent a message that she felt unwell and had gone to bed. Madame de Rosemonde wished to go up to her bed-room; but the cunning invalid pretended a headache which did not allow her to see anybody. You may imagine that I did not sit up late after supper and that I too had my headache. Having retired to my room, I wrote a long letter to complain of this harshness and I went to bed with the idea of delivering it this morning. I slept badly, as you may perceive by the date of this letter. I got up and re-read my epistle. I saw that I had not watched myself enough; that I showed more ardour than love in it and more ill-humour than sadness. I ought to re-write it; but then, I should have to be calmer to do so.

I see it is dawn, and I hope its accompanying freshness will bring me sleep. I am going back to bed; and, whatever may be this woman's domination, I promise you not to be so absorbed with her that I have not plenty of time left to think of you. Good-bye, my fair friend.

From . . . , 21st of August, 17—. 4 o'clock in the morning.

LETTER XXIV

The Vicomte de Valmont to Madame de Tourvel

Ah! For pity's sake, Madame, deign to calm the distemper of my soul; deign to inform me what I am to hope or fear. Placed between an excess of happiness or of misfortune, uncertainty is a cruel torture. Why did I speak to you? Why could I not resist the dominating charm which delivered up my thoughts to you? Content to adore you in silence, I at least enjoyed my love; that pure sentiment, which was not then troubled by the image of your grief, sufficed for my felicity; but that source of happiness has become a source of despair since I saw your tears flow, since I heard that cruel "*Ah! unhappy woman!*" Those words will long continue to re-echo in my heart. Through what fatality is it that the softest of sentiments is only able to inspire you with fear? But what is this apprehension? Ah! not that of sharing it; I have misunderstood your heart, it is not made for love; mine, which you constantly depreciate, is alone sensitive, yours does not even feel pity. If it were not so, you would not have refused a word of consolation to a wretch who confessed to you his sufferings; you would not have avoided his gaze when he has no pleasure but to look at you; you would not have played cruelly with his uneasiness by announcing that you were ill, without allowing him to enquire after your state of health; you would have felt that this very night, which for you was merely twelve hours of repose, would be a century of pain for him.

Tell me, how have I deserved this cruel harshness? I am not afraid to refer to you as judge; what have I done, but yield to an involuntary sentiment; inspired by beauty and justified by virtue, always restrained by respect, the innocent admission of which was the result of confidence and not of hope; will you betray the confidence which you yourself seem to allow me and to which I yielded myself unreservedly? No, I cannot believe it; that would be imputing a fault to you, and my heart revolts at the mere idea of finding one in you; I disavow my complaints; I may have written, but would not have thought them. Ah! let me think you perfect—it is the sole pleasure left me. Prove to me that you are so by granting me your generous aid. What wretch have you succoured who needed it as much as I? Do not abandon me in the delirium into which you have plunged me; lend me your good sense since you have stolen my own; when you have corrected me, enlighten me in order to finish your work.

I do not wish to deceive you, you will not succeed in extinguishing my love; but you will teach me to control it; by guiding my steps, by dictating my speech, you will at least save me from the terrible misfortune of displeasing you. I beg you to dissipate this despairing thought; tell me that you forgive me, that you pity me; assure me of your indulgence. You will never grant me all I desire; but I ask for what I need—will you refuse it?

Farewell, Madame; receive graciously the homage of my feelings; it does not diminish my respect.

From . . . , 20th of August, 17—.

LETTER XXV

The Vicomte de Valmont to the Marquise de Merteuil

Here is today's bulletin.

At eleven o'clock I went to Madame de Rosemonde's apartment; and, under her protection, I was taken to see the feigned invalid, who was still in bed. Her eyes were very tired; I hope she slept as badly as I did. I seized a moment, when Madame de Rosemonde had turned away to deliver my letter; she refused to take it; but I left it on the bed and very virtuously went over to my old aunt's armchair for she wanted to be near her dear child—the letter had to be concealed to avoid a scandal. The invalid said awkwardly that she thought she was a little feverish. Madame de Rosemonde called on me to feel her pulse and boasted of my medical knowledge. My fair one therefore had the double disappointment of being obliged to give me her arm, and of feeling that her little lie would be discovered. I took her hand and clasped it in one of mine, while with the other I wandered over her cool dimpled arm; the malicious creature made no response whatever which made me say as I let her go: "There is not even the least excitement." I guessed her looks must be severe and, to punish her, I did not seek them; a moment after, she said she wished to rise and we left her alone. She appeared at dinner, which was a gloomy meal; she announced she would not go walking, which was to tell me I should not have a chance of speaking to her. I felt that was the moment for a sigh and a pained glance; no doubt she expected it, for it was the only moment in the day when I succeeded in catching her eye. Modest as she is, she has her little wiles like the rest. I found an opportunity to ask her if she would have the goodness to tell me what was my fate, and I was a little astonished to hear her reply: "Yes, Monsieur I have written to you." I was very anxious to have this letter; but either from wiles again, or from clumsiness or from timidity, she only handed it to me this evening just before she went to bed. I send you her letter as well as the rough draft of my own; read them and judge; notice with what obvious falsity she asserts she is not in love, when I am sure of the contrary; and then she will complain if I am unfaithful to her afterwards, when she is not afraid to be faithless to me beforehand! My fair friend, the most skilful man can do no more than keep himself on a level with the most sincere woman; and yet I must feign to believe all this nonsense, and weary myself out with despairing, because it pleases the lady to play at cruelty!

How can one avoid being avenged on such baseness! Ah! patience . . . but good-bye, I have still a lot to write.

Apropos, you will send me back the cruel one's letter; it may be that later she will want to set a value on these trifles, and I must be prepared.

I have not said anything about the Volanges girl; we will talk about her at the first opportunity.

From the *Château*, 22nd of August, 17.

LETTER XXVI

Madame de Tourvel to the Vicomte de Valmont

Indeed, Monsieur, you would have had no letter from me if my foolish conduct of yesterday evening did not compel me to enter into explanations with you. Yes, I wept, I admit it; and perhaps those two words which you are so careful to quote to me did escape me. Tears and words, you noticed them all; and therefore all must be explained to you.

I am accustomed to inspire none but virtuous sentiments, to hear no speech save that which can be listened to without blushing; consequently to enjoy a feeling of confidence which I dare to say I deserve and therefore I do not know how to dissimulate or to combat the impressions I feel. The astonishment and embarrassment into which I was thrown by your behaviour, a certain fear inspired by a situation which ought never to have occurred to me, perhaps the revolting idea of seeing myself confounded with women you depised and treated as lightly as they—all these causes provoked my tears and may have made me say (rightly, I think) that I was unhappy. This expression, which you thought so strong, would be far too weak if my tears and words had had another motive; if instead of disapproving sentiments which must offend me I could have feared I might share them.

No, Monsieur, I do not have this fear; if I had, I should fly a hundred miles from you; I should bewail in a desert the misfortune of having known you. Perhaps, in spite of my certainty that I do not love you, that I shall never love you, perhaps I should have done better to follow the advice of my friends and not allow you to come near me.

I thought—and this is my one error—I thought you would respect a virtuous woman who asked no more than to find you so and to do you justice; who was defending you while you outraged her by your criminal desires. You do not know me; no, Monsieur, you do not know me. If you did, you would not have supposed you could make your errors into a right; because you said to me words I ought not to have heard, you would not have thought yourself authorised to write me a letter I ought not to have read—and you ask me to "guide your steps, to dictate your speech!" Monsieur, silence and oblivion, that is the advice I must give you, which you must follow; then indeed you will have a right to my indulgence and it will depend upon you to obtain even my gratitude . . . And yet, no, I will ask nothing of him who has not respected me; I will not give a mark of confidence to one who has abused my frankness.

You compel me to fear you, perhaps to hate you; I did not wish to; I wished to see in you only the nephew of my most respectable friend; I opposed the voice of friendship to the accusations of public fame. You have destroyed everything; and, I foresee, you will not wish to repair anything.

I must tell you, Monsieur, that your feelings are an offence to me, that the admission of them is an outrage, and above all that, far from my coming one day to share them, you would force me never to see you again, if you did not keep a silence in this matter such as I feel I have a right to expect and even to demand from you. I enclose with this letter the one you wrote me and I hope you will be good enough to return me this one; I should indeed be pained if there existed any trace of an event which ought never to have taken place.

I have the honour to be, etc.

From . . . , 21st of August, 17—.

LETTER XXVII

Cécile Volanges to the Marquise de Merteuil

Ah! How kind you are, Madame! How well you realised that it would be easier for me to write than to speak to you! And then, what I have to say to you is very difficult; but you are my friend, are you not? Oh! yes, my most kind friend! I will try not to be afraid; and then I need so much both you and your advice! I am in great trouble, it seems to me that every one guesses what I am thinking; and especially when *he* is there, I blush as soon as I am looked at. Yesterday, when you saw me crying, I wanted to speak to you, and then something or other prevented me; and when you asked me what was the matter, the tears came in spite of me. I could not have said a word. But for you Mamma would have noticed it and then what would have happened to me? And this is how I spend my life, especially the last four days!

That was the day, Madame, yes, I am going to tell you about it, that was the day the Chevalier Danceny wrote to me. Oh! I assure you that when I found his letter I did not know at all what it could be; but, to tell you the truth, I cannot say that I had no pleasure in reading it; I would rather be unhappy all my life than that he should not have written to me. But I knew very well that I ought not to write to him and I can assure you that I told him I was very angry about it; but he said it was stronger than he was, and I can believe him; for I had made up my mind not to answer him and yet I could not prevent myself. Oh! I only wrote to him once and even then it was partly to tell him not to write to me again; but in spite of that he keeps on writing; and as I do not answer, I can see he is sad, and that grieves me still more; and so I don't know what to do or what will become of me and I am very much to be pitied.

Tell me, Madame, I beg you, would it be very wrong to answer him from time to time? Only until he can manage not to write to me any more himself and to be as we were before; because for my part if this goes on I do not know what will happen to me. Why, when I read his last letter, I cried as if I would never stop; and I am quite sure if I do not answer him again, it will be very sad for us both.

I send you his letter too, or rather a copy, and you can judge; you will see he does not ask for anything wrong. However, if you think it ought not to be, I promise you not to do it; but I believe you will think as I do, that there is nothing wrong in it.

And while I am about it, Madame, allow me to ask you another question; I have been told that it is wrong to love anybody—but why is that? What makes me ask you is that the Chevalier Danceny says that it is not wrong at all and that almost everybody loves; if that is true, I do not see why I should be the only one not to; or is it only wrong for girls? Because I heard Mamma herself say that Madame D. . . . loved M. . . . and she did not speak of it as a thing which was so very bad; and yet I am sure she would be angry with me if she even suspected my friendship for Monsieur Danceny. Mamma still treats me like a child; and she tells me nothing. When she took me out of the convent, I thought it was to marry me; but now I don't think so; not that I care, I assure you; but you, who are so much her friend, perhaps know about it and if you know, I hope you will tell me.

This is a very long letter, Madame, but since you have allowed me to write to you I have taken advantage of it to tell you everything and I rely on your friendship.

I have the honour to be, etc.

Paris, 23rd of August, 17—.

LETTER XXVIII

The Chevalier Danceny to Cécile Volanges

And so, Mademoiselle, you still refuse to answer me! Nothing can bend you; and each day carries away with it the hopes it brought! What is the friendship you consented should exist between us if it is not even powerful enough to make you sensible of my suffering; if it leaves you calm and cold while I endure the torments of a fire I cannot extinguish; if, so far from inspiring you with confidence, it does not even suffice to awaken your pity? What! your friend suffers, and you do nothing to help him! He asks only a word from you and you refuse it to him! And you want him to be contented with a sentiment so weak that you are afraid to repeat your assurance of it!

Yesterday you said you wished not to be ungrateful; ah! believe me, Mademoiselle, to wish to repay love with friendship is not to fear ingratitude, it is only fearing to seem ungrateful. But I dare no longer occupy you with feelings which can only be troublesome to you if they do not affect you; I must shut them up in my own heart until I learn to overcome them. I know how painful this task will be; I do not conceal from myself the fact that I shall need all my strength; I shall try every means—there is one which will wound my heart deepest, and that will be to remind myself frequently that yours is insensible. I shall even try to see you less often and I am already considering how to find a plausible excuse.

What! Must I abandon the sweet custom of seeing you each day! Ah! at least I shall never cease to regret it. An endless misery will be the reward of a most tender love; and you will have willed it so, it will be your work! Never—I see it—shall I recapture the happiness I lose to-day; you only were made for my heart; with what pleasure would I swear to live only for you! But you do not desire to accept that oath; your silence shows me plainly that your heart makes no plea for me; it is at once the clearest proof of your indifference and the cruellest manner of telling me. Farewell, Mademoiselle.

I dare not flatter myself with hopes of an answer; love would have written eagerly, friendship with pleasure, even pity with indulgence; but pity, friendship and love are alike strangers to your heart.

Paris . . . , 23rd of August, 17—.

LETTER XXIX

Cécile Volanges to Sophic Carnay

I told you, Sophie, that there are cases when one can write; and I assure you I greatly regret having followed your advice, because it gave so much pain to the Chevalier Danceny and to me. The proof that I was right is that Madame de Merteuil, a woman who must surely know, ended by agreeing with me. I told her everything. At first she said what you did; but when I explained everything to her she admitted it was quite a different matter; she only demands that I show her all my letters and all those from the Chevalier Danceny, to be certain I only say what is proper; and so for the present I am content. Ah! how much I love Madame de Merteuil! She is so kind! And such a respectable woman. So nothing can be said.

How I shall write to Monsieur Danceny and how pleased he will be! He will be more pleased than he imagines; for up to now I have only spoken to him of my "friendship" and he always wanted me to say my "love." I think it came to the same thing; but still I was afraid to, and he wanted it. I spoke of it to Madame de Merteuil; she said I was right and that one should never admit loving someone until it was impossible to prevent it any longer. I am sure I shall not be able to prevent myself much longer; after all it is the same thing, and will please him more.

Madame de Merteuil said too that she would lend me books

which speak about all this and will teach me how to conduct myself and how to write better than I do; for, you see, she tells me all my faults, which is a proof of how much she loves me; she only advised me not to say anything to Mamma about these books because it would look as if she had neglected my education and that might make her angry. Oh! I shall say nothing to her about them.

But it is very extraordinary that a woman who is scarcely related to me should take more care of me than my mother! It is very fortunate for me that I know her!

She also asked Mamma to allow her to take me to-morrow to her box at the Opera; she says we shall be all alone there, and we will talk the whole time without any fear of being overhead; I shall like that much more than the Opera. We shall also talk about my marriage; for she tells me that it was quite true I was to be married; but we have not been able to say any more about it. Now, is it not really very surprising that Mamma says nothing at all about it?

Good-bye, dear Sophie, I am going to write to the Chevalier Danceny. Oh! I am so glad!

From . . . , 24th of August, 17—.

LETTER XXX

Cécile Volanges to the Chevalier Danceny

At last I agree to write to you, Monsieur, to assure you of my friendship, of my *love*, since otherwise you would be unhappy. You say I am not kind-hearted; I assure you that you are wrong and I hope you will not think so any more. If you were hurt because I did not write to you, do you think I was not pained too? But the reason was that nothing in the world could make me do something I thought was wrong; and I would not even have admitted my love, if I could have prevented myself; but your sadness hurt me too much. I hope now you will not be sad any more and that we shall be very happy.

I expect to have the pleasure of seeing you this evening and that you will come early; it will never be as early as I desire. Mamma dines at home and I think she will invite you to stay; I hope you will not be engaged as you were the day before yesterday. Was it very pleasant, the dinner you went to? For you left very early. But we will not speak of that any more; now you know I love you, I hope you will stay with me as much as you can; for I am only happy when I am with you and I should like you to be the same.

I am very sorry you are still sad, but it is not my fault. I shall ask for my harp as soon as you arrive so that you get my letter at once. I cannot do any more.

Good-bye, Monsieur. I love you with all my heart; the oftener I tell you so, the happier I am; I hope you will be so too.

From . . . , 24th of August, 17—.

LETTER XXXI

The Chevalier Danceny to Cécile Volanges

Yes, we shall indeed be happy. My happiness is certain, since I am loved by you; your happiness will never end if it lasts as long as the love you have inspired in me. Ah! you love me, you are no longer afraid to assure me of your *love*! "the oftener you tell me so, the happier you are!" After I had read that charming "I love you," written by your hand, I heard the same admission from your lovely mouth. I saw those charming eyes gaze upon me, embellished by an expression of tenderness. I received your vow to live only for me. Ah! accept my vow to devote my whole life to your happiness; accept it and be sure I shall never betray it.

What a happy day we spent yesterday! Ah! Why is it that Madame de Merteuil has not secrets to tell your Mamma every day? Why must the idea of the constraint which awaits us mingle with the delicious memory which possesses me? Why can I not forever hold that pretty hand which wrote to me "I love you!" cover it with kisses, and thus avenge myself for the refusal of a greater favour!

Tell me, my Cécile, when your Mamma came back; when we were forced by her presence to exchange only indifferent looks; when you could no longer console me by the assurance of your love for your refusal to give me proofs of it; did you feel no regret? Did you not say to yourself: "A kiss would have made him happier and I have denied him this happiness?" Promise me, my dearest, that you will be less severe next time. With the aid of this promise I shall find courage to endure the difficulties which circumstances are preparing for us; and cruel privations will at least be softened by the certainly that you share my affliction.

Good-bye, my charming Cécile; the hour has come when I am to visit you. It would be impossible for me to leave you if it were not to go and see you again. Good-bye, you whom I love so much! You whom I shall ever love more!

From . . . , 24th of August, 17—.

LETTER XXXII

Madame de Volanges to Madame de Tourvel

You wish me, Madame, to believe in the virtues of Monsieur de Valmont? I confess I cannot bring myself to do so, and it would be as difficult for me to consider him virtuous from the one fact which you tell me, as it would be to consider as vicious a man of recognised virtue about whom I learned one lapse. Humanity is not perfect in any type, no more in evil than in good. The wicked man has his virtues, as the good man has his weaknesses. It seems to me the more necessary to believe this truth since from it arises the necessity of indulgence for the wicked as well as for the good; since it saves the latter from pride and the former from discouragement. No doubt you will think I am now practising very ill the indulgence I preached; but I see in indulgence only a dangerous weakness when it leads us to treat the good and the wicked man in the same way.

I will not allow myself to scrutinise the motives for Monsieur de Valmont's action; I like to think they were as praiseworthy as it; but has he any the less spent his life in bringing into families confusion, disorder and scandal? Hearken, if you will, to the voice of the wretch he has succoured; but do not let it prevent you from hearing the cries of a hundred victims he has sacrificed. Even if he were, as you say, merely an example of the danger of acquaintance, would he himself be any the less a dangerous acquaintance? You suppose him capable of reformation? Let us go further; let us suppose this miracle had happened. Would not public opinion against him still remain, and does not that suffice to govern your conduct? God alone can absolve at the moment of repentance; He reads our hearts: but men can only judge of thoughts by actions, and no one who has lost the esteem of others has a right to complain of necessary suspicion, which renders this loss so difficult to repair. Above all, remember, my young friend, that sometimes, to lose this esteem, one need only appear to attach too little value to it; and do not call this severity injustice; for, beside the fact that we have a right to think people do not renounce this valuable advantage when they have a right to it, those who are no longer withheld by this powerful curb are indeed the nearer to acting ill. Such, however, would be the aspect in which you would appear through an intimate acquaintance with Monsieur de Valmont, however innocent it might be.

Alarmed by the warmth with which you defend him, I hasten to forestall the objections I foresee. You will adduce Madame de Merteuil, to whom this acquaintance has been pardoned; you will ask me why I receive him in my house; you will tell me that far from being rejected by virtuous people, he is admitted into, even sought by, what is called good company. I think I can answer everything.

First of all, Madame de Merteuil, who is indeed most estimable, has perhaps no fault but too great a confidence in her own strength; she is a skilful driver who enjoys guiding her chariot among rocks and precipices, and only her success is her justification; it is right to praise her, it would be imprudent to follow her; she herself admits it and reproaches herself. The more she has been, the more severe her principles have become; and I am bold to assure you that she would think as I do.

As to what concerns me, I will not justify myself any more than others. No doubt I receive Monsieur de Valmont and he is received everywhere; it is one more inconsistency to add to the thousand others which govern society. You know as well as I do that we spend our lives in noticing them, in complaining of them and submitting to them. Monsieur de Valmont, with his ancient name, his wealth, and many amiable qualities, early recognised that to dominate society all he needed was to wield praise and ridicule with equal skill. No one possesses this double talent to the extent he does; with the one he seduces, with the other he makes himself feared. He is not esteemed; but he is flattered. Such is his existence in the midst of a world which, with more prudence than courage, prefers to treat him with deference rather than to combat him.

But neither Madame de Merteuil nor any other woman would dare to shut herself up in the country almost alone with such a man. It was reserved for the most prudent and most modest of women to set the example of this inconsistency—forgive the word, it slipped out from friendship. My dear friend, your very virtue betrays you, through the confidence it inspires in you. Remember that your judges on the one hand will be frivolous people who will not believe in a virtue whose model they do not find in themselves, and, on the other, malignant people who will feign not to believe in it to punish you for having had it. Consider that at this moment you are doing what several men would not dare to risk. Indeed, I see that among the young men (of whom Monsieur de Valmont is but too much the oracle) the most prudent are afraid to appear too intimate with him; and you are not afraid! Ah! recover yourself, I beg you . . . If my reasons do not suffice to persuade you, yield to my friendship; it is that which causes me to renew my entreaties, which justifies them. You will think it severe, and I hope it may be unnecessary; but I prefer that you should have to complain rather of its solicitude than of its negligence.

From . . . , 24th of August, 17—.

LETTER XXXIII

The Marquise de Merteuil to the Vicomte de Valmont

As soon as you are afraid of succeeding, my dear Vicomte, as soon as your object is to furnish arms against yourself, and you are less anxious to triumph than to combat, I have nothing more to say. Your conduct is a masterpiece of prudence. It would be a masterpiece of folly on the contrary assumption; and to tell you the truth, I am afraid you are deluding yourself.

I do not reproach you for not having profited by the moment. On the one hand, I do not quite see that it had arrived; and on the other I know too well—whatever they may say—that an occasion lost may be found again, while there is no recovery from too hasty a step.

But your real blunder is to have allowed yourself to start writing. I defy you now to foresee where that may lead you. Do you hope perhaps to prove to this woman that she ought to yield herself? It seems to me that could only be a truth of sentiment, not of demonstration; and to make her accept it you must move her and not reason with her; but how can it help you to move her by letter, since you will not be at hand to profit by it? Suppose your fine phrases should produce the intoxication of love, do you flatter yourself it will last so long that reflection will not have time to prevent her from admitting it? Remember what is needed for writing a letter, what takes place before it is delivered; and then see if a woman with principles like your devotee can desire for so long what she tries never to desire at all. This course may succeed with children who, when they write "I love you," do not know they are saying "I yield myself." But the rationalising virtue of Madame de Tourvel seems to me to understand perfectly the value of phrases. So, in spite of the advantage you obtained over her in conversation, she beats you in her letter. And then, do you realise what is happening? The mere fact of debating makes her unwilling to yield. By dint of looking for good reasons she finds them; she expresses them; and then she will hold to them not so much because they are good as not to contradict herself.

And then, I am surprised you have not noticed that there is nothing so difficult in love as to write what one does not feel. I mean, to write it in a credible way. It is not that the same words are not employed; but they are not arranged in the same way, or rather they are arranged—and that is enough. Re-read your letter; there is an order in it which betrays you at every phrase. I am ready to believe that your Madame de Tourvel is sufficiently inexperienced not to notice it; but what does that matter? The effect fails none the less. That is the defect of novels; the author lashes his sides to warm himself up, and the reader remains cold. *Héloise* is the one exception; and in spite of the author's talents this observation has always made me think that its subject was true. It is not the same in speaking. Practice in exerting one's speech gives it feeling; facility of tears adds to it; the expression of desire in the eyes may be mistaken for that of tenderness; and then a less coherent language more easily introduces that air of uncertainty and disorder which is the real eloquence of love;

above all the presence of the loved person prevents reflection and makes us women wish to be overcome.

Take my advice, Vicomte; you are asked not to write again; take advantage of it to repair your mistake and wait for an opportunity to speak. Do you know, this woman has more strength than I thought? Her defence holds out well; and but for the length of her letter and the pretext she gives you to resume by her phrase of gratitude, she would not have betrayed herself at all.

What seems to me still to promise you success is that she uses too many influences at once; I foresee she will exhaust them all in defending the word and will have none left to defend the thing.

I send you back your two letters; if you are wise they will be the last until after the happy moment. If it were not so late, I should speak to you of the Volanges girl who is advancing rapidly; I am very pleased with her. I think I shall have finished before you, and you ought to be very glad of it. Good-bye for to-day.

From . . . , 24th of August, 17—.

LETTER XXXIV

The Vicomte de Valmont to the Marquise de Merteuil

You speak admirably, my fair friend; but why take so much trouble to prove what every one knows? To advance rapidly in love it is better to speak than to write; that, I think, is the whole of your letter. Why! These are the simplest elements in the art of seduction. I will only point out that you make but one exception to this principle, and there are two. To children who follow this course from timidity and yield from ignorance, must be added the blue-stockings who start on it from conceit and who are led into the snare by vanity. For instance, I am quite sure that the Comtesse de B . . . who readily answered my first letter was no more in love with me then than I was with her; she only saw an opportunity of writing on a subject which would show her wit to good advantage.

However that may be, a lawyer would tell you that in this case the principle does not apply. You suppose I have a choice between writing and speaking, and I have not. Since the occurrence of the nineteenth my cruel one, who is on the defensive, has avoided any meeting with a skill which outwits my own. It has reached a point where if it continues she will force me to think seriously of means to regain this advantage; for assuredly I do not mean to be beaten by her in any way. Even my letters are the subject of a little war; not content with leaving them unanswered, she refuses to take them. Each one needs a new ruse which is not always successful.

You remember in what a simple way I delivered the first one; the second was just as easy. She had asked me to return her letter; I gave her mine instead, without her having the least suspicion of it. But either from annoyance at having been caught, or from caprice, or perhaps even from virtue—for she will at last force me to believe in it—she obstinately refused the third. However I hope that the embarrassment which nearly involved her as a result of this refusal will correct her for the future.

I was not very surprised when she refused to take this letter, which I merely offered her; that would have been granting something and I expect a longer defence. After this attempt, which was simply an experiment made by the way, I put my letter in an envelope and, choosing the moment of her toilet, when Madame de Rosemonde and the waiting-woman were present, I sent it to her by my man-servant with orders to say to her that this was the paper she had asked for. I had foreseen that she would shrink from the scandal of explanation which would be necessitated by a refusal; she took the letter, and my ambassador, who had been told to watch her face and is no bad observer, only noticed a slight blush and more embarrassment than anger.

I was congratulating myself, you may be sure, on the idea that she would either keep my letter or that if she wished to return it she would be compelled to be alone with me; which would give me an opportunity to speak to her. About an hour afterwards one of her servants came to my room and delivered to me, from his mistress, a packet shaped differently from mine, on whose envelope I recognised the much-desired handwriting. I opened it hastily . . . It was my own letter unopened and merely folded in two. I suspect that her fear lest I should be not so scrupulous as she in the matter of scandal made her employ this diabolical ruse.

You know me; there is no need for me to describe my rage. However, it was necessary to be calm and to think of some new method. This is the only one I could devise.

Every morning someone goes from here to get the letters at the post office which is about three quarters of a mile away; a closed box rather like a trunk is used for this purpose, and the post master has one key and Madame de Rosemonde the other. Every one puts his letters in at any time during the day; in the evening they are taken to the post office and the next morning someone goes for the letters which have come in. All the servants, visitors' or others, discharge this task. It was not my man's turn; but he undertook to go on the pretext that he had something to do in the neighbourhood.

Meanwhile I wrote my letter. I disguised my handwriting for the address and imitated fairly well on the envelope the postmark of Dijon. I chose this town because, since I was asking for the same rights as the husband, I thought it more amusing to write from the same place, and also because my fair one had talked all day of her desire to receive letters from Dijon. I thought it only just to procure her that pleasure.

These precautions once taken, it was easy to add my letter to the others. By this expedient I had the advantage of being able to watch its reception; for the custom here is to meet for breakfast and to await the arrival of the letters before separating. At last they arrived.

Madame de Rosemonde opened the box. "From Dijon," said she, giving the letter to Madame de Tourvel. "It is not my husband's handwriting," said she in an anxious voice, as she broke the seal quickly; the first glance showed her what it was; she changed countenance to such an extent that Madame de Rosemonde noticed it and said to her: "What is the matter?" I went up to her too, saying: "Is this letter so very terrible then?" The timid devotee dared not raise her eyes, said not a word, and to hide her embarrassment, pretended to run over the letter, although she was in no state to read. I was enjoying her distress and, not being sorry to tease her a little, I added: "Your calmer looks make me hope that this letter gives you more astonishment than pain." Her anger then inspired her better than her prudence had done. "It contains," she replied, "things which offend me and which I am astounded anyone should have dared to write to me." "And who is it?" interrupted Madame de Rosemonde. "It is not signed," replied the enraged fair one, "but both the letter and its writer inspire me with the same contempt. I shall be obliged if you will not mention it to me again." So saying, she tore up the audacious letter, put the pieces in her pocket, rose, and left the room.

In spite of her anger, she has none the less had my letter; and I rely on her curiosity at least to read it in full.

The details of the day would take me too far. I add to this the rough draft of my two letters; you will know as much as I do. If you wish to keep informed of this correspondence you must get used to deciphering my notes; nothing in the world would induce me to endure the boredom of re-copying them. Good-bye, my fair friend.

From . . . , 25th of August, 17—.

LETTER XXXV

Vicomte de Valmont to Madame de Tourvel

I must obey you, Madame, I must prove to you that in spite of all the faults you are pleased to credit me with I still retain sufficient delicacy not to permit myself a complaint and sufficient courage to burden myself with the most painful of sacrifices. You impose upon me silence and oblivion! Well! I will force my love to be silent and I will forget, if it is possible, the cruel manner in which you received it. Of course my wish to please you did not give me the right to do so and I admit that my need for your indulgence constituted no claim to obtaining it; but you consider my love an outrage and you forget that if it were a fault you are at once its cause and its excuse. You forget also that I was accustomed to show you my soul even when such confidence might harm me and that it was no longer possible to hide from you the emotions which move me; and so that which was the result of my good faith you consider the fruit of presumption. As a reward for the most tender, the most respectful, the most sincere love, you cast me far from you. You even speak of your hatred . . . Who else would not complain of such treatment? I alone am submissive; I endure everything without a murmur; you strike and I adore. The inconceivable power you have over me makes you the absolute mistress of my feelings; and if my love alone resists you, if you cannot destroy it, that is because it is your work and not mine.

I do not ask a return which I never flattered myself I should obtain. I do not expect even that pity which I might have hoped from the interest you have sometimes shown in me. But I confess I think I may appeal to your justice.

You inform me, Madame, that an attempt has been made to harm me in your estimation. If you had followed the advice of your friends you would not even have allowed me to approach you—those are your words. And who are these officious friends? No doubt such severe people, of so rigid a virtue, will consent to be named; no doubt they would not hide themselves in an obscurity which would make them resemble vile calumniators; and I shall learn both their names and their censures. Remember, Madame, I have a right to know both since by them you have judged me. A criminal is not condemned without being told his crimes and the names of his accusers. I ask no other favour and I undertake beforehand to justify myself, to force them to withdraw their statements.

If I have been perhaps too contemptuous of the vain clamour of a public for which I care little, it is different with your esteem; and when I am consecrating my life to the task of meriting it, I shall not allow myself to be deprived of it with impunity. It becomes the more precious to me since I must owe to it the request you are afraid to ask of me, which would give me, you say, "a right to your gratitude." Ah! far from claiming your gratitude I should think I owed it to you if you would grant me an opportunity to be agreeable to you. Begin by doing me greater justice, and do not leave me ignorant of what you desire from me. If I could guess it, I would spare you the trouble of saying it. To the pleasure of seeing you, add the happiness of serving you and I shall congratulate myself upon your indulgence. What then can restrain you? Not, I hope, the fear of a refusal? I feel I could not pardon you that. There was no refusal in my not returning your letter. More than you, I desire that it should not be necessary to me; but I was accustomed to believe your spirit was so gentle and it is only in that letter that I can find you such as you would wish to appear. While I form the hope that you will be moved, I see you would fly a hundred miles from me rather than consent; when everything in you increases and justifies my love, it is your letter which repeats to me that my love outrages you; when I look at you and that love seems to me the supreme good, I need to read you to feel it is only a dreadful torment. You can now perceive that my greatest happiness would be to be able to return you that fatal letter; to ask me for it again would be authorising me to cease to believe what it contains; I hope you do not doubt my eagerness to return it to you.

From . . . , 21st of August, 17—.

LETTER XXXVI

The Vicomte de Valmont to Madame de Tourvel

(Postmark of Dijon).

Your severity increases every day, Madame, and if I may be bold to say so, you seem to be more afraid of being indulgent than of being unjust. After having condemned me without hearing me, you must have felt indeed that it would be easier for you not to read my arguments than to reply to them. You refuse my letters with obstinacy; you send them back to me with contempt. And you finally force me to a trick at the very moment when my sole object is to convince you of my good faith. The necessity of defending myself which you have forced upon me no doubt will suffice to excuse the means. Convinced moreover by the sincerity of my feelings that it would suffice to explain to you in order to justify them in your eyes, I felt I might allow myself this slight subterfuge. I even dare to think that you will forgive me for it and that you will not be very surprised that love is more ingenious in putting itself forward than indifference is in putting it aside.

Madame, allow my heart to reveal itself completely to you. It belongs to you and you have a right to know it.

When I came to Madame de Rosemonde's house I was very far from foreseeing the fate which awaited me there. I did not know you were there; and I will add, with my accustomed sincerity, that if I had known it my self-confidence would not have been disturbed—not that I do not render your beauty the justice it cannot be refused, but as I was accustomed to experience nothing but desires and to yield only to those encouraged by hope, I was ignorant of the tortures of love.

You yourself witnessed the entreaties of Madame de Rosemonde to retain me here for some time. I had already passed a day with you; yet I only yielded—or at least I only thought I yielded—to the natural and legitimate pleasure of showing my regard for a respectable relative. The kind of life led here was certainly very different from that to which I was accustomed. It was easy for me to conform to it; and, without seeking to discover the cause of the change which was taking place in me, I still attributed it solely to that facility of character of which I think I have already spoken to you.

Unfortunately (and why must it be a misfortune?) as I came to know you better I soon recognised that the enchanting face, which alone had struck me, was the least of your advantages; your heavenly soul surprised, seduced mine. I admired beauty, I adored virtue. Without hoping that I might possess you I pondered on the means to deserve you. By asking your indulgence for the past, I aspired to your approbation in the future; I sought for it in your speech, I watched for it in your looks; in those looks which darted a poison all the more dangerous in that it was cast out undesignedly, and received without suspicion.

And then I knew love. But how far I was from complaining! Resolved to bury it in eternal silence I gave myself up without fear, as without reserve, to that delicious emotion. Each day increased its power. Soon the pleasure of seeing you became a necessity. Were you absent for a moment? My heart shrank with sadness; at the sound which announced your return, it palpitated with joy. I existed only through you and for you. And yet, I call you to witness, never in the gaiety of playful games or in the interest of a serious conversation did there escape from me a word which might betray the secret of my heart.

At last there came a day from which my misfortune was to

begin; and through some inscrutable destiny a good action was the beginning. Yes, Madame, it was in the midst of the wretches I had succoured that, yielding yourself to that precious sensibility which embellishes beauty itself and adds a value to virtue, you completely unsettled a heart which was already intoxicated by an excess of love. You remember perhaps how preoccupied I was during our return! Alas! I was trying to overcome an inclination which I felt was becoming stronger than I.

It was when I had exhausted my strength in this unequal struggle that a chance I could not have foreseen left us alone together. There I succumbed, I admit it. My overwrought heart could not restrain its words or its tears. But is that a crime? And if it is one, is it not sufficiently punished by the frightful torments to which I am abandoned?

Devoured by a hopeless love, I implore your pity and find nothing but your hatred; with no other happiness than that of seeing you my eyes seek you out in spite of myself and I tremble to meet your looks. In the cruel condition to which you have reduced me, I spend the days in hiding my pain and the nights in abandoning myself to it; while you in your peace and tranquillity know nothing of these torments except that you caused them and congratulate yourself on doing so. Yet it is you who complain and I who make excuses.

Here, Madame, is a faithful account of what you call my faults, which might more justly perhaps be called my misfortunes. A pure and sincere love, a respect which has never failed, a perfect submission, such are the feelings you have inspired in me. I should not have been afraid to offer them to the divinity itself. O you, you who are its fairest work, imitate it in its indulgence! Think of my cruel distress; think above all that since you have placed me between despair and supreme happiness, the first word you pronounce will decide my fate forever.

From . . . , 23rd of August, 17—.

LETTER XXXVII

Madame de Tourvel to Madame de Volanges

I submit myself, Madame, to the advice your friendship gives me. Accustomed to defer to your opinion in everything I have come to think it is always founded upon reason. I will even admit that Monsieur de Valmont must indeed be infinitely dangerous if he can at the same time feign to be what he appears here and remain as you describe him. However this may be, since you demand it, I will send him away from me; at least I will do my best; for often those things which in essence should be the most simple, become difficult in form.

It still seems to me impractical to make this request of his aunt; it would be equally impolite to her and to him. On the other hand I should feel some repugnance if I took the decision to go away myself; for in addition to the reasons I have already mentioned relative to Monsieur de Tourvel, if my departure provoked Monsieur de Valmont (as is possible), would it not be easy for him to follow me to Paris? And his return, of which I should be, or at least appear to be, the cause, might seem stranger than a meeting in the country at the house of a person who is known to be his relative and my friend.

My only resource then is to persuade him to leave of his own accord. I feel this is a difficult proposition to make; but since he seems to desire to prove to me that he is more virtuous than is supposed, I do not despair of success. I shall not even be sorry to attempt it and to have an opportunity of judging whether, as he says so often, really virtuous women never have and never will have reason to complain of his actions. If he goes, as I desire, it will indeed be out of regard for me; for I cannot doubt but that he intends to pass a large part of the autumn here. If he refuses my request and persists in remaining, I shall still have time to go myself and I promise you I will.

That, I think, Madame, is all your friendship exacted of me; I hasten to satisfy it, and to prove to you that in spite of the *warmth* I may have showed in defending Monsieur de Valmont, I am none the less disposed, not only to listen to, but to follow, the advice of my friends.

I have the honour to be, etc.

From . . . , 25th of August, 17—.

LETTER XXXVIII

The Marquise de Merteuil to the Vicomte de Valmont

Your enormous packet has just arrived, my dear Vicomte. If its date is correct, I ought to have received it twenty-four hours sooner; however that may be, if I took the time to read it I should have none left in which to answer it. So I prefer to acknowledge receipt of it—and to talk of something else. It is not that I have anything to tell you on my own account; the autumn hardly leaves in Paris one man with a human visage; and so for the last month I have been desperately sober and anyone but my Chevalier would be wearied by the proofs of my constancy. Having nothing to do, I amuse myself with the Volanges girl; and it is about her that I wish to speak to you.

Do you know you have lost more than you thought by not undertaking this child? She is really delicious! She has neither character nor principles; so you may judge how charming and facile her company would be. I do not think that she would ever shine by sentiment but everything about her presages the keenest sensations. Without wit and without finesse, she has yet a certain natural duplicity (if I may say so) which sometimes surprises even me and which will be all the more successful since her face is the image of candour and ingenuousness. She is naturally very caressing, and I sometimes amuse myself with her; she grows excited with incredible facility; and she is all the more amusing because she knows nothing, absolutely nothing, of what she so much wishes to know. She has the most diverting fits of impatience; she laughs, is piqued, weeps, and then begs me to tell her with a truly seductive good faith. Really, I am almost jealous of the man for whom this pleasure is reserved.

I do not know if I told you that for four or five days I have had the honour to be her confident. You can guess that at first I pretended to be severe; but as soon as I saw that she thought she had convinced me with her bad reasons, I pretended to accept them as good, and she is intimately persuaded that she owes this success to her eloquence; this precaution was necessary to avoid compromising myself. I have allowed her to write and say "I love"; and the same day, without her guessing it, I arranged her an interview alone with Danceny. But imagine, he is still such a fool that he did not even get a kiss from her. Yet the lad writes very pretty verses! Heavens! How silly these wits are! He is silly to an extent which embrasses me; for after all I cannot be his guide.

You would be very useful to me at this juncture. You know Danceny sufficiently well to obtain his confidence, and if he once gave it to you, we should progress rapidly. Hurry on with your Madame de Tourvel, for I do not want Gercourt to escape; I spoke of him yesterday to the little girl and painted him in such colours that she could not hate him more if she had been married to him for ten years. I gave her a long lecture on conjugal fidelity; nothing could equal my severity on this point. Thus, on the one hand, I re-established in her eyes my reputation for virtue which might be destroyed by too much condescension; and on the other hand I increased in her the hatred with which I desire to gratify her husband. And finally I hope that by making her think that it is not permitted to yield to love except in the little time she has left before marriage, she will decide the quicker not to lose any of it. Good-bye, Vicomte; I must go to my toilet and will read your volume.

From . . . , 27th of August, 17—.

LETTER XXXIX

Cécile Volanges to Sophie Carnay

I am sad and uneasy, my dear Sophie, I cried almost all night. It is not that I am not very happy at the moment, but I foresee that it will not last.

Yesterday I went to the Opera with Madame de Merteuil; we talked a lot about my marriage and what I found out about it was not at all nice. It is M. le Comte de Gercourt I am to marry and it is to be in October. He is rich, he is a man of quality, he is colonel of the . . . Regiment. So far so good. But first of all he is old; consider, he is at least thirty-six! And then Madame de Merteuil says he is melancholy and severe and that she is afraid I shall not be happy with him. I could see quite well she was sure I should not be, and that she did not wish to say so, in order not to distress me. She talked to me almost all the evening about the duties of wives to husbands; she admits that M. de Gercourt is not in the least agreeable, and yet she says I must love him. Did she not tell me also that once I am married I must no longer love the Chevalier Danceny? As if that were possible! Oh! I assure you I shall always love him. Let M. de Gercourt look after himself, he was none of my seeking. At present he is in Corsica, far away from here; I wish he would stay there ten years. If I were not afraid of going back to the convent I should tell Mamma this is not the husband I want; but that would be even worse. I am very perplexed. I feel I have never loved M. Danceny so much as I do now; and when I think that I have only one month left to be as I am, the tears come into my eyes at once; my only consolation is the friendship of Madame de Merteuil—she is so kind-hearted! She shares all my griefs with me and then she is so agreeable that when I am with her I hardly think about them. Besides she is very useful to me; for the little I know I have learnt from her; and she is so kind that I tell her everything I think without being at all bashful. When she thinks it is not right, she scolds me sometimes; but it is very gently, and then I kiss her with all my heart until she is no longer vexed. At least I can love her as much as I like without there being anything wrong in it, and it gives me a great deal of pleasure. However, we have agreed that I am to appear not to love her so much in front of other people, especially in front of Mamma, so that she will not suspect anything about the Chevalier Danceny. I assure you that if I could always live as I do now, I think I should be very happy. There is only that horrid M. de Gercourt . . . But I will not speak of him again; for it would make me sad once more. Instead of that I shall write to the Chevalier Danceny; I shall only speak to him about my love but not about my griefs, for I do not wish to distress him.

Good-bye, my dear. You see it would be wrong for you to complain and that although I am so *busy*, as you put it, I still have time to love you and to write to you.[1]

From . . . , 17th of August, 17—.

LETTER XL

The Vicomte de Valmont to the Marquise de Merteuil

It is not enough for my cruel one that she does not answer my letters, that she refuses to receive them; she wants to deprive me of the sight of her, she insists that I go away. What will surprise you more is that I am submitting to such harshness. You will find fault with me. But I did not think I ought to lose an opportunity of allowing myself to be given an order; being convinced on the one hand that she who commands commits herself, and on the other hand that the illusory authority we appear to let women take is one of the snares they avoid with the most difficulty. Moreover the skill she has shown in avoiding being alone with me placed me in a dangerous situation, which I felt I ought to get out of at any cost; for as I was continually near her without being able to occupy her with my love I had reason to fear that at last she would grow used to seeing me without disquiet—a state of mind which you know is difficult to alter.

For the rest you can guess I did not yield without conditions. I was even careful enough to make one it was impossible to grant; so that I can either keep my word or break it, and at the same time begin a discussion either by word of mouth or by letter at a time when my fair one is better pleased with me and when she needs me to be so with her—without reckoning that I should be very clumsy if I found no means of obtaining some satisfaction for desisting from this claim, however indefensible it may be.

After having explained my reasons in this long preamble, I begin the history of the last two days. As corroborating documents I shall add my fair one's letter and my reply. You will admit that few historians are so precise as I am.

You will remember the effect produced by my letter from Dijon the day before yesterday; the rest of the day was very stormy. The pretty prude only appeared at dinner time and announced a bad headache—a pretext by which she tried to conceal one of the most violent fits of ill humour a woman could have. Her face was positively changed by it; the expression of gentleness you are familiar with in her was changed into a rebellious look which gave it a new beauty. I have promised myself to make use of this discovery later and sometimes to replace the tender mistress by the rebellious mistress.

I foresaw that the time after dinner would be dull and, to avoid the boredom of it, I pretended I had letters to write and went to my room. I returned to the drawing-room about six o'clock; Madame de Rosemonde proposed that we should go out, which was agreed upon. But at the moment when she was getting into the carriage, the pretended invalid by an infernal piece of malice pretended in her turn (and perhaps to avenge herself for my absence) that she had an increase of pain, and without any pity forced me to endure a *tête a tête* with my old aunt. I do not know if my imprecations against this female demon were heard, but when we got back we found she had gone to bed.

Next morning at breakfast she was a different woman. Her natural gentleness had returned and I had good reason to think I was forgiven. Breakfast was scarcely over when the gentle creature rose with an indolent air and went out into the park; I followed her, as you may suppose. "What has caused this desire for walking?" said I as I met her. "I have done a lot of writing this morning," she replied, "and my head is a little weary." "I am not so fortunate," I replied, "as to have to reproach myself with this fatigue?" "I have indeed written to you," she answered again, "but I hesitate to give you my letter. It contains a request and you have not accustomed me to hope for its success." "Ah! I swear that if it is possible . . ." "Nothing is easier," she interrupted, "and although you ought perhaps to grant it as a matter of justice, I consent to take it as a grace." Saying these words, she handed me her letter; as I took it I took her hand too, which she withdrew but without anger and with more embarrassment than vivacity. "It is hotter than I thought," she said, "I must go back." And she took the path to the *Château*. I made vain efforts to persuade her to continue her walk and I was forced to remind myself that we might be seen in order to use nothing more than eloquence. She returned without speaking a word and I saw plainly that this pretended walk had no other purpose than that of giving me her letter. She went to her room as soon as we got back and I went to mine to read the epistle, which you will do well to read also, as well as my reply before going any further . . .

LETTER XLI

Madame de Tourvel to the Vicomte de Valmont

It seems from your conduct towards me, Monsieur, that you are every day seeking to increase the reasons I have to complain of you. Your obstinacy in wishing to occupy me ceaslessly with a sentiment which I ought not and do not wish to listen to; the way in which you have dared to abuse my good faith or my timidity in order to send me your letters; above all the method, which I dare to say was indelicate, you made use of to send me the last, without even fearing the effect of a surprise which might have compromised me; all these ought to have given rise on my part to reproaches as sharp as they are justly deserved. However, instead of dwelling upon these injuries, I merely make a request as simple as it is just; and if I obtain it from you, I agree that all shall be forgiven.

You have said yourself, Monsieur, that I need not fear a refusal; and although, by an inconsistancy which is peculiar to you, this very phrase was followed by the only refusal you could give me,[1] I am ready to believe that to-day you will none the less keep the word formally given so short a time ago.

It is my desire than that you will be good enough to go away from me; to leave this *Château*, where a longer stay on your part could only expose me further to the judgment of a public always quick to think ill of others, a public whom you have but too much accustomed to fix their eyes on the women who admit you into their society.

Although I had long been warned of this danger by my friends, I neglected, I even opposed their advice as long as your conduct towards me led me to believe that you did not wish to confound me with that crowd of women who all have reason to complain of you. Now that you treat me like them, that I can no longer remain ignorant of it, I owe to the public, to my friends, to myself to follow this imperative course. I might add here that you will gain nothing by refusing my request, for I am determined to leave myself if you persist in remaining; but I do not wish to lessen my obligation to you for this complaisance and I wish you to know that by forcing me to leave here you would disorganise my plans. Prove to me, as you have so often told me, Monsieur, that virtuous women will never have reason to complain of you; prove to me at least that when you have done them an injury you can repair it.

If I thought there was any need to justify my request to you, it would be enough to say to you that you have passed your life in making it necessary and that nevertheless it is not my fault if I have to make it. But do not let us recall events I wish to forget, which force me to judge you sternly at a moment when I give you an opportunity to deserve my gratitude. Good-bye, Monsieur. Your conduct will show me with what sentiments I should be, for life, your most humble, etc.

From . . . , 25th of August, 17—.

LETTER XLII

The Vicomte de Valmont to Madame de Tourvel

However harsh the conditions you impose upon me, Madame, I do not refuse to carry them out. I feel it would be impossible for me to thwart any of your desires. Once agreed on this point, I dare to flatter myself that in my turn you will allow me to make a few requests, much more easy to grant than yours, which I only desire to obtain by my perfect submission to your will.

One, which I hope will be pleaded for by your justice, is to be good enough to tell me the names of those who have accused me to you; it seems to me they have done me enough wrong for me to have a right to know them; the other, which I expect from your indulgence, is to be good enough to allow me sometimes to renew the homage of a love which more than ever now will deserve your pity.

Remember, Madame, that I hasten to obey you even when I can only do so at the expense of my own happiness; I will say more, in spite of my persuasion that you only desire my departure to save yourself from the sight (which is always painful), of the object of your injustice.

Acknowledge, Madame, that you do not so much fear a public which is too much accustomed to respect you to dare to form a disadvantageous judgement of you, as you are inconvenienced by the presence of a man whom it is easier to punish than to blame. You send me away from you as people avert their gaze from an unfortunate they do not wish to aid.

But when absence is about to increase my anguish, to whom but to you can I address my complaints? From whom else can I expect the consolations which will be so necessary to me? Will you refuse them to me, when you alone are the cause of my sufferings?

No doubt you will not be surprised that before I go I should desire to justify to you the sentiments you have inspired in me; and also that I lack the courage to go away if I do not receive the order from your mouth.

This double reason makes me ask you for a brief interview. It is useless for us to try to replace it by letters; we write volumes and explain badly what could be perfectly understood in a quarter of an hour's conversation. You will easily find time to grant it to me; for, however anxious I am to obey you, you know that Madame de Rosemonde is aware of my plan to spend part of the autumn with her and I must at least wait for a letter to be able to form a pretext for leaving.

Good-bye, Madame; never has it been so difficult for me to write this word as now when it calls to my mind the idea of our separation. If you could imagine how it makes me suffer I dare to think that you would feel some gratitude to me for my docility. At least accept, with more indulgence, the assurance and homage of the tenderest and most respectful love.

From . . . , 26th of August, 17—.

CONTINUATION OF LETTER XL

The Vicomte de Valmont to the Marquise de Merteuil

And now let us discuss matters, my fair friend. You will agree with me that the scrupulous, the honest Madame de Tourvel, cannot grant me the first of my requests and betray the confidence of her friends, by telling me the names of my accusers; so by promising everything on this condition, I commit myself to nothing. But you will agree also that her refusal of this will become a claim to obtaining all the rest; and that, by going away, I then gain the advantage of starting a regular correspondence with her by her own permission; I make little account of the meeting I have asked from her which has practically no other object but to accustom her beforehand not to refuse me others when they will be really necessary to me.

The only thing remaining for me to do before I leave is to find out who are the people who have busied themselves in harming me in her opinion. I presume it is her pedant of a husband; I wish it were; in addition to the fact that a conjugal interdiction is a spur to desire, I should be certain that from the moment my fair one consents to write to me I shall have nothing more to fear from her husband, since she will already be compelled to deceive him.

But if she has a woman friend sufficiently intimate to possess her confidence, and if this friend is against me, it seems to me necessary to set them at variance, and I expect to succeed in this; but first of all I must know who it is.

I thought yesterday I should find it out; but this woman does nothing like other women. We were in her room at the moment when she was informed that dinner was served. She was just finishing her toilet and as she hurried and made apologies, I noticed she had left the key in her writing-table; and I knew it was her habit not to remove the key of her apartment. I was thinking of it during dinner when I heard her maid come down; I decided what to do at once; I pretended my nose was bleeding and left the room. I rushed to the writing-table; but I found all the drawers open and not a piece of written paper. Yet there is no need to burn them at this time of the year. What does she do with the letters she receives? For she often receives them. I neglected nothing; everything was open and I looked everywhere; but I found nothing except that I convinced myself that this precious hoard remains in her pocket.

How am I to get them out? Ever since yesterday I have been trying in vain to find the way; yet I cannot overcome my desire for them. I regret I do not possess the talents of a pickpocket. Ought it not indeed to make part of the education of a man who dabbles in intrigues?[1] Would it not be amusing to steal a rival's letter or portrait, or to take from a prude's pocket the evidence to unmask her? But our parents never foresee anything; and though I foresee everything I only perceive that I am clumsy, without being able to remedy it.

In any case I came back to table very much out of humour. However, my fair one calmed my ill humour a little by the air of interest which my feigned indisposition gave her; and I did not fail to assure her that for some time I had undergone violent agitations which had harmed my health. Persuaded as she is that she is the cause of them, ought she not in conscience to labour to calm them? But, although she is pious, she is not very charitable; she refuses all amorous alms and it seems to me that this refusal suffices to justify my stealing them. But good-bye; the whole time I am talking to you I am thinking of nothing but those accursed letters.

From . . . , 27th of August, 17—.

LETTER XLIII

Madame de Tourvel to the Vicomte de Valmont

Why do you attempt, Monsieur, to lessen my gratitude? Why do you wish to obey me only by half and as it were to haggle over a right proceeding? It is not sufficient for you, then, that I should feel its cost? You not only ask a great deal; you ask impossible things. If indeed my friends have spoken to me of you, they can only have done so out of interest in me; even if they were wrong, their intention was none the less good and you propose to me to acknowledge this mark of attachment on their part by giving you their secret! I was wrong to speak to you of it and you make me feel so at this moment. What would have been merely candour with any one else becomes thoughtlessness with you, and would lead me to a base action if I yielded to your request, I appeal to you, to your honour; did you believe me capable of this action? Ought you to have suggested it to me? Certainly not; and I am sure that when you have thought it over you will not repeat this request.

The request you make me that you shall write to me is no easier to grant; and if you have any wish to be just you will not blame me. I do not wish to offend you; but with the reputation you have acquired (which, by your own admission, you deserve at least in part), what woman could admit she was in correspondence with you! And what virtuous woman can make up her mind to do what she feels she will be compelled to hide?

Still, if I were assured that your letters would be such that I should never have to complain of them, that I could always justify myself in my own eyes for having received them, perhaps then the desire of proving to you that I am guided by reason and not by hatred would make me pass over these powerful considerations and do far more than I ought to do, by allowing you to write to me sometimes. If indeed you desire it as much as you tell me, you will willingly submit to the one condition which could make me consent to it; and if you have any gratitude for what I am doing for you now you will delay your departure no further.

Allow me to point out to you in this connection that you received a letter this morning and did not make use of it to announce your departure to Madame de Rosemonde, as you had promised me. I hope that now nothing will prevent you from keeping your word. Above all I desire you will not expect from me, in exchange, the meeting you ask for, to which I absolutely will not lend myself; and that, instead of the order which you pretend is necessary to you, you will content yourself with the entreaty I make you again.

Farewell, Monsieur.

From . . . , 27th August, 17—.

LETTER XLIV

The Vicomte de Valmont to the Marquise de Merteuil

Share my joy, my fair friend; I am loved; I have triumphed over this rebellious heart. It is in vain for it to dissimulate longer; my fortunate skill has surprised its secret. Thanks to my active exertions, I know everything I want to know; since night, the happy night of yesterday, I am back in my element; I have resumed my whole existence; I have unveiled a double mystery of love and iniquity; I shall enjoy the one, I shall avenge myself upon the other; I shall fly from pleasures to pleasures. The mere idea of it so transports me that I have some difficulty in recalling my prudence, that I find it difficult perhaps to put any order into the account I have to give you. However, let me try.

Yesterday, after I had written you my letter, I received one from the charming devotee. I send it to you; you will see that she gives me, as adroitly as she can, the permission to write to her; but she urges me to leave and I felt I could not put off going much longer without doing myself harm.

Tormented, however, with the desire to know who could have written against me, I was still uncertain what course to take. I tried to win over the waiting-maid and I desired her to let me have her mistress's pockets, which she could easily get hold of at night and replace in the morning, without creating the least suspicion. I offered ten louis for this slight service; but I found only a scrupulous or timid prude, who was moved neither by my money nor my eloquence. I was still lecturing her when the supper bell rang. I had to leave her, only too happy that she consented to promise secrecy, upon which as you may suppose I did not rely.

I was never more out of humour. I felt I had compromised myself; and the whole evening I reproached myself for this imprudent step.

I went to my apartment, not without uneasiness, and spoke to my man who, in his character of the successful lover, ought to have some influence. I wished him either to get this girl to do what I had asked of her or at least to make certain of her discretion; but he who is usually uncertain of nothing seemed uncertain of the success of this negotiation, and made me an observation on this subject which astonished me by its profoundity.

"You know better than I, Monsieur," said he, "that to lie with a girl is only to make her do what pleases her; there is often a great distance between that and making her do what we want."

The rascal's good sense sometimes terrifies me.[1]

"I can rely on her the less," he added, "since I have reason to believe that she has a lover and that I only owe her to the inactivity of the country. So, were it not for my zeal in your service, Monsieur, I should only have had her once." (This fellow is a real treasure!) "As to secrecy," he went on, "of what use will it be to make her promise since she will risk nothing by betraying us? To speak of it to her again would only better show her that it is important and so make her more anxious to please her mistress."

The more true these remarks were, the more my embarrassment was increased. Luckily the fellow was in a mood to chatter;

and since I needed him I let him talk. As he related his affair with this girl, he informed me that since the room she occupies is only separated from her mistress's by a thin partition, which might allow a suspicious noise to be heard, they came together each night in his room. Immediately I formed my plan, communicated it to him, and we carried it out with success.

I waited until two o'clock in the morning; then, as we had agreed, I went to the room where their meeting took place, carrying a light with me under the pretext that I had rung several times without getting an answer. My confident, who plays his parts perfectly, gave a little scene of surprise, despair and excuses which I closed by sending him to heat me some water I feigned to need; while the scrupulous chamber-maid was the more shamefaced because the rascal, who had wanted to improve upon my plan, had persuaded her to a lack of costume which the season admitted of but did not excuse.

Since I felt that the more this girl was humiliated the more easily I could make use of her, I did not allow her to alter either her posture or her dress; and after I had bade my valet wait for me in my room, I sat down beside her on the bed which was in great disorder and began my conversation. I needed to preserve all the domination over her which the occasion gave me; I therefore maintained a calm which would have done honour to the continence of Scipio; and without taking the slightest liberty with her, which her freshness and the opportunity appeared to give her a right to hope for, I talked to her about business as calmly as I could have done with a solicitor.

My conditions were that I would faithfully keep her secret provided that the next day at about the same hour she would submit to me her mistress's pockets. "Moreover," I added, "I offered you ten louis yesterday; I promise them to you again to-day. I do not wish to abuse your situation." Everything was granted, as you may believe; then I retired and permitted the happy couple to make up for lost time.

I spent my own time in sleeping; and on awaking, since I wished to have an excuse for not replying to my fair one's letter before I had gone through her papers (which I could not do until the following night), I decided to go shooting, where I remained almost all day.

At my return I was received rather coldly. I have reason to think that she was a little piqued by my lack of anxiety to make use of the time which remained; especially after the kinder letter she had written me. I suppose this, because when Madame de Rosemonde reproached me a little for this long absence, my fair one replied rather tartly: "Ah! Do not let us reproach Monsieur de Valmont for giving himself up to the sole pleasure he can find here." I complained of this injustice, and I made use of it to assert that I enjoyed the company of these ladies so much I sacrified to it a very interesting letter I had to write. I added that, not having been able to sleep for several nights, I had wished to try whether weariness would give it me; and my looks explained sufficiently both the subject of my letter and the cause of my insomnia. I was careful to be gently melancholy all evening which seemed to me to succeed quite well; and under this I hid the impatience I was in for the arrival of the hour which would yield me the secret she persisted in hiding from me. At last we separated and some time after the faithful waiting-woman came to bring me the price agreed upon for my discretion.

Once master of this treasure, I proceeded to go through it with my accustomed prudence; for it was important to put everything in its place. I came first of all upon two letters from the husband, a confused mixture of law-suit details and tirades of conjugal love, which I had the patience to read right through without finding a word referring to myself. I put them back with annoyance; but this was mollified by my finding under my hand the pieces of my famous letter from Dijon carefully put together. Fortunately I took it into my head to run over it. Imagine my joy at perceiving very distinct traces of my adorable devotee's tears!

I confess it, I yielded to the sentiment of a young man, and kissed this letter with a rapture of which I thought myself incapable. I continued my happy search; I found all my letters arranged in the order of their date; and what surprised me still more agreeably, was to find the first of all, which I thought had been returned to me by an ingrate, faithfully copied by her hand and in a changed trembling handwriting which sufficiently showed the soft agitation of her heart during this occupation.

Up till then I was entirely given over to love; but soon it gave place to fury. Who do you think it is tries to ruin me with the woman I adore? What fury do you suppose is so malicious as to contrive such a perfidy? You know her; she is your friend, your relative; it is Madame de Volanges. You cannot imagine what a tissue of horrors that infernal shrew has written her about me. It is she and she alone who has disturbed the confidence of this angelic woman; it is by her designs, her pernicious warnings, that I am forced to go away; it is to her I am sacrificed. Ah! Her daughter must certainly be seduced, but that is not enough, I must ruin her; and since the age of this accursed woman protects her from my attacks, I must strike her through the object of her affection.

She wishes me to return to Paris! She forces me to do so. So be it, I will return, but she shall groan for my return. I am sorry Danceny is the hero of this adventure; he has a foundation of virtue which will impede us; however he is in love and I often see him; perhaps some use can be made of this. But I am forgetting myself in my anger and that I owe you an account of what happened to-day. Let us go back to it.

This morning I saw my delicate prude. Never had I thought her so beautiful. That was inevitable: a woman's fairest moment, the only one in which she can produce that intoxication of the soul, which is always talked of and so rarely experienced, is the moment when we are certain of her love but not of her favours; and that is precisely the position in which I am. Perhaps also the idea that I was about to be deprived of the pleasure of seeing her served to embellish her. At length, when the post arrived I was handed your letter of the twenty-seventh; and while I read it I still hesitated as to whether I should keep my word; but I met my fair one's eyes and it would have been impossible for me to refuse her anything.

I therefore announced my departure. A minute later Madame de Rosemonde left us alone; but I was still four paces from the timid person when she rose with an air of terror, saying: "Leave me, leave me, Monsieur, in the name of God, leave me." This fervent prayer which revealed her emotion could not but animate me the more. Already I was beside her, I held her hand which she clasped in a most touching way; I was beginning a series of tender complaints when some hostile demon brought back Madame de Rosemonde. The timid devotee, who indeed had some reason to be fearful, took advantage of it and retired.

Nevertheless, I offered her my hand and she accepted it; and auguring well from this mildness, which she had not showed me for a long time, I attempted to press her hand as I recommenced my complaints. First of all she tried to withdraw it; but at my warmer entreaty she yielded with a comparatively good grace, but without replying either to this gesture or to my speeches. When we reached the door of her apartment I tried to kiss her hand before leaving her. The defence was at first thorough; but a "remember I am going" spoken very tenderly made it awkward and inadequate. Scarcely was the kiss given when the hand recovered enough strength to escape and the fair one entered her apartment, where her waiting-woman was. Here ends my story.

Since I presume you will be at the house of the Maréchale de . . . to-morrow, where assuredly I shall not go to find you; since I suppose also that at our first interview we shall have more than one matter to discuss, notably that of the Volanges girl which I have not lost sight of; I am sending this letter on ahead of me, and long as it is, I shall only close it at the moment it goes to post; but in my present condition everything may depend upon an opportunity; and I leave you, to go and watch for it.

P.S. Eight o'clock in the evening.

Nothing new; not the smallest moment of freedom; care even to avoid it. However, as much sadness as decency permitted, at least. Another event which cannot be unimportant, is that I am entrusted with an invitation from Madame de Rosemonde to Madame de Volanges to come and spend some time with her in the country.

Good-bye, my fair friend; to-morrow or the day after at latest.

From . . . , 28th of August, 17—.

LETTER XLV

Madame de Tourvel to Madame de Volanges

Monsieur de Valmont left this morning, Madame; you seemed to desire this departure so much that I felt I ought to inform you. Madame de Rosemonde misses her nephew very much and it must be admitted that his society is agreeable; she passed the whole morning talking to me with her accustomed sensibility; she never ceased praising him. I felt I owed her the gratification of listening without contradicting her, the more so since it must be admitted she is right in many respects. I felt moreover that I had to reproach myself with being the cause of the separation and I do not hope to be able to make up to her for the pleasure of which I have deprived her. You know I am not naturally gay and the kind of life we shall lead here will not increase my gaiety.

If I had not been led by your advice, I should fear I had acted somewhat inconsiderately; for I was really hurt by the grief of my respectable friend; it touched me so much that I would gladly have mingled my tears with hers.

We now live in the hope that you will accept the invitation

which Monsieur de Valmont brings you from Madame de Rosemonde, to come and spend some time with her. I hope you have no doubt of the pleasure I shall have in seeing you again; and indeed you owe us this compensation. I shall be very happy to have this opportunity of making an earlier acquaintance with Mademoiselle de Volanges and of being in a position to convince you more and more of my respectful sentiments, etc.

From . . . , 29th of August, 17—.

LETTER XLVI

The Chevalier Danceny to Cécile Volanges

What has happened to you, my adorable Cécile? What can have caused so sudden and so cruel a change in you? What has become of your vows that you would never change? But yesterday, you repeated them with so much pleasure! What can have made you forget them to-day? However much I scrutinize myself I cannot find the cause in me and it is dreadful to me to have to look for it in you. Ah! You cannot be either fickle or deceitful; and even in this moment of despair my soul shall not be withered by an insulting suspicion. And yet by what fatality is it that you are no longer the same? No, cruel one, you are no longer the same! The tender Cécile, the Cécile I adore, whose vows I have received, would not have avoided my glances, would not have thwarted the lucky chance which placed me beside her; or if some reason which I cannot conceive had forced her to treat me so harshly she would at least not have disdained to tell me of it.

Ah! You do not know, you will never know, my Cécile, how much you have made me suffer to-day, how much I suffer at this moment. Do you think I can live if I am not loved by you? And yet, when I begged a word from you, one word, to dissipate my fears, instead of replying you pretended a fear that we should be overheard; and you created that obstacle, which up till then did not exist, by the place you chose in the company. When forced to leave you I asked you at what hour I might see you to-morrow, and you feigned not to know, so that Madame de Volanges had to tell me. And so the moment I always desire so much, the moment which brings me back to you, to-morrow will create nothing but uneasiness for me; and the pleasure of seeing you, hitherto so dear to my heart, will be replaced by the fear of being wearisome to you.

Already this fear checks me, I feel it, and I dare not speak to you of my love. That "I love you" which I delighted to repeat when I could hear it in my turn, that sweet phrase which sufficed for my felicity, now offers me, if you are changed, nothing but the image of an eternal despair. Yet I cannot believe that this talisman of love has lost all its power and I still attempt to use it.[1] Yes, my Cécile, I love you. Repeat with me that expression of my happiness, remember that you have accustomed me to hear it and that by depriving me of it you condemn me to a torment which, like my love, will end only with my life.

From . . . , 29th of August, 17—.

LETTER XLVII

The Vicomte de Valmont to the Marquise de Merteuil

I shall not be able to see you to-day, my fair friend, for the following reasons which I beg you to accept with indulgence.

Instead of returning directly yesterday I stopped with the Comtesse de . . ., whose *Château* was almost on my road, and from whom I asked a dinner. I did not reach Paris until seven o'clock and I went to the Opera where I hoped I might find you.

After the Opera I went to see my ladies of the green-room; there I found my former love, Emilie, surrounded by a retinue of admirers, women as well as men, whom she had invited to supper that evening at P . . . I had no sooner entered the circle than I was invited to the supper by general acclamation. I was also invited by a short, fat, little creature who jabbered an invitation in Dutch French, whom I perceived to be the true hero of the feast. I accepted.

I learned on the way that the house to which we were going was the price agreed upon for Emilie's favours to this grotesque creature and that the supper was a positive wedding-feast. The little man could not contain himself with joy in expectation of the happiness he was about to receive; he seemed to me to feel so much satisfaction that he made me wish to disturb it; which I did accordingly.

The only difficulty I met with was in persuading Emilie who was made a little scrupulous by the Burgermaster's wealth. However, after some demurring, she consented to my plan, which was to fill up this little beer-tub with wine and thus put him out of action for the whole night.

The sublime conception we had of a Dutch drinker caused us to employ all the known methods. We succeeded so well that at dessert he had not even the strength to hold his glass; but the charitable Emilie and I vied with each other in filling him up. At last he fell under the table in such a state of intoxication that it will last at least a week. We then decided to send him back to Paris; and since he had not kept his carriage I had him put into mine and remained in his place. I then received the congratulations of the company who retired soon after, leaving me master of the battle-field. This freak, and possibly my long retirement, made me think Emilie so desirable that I have promised to remain with her until the Dutchman's resurrection.

This complaisance on my part is the reward for one she has just granted me, that of acting as a desk for me to write to my fair devotee; I thought it amusing to send her a letter written from the bed and almost in the arms of a girl, interrupted even for a complete infidelity, in which letter I give her an exact description of my situation and my conduct. Emilie, to whom I read the epistle, laughed extravagantly, and I hope you too will laugh.

Since my letter must have the Paris postmark I am sending it to you; I have left it open. Be good enough to read it, to seal it, and to send it to the post. Be careful not to use your own seal nor any amorous emblem, simply a head. Good-bye, my fair friend.

P.S. I re-open my letter; I have persuaded Emilie to go to the Italiens . . . I shall make use of this time to come to see you. I shall be with you at six o'clock at the latest; and if it suits you, we will go together to Madame de Volanges at seven. It will be decent not to delay the invitation I have to give her from Madame de Rosemonde; moreover I shall be very glad to see the Volanges girl. Good-bye, most fair lady. I hope to have so much pleasure in embracing you that the Chevalier may be jealous of it!

From P . . . , 30th August, 17—.

LETTER XLVIII

The Vicomte de Valmont to Madame de Tourvel

(Postmark of Paris)

After a restless night, during which I have not closed my eyes; after having been ceaselessly either in the agitation of a devouring ardour or in the complete annihilation of all the faculties of my soul; I seek from you, Madame, the calm I need but do not hope to enjoy yet. Indeed the position I am in as I write to you makes me understand more than ever the irresistible power of love; I can scarcely preserve sufficient control over myself to put my ideas into some order; and already I perceive that I shall not finish this letter without being forced to interrupt it. Ah! May I not hope that some day you will share the perturbation I feel at this moment? Yet I dare to think that if you really knew it you would not be entirely insensible to it. Believe me, Madame, cold tranquillity, the soul's sleep, the image of death, do not lead to happiness; only the active passions can lead to it; and, in spite of the torments you make me endure, I think I may boldly assure you that at this moment I am happier than you. In vain do you overwhelm me with your discouraging severities; they do not prevent me from abandoning myself wholly to love and from forgetting in the delirium it causes me the despair to which you surrender me. It is thus I mean to avenge myself for the exile to which you condemn me. Never did I have so much pleasure in writing to you; never in that occupation did I feel so soft and yet so keen an emotion. Everything seems to increase my raptures; the air I breathe is filled with voluptuousness; the very table upon which I write to you, which for the first time is devoted to that use, becomes for me the sacred altar of love; how much it will be embellished in my eyes! I shall have traced upon it the vow to love you forever! I beg you to pardon the disorder of my senses. Perhaps I ought to abandon myself less to raptures you do not share; I must leave you a moment to dispel an ecstasy which increases every instant, which becomes stronger than I am.

I return to you, Madame, and without doubt I shall always return with the same eagerness. Yet the feeling of happiness has fled far from me; it has given place to that of cruel privations. Of what use is it for me to speak to you of my sentiments if I seek vainly the means to convince you? After so many repeated efforts, confidence and strength abandon me together. If I recall once more the pleasures of love it is but to feel more keenly the regret at being deprived of them. I see no resource save in your indulgence and at this moment I feel but too well how much I need that indulgence to hope to obtain it. However, never was my love more respectful, never ought it to offend you less; it is such, I dare to say, that the most austere virtue need not fear it; but I fear myself that I am occupying you too long with the distress I feel. Although I know that the object which causes it does not share that distress, I must not abuse its favours; and it would be doing so to employ more time in recalling this painful image to you. I shall only beg you to reply to me and never to doubt the truth of my sentiments.

Written from P . . . , dated from Paris, 30th of August, 17—.

LETTER XLIX

Cécile Volanges to the Chevalier Danceny

Without being either fickle or deceitful, Monsieur, it is sufficient for me to be enlightened as to my conduct, to feel the necessity of altering it; I have promised this sacrifice to God, until I can offer Him also that of my sentiments to you which your religious condition renders still more criminal. I feel indeed that this will give me great pain and I will not hide from you the fact that since the day before yesterday I have cried every time I have thought of you. But I hope that God will grant me the grace of sufficient strength to forget you, as I pray Him morning and evening. I expect from your friendship and from your honour that you will not seek to disturb me in the good resolution which has been inspired in me and in which I am trying to remain. Consequently I beg you will be kind enough to write to me no more, especially as I warn you that I shall not answer you again and that you would force me to inform Mamma of all that has happened; which would entirely deprive me of the pleasure of seeing you.

None the less I still have all the attachment for you that may be without there being anything wrong in it; and it is indeed with all my soul that I wish you every kind of happiness. I feel that you will not love me so much now and that perhaps you will soon love another more than me. But that will be one penance the more for the fault I have committed by giving you my heart, which I ought to give only to God and to my husband, when I have one. I hope that the divine mercy will have pity on my weakness and send me no grief that I cannot endure.

Farewell, Monsieur; I can assure you that if I were permitted to love anyone I should never love anybody but you. But that is all I can say to you and perhaps even that is more than I should say.

From . . . , 31st of August, 17—.

LETTER L

Madame de Tourvel to the Vicomte de Valmont

Is it thus, Monsieur, that you carry out the conditions under which I have consented sometimes to receive your letters? And can I "have no reason to complain" when you speak to me of nothing but a sentiment to which I should fear to yield myself even if I could do so without failing in all my duties?

Moreover, did I need fresh reasons for retaining this salutary fear, it seems to me that I could find them in your last letter. Indeed, at the very moment when you seem to make love's apology, what do you do but show its dreadful storms? Who can wish for happiness that is bought at the price of reason, whose fleeting pleasures are at least followed by regret, if not remorse?

And you, upon whom the effect of this dangerous delirium must be lessened by habit, are yet obliged to admit that it often becomes stronger than you; and are you not the first to complain of the involuntary distress it causes you? What fearful ravages would it not make upon a fresh and sensitive heart, which would yet increase its power by the extent of the sacrifices such a heart would be compelled to make it?

You believe, Monsieur, or you feign to believe, that love leads to happiness; and I am so convinced it would render me unhappy that I wish never to hear its name mentioned. It seems to me that one's tranquillity is disturbed merely by speaking of it; and it is as much from choice as from a sense of duty that I beg you to keep silent upon this point.

After all, this request must now be very easy for you to grant me. Now you are back in Paris you will find plenty of opportunities to forget a sentiment whose origin was perhaps only due to your habit of occupying yourself with such objects, and whose strength was merely due to the inactivity of the country. Are you not in the very place where you formerly saw me with so much indifference? Can you make a step without meeting an instance of your facility in changing? And are you not surrounded by women who are all more amiable than I and have more rights to your homage? I do not possess the vanity of which my sex is accused; still less have I that false modesty which is only a refinement of pride; and it is in good faith that I tell you I know I have very few means of pleasing; did I possess them all, I should not think them sufficient to secure your constancy. To request you to concern yourself with me no more is therefore only to beg you to do to-day what you have already done and what you will surely do again in a little time, even if I should beg you to do the contrary.

This truth, which I do not lose sight of, would of itself be a sufficiently strong reason for my not wishing to listen to you. I have a thousand others; but without entering upon a long discussion I limit myself to asking you (as I have done already) to occupy me no longer with a sentiment I ought not to listen to and to which I ought still less to respond.

From . . . , 1st of September, 17—.

END OF PART ONE

PART II

LETTER LI

The Marquise de Merteuil to the Vicomte de Valmont

Really, Vicomte, you are insupportable. You treat me as carelessly as if I were your mistress. Do you know, I shall grow angry, and that I am now in a frightful temper? What! You are to see Danceny to-morrow morning; you know how important it is that I should speak to you before this interview; and without troubling yourself any further, you let me wait for you all day long and run off I know not where. Owing to you I was *indecently* late in reaching Madame de Volanges's and all the old women said I was *wonderful*. I had to flatter them the whole evening to appease them; for old women must not be angered—they make young women's reputations.

It is now one o'clock in the morning and instead of going to bed, as I am dying to do, I must write you a long letter which will increase my drowsiness by the boredom it will cause me. It is very lucky for you that I have no time to scold you more. And do not let that make you believe I have forgiven you; I am simply in a hurry. Now listen to me; I must hurry.

If you are at all skilful, you ought to secure Danceny's confidence to-morrow. This is a favourable moment for confidence; he is unhappy. The little girl has been to confession; she told everything, like a child; and since then she has been so tormented by fear of the devil that she wishes to break off altogether. She told me all her little scruples with a vehemence which showed me how excited she was. She let me see the letter breaking things off, and it is a positive sermon. She prattled a whole hour to me without speaking one word of common sense. None the less she embarrassed me; for as you may suppose I could not risk confiding myself to such a reckless creature.

Nevertheless, I saw through all her chatter that she does not love Danceny any the less; I even noticed one of those expedients (which never fail in love), by which the little girl rather amusingly deceives herself. Tormented by the desire to think about her lover and by the fear of damning herself by thinking about him, she has devised the plan of praying God to make her forget him; and since she renews this prayer at every moment of the day she finds a way of constantly thinking about him.

With someone more *accustomed* than Danceny this little event might perhaps be more favourable than hurtful; but the young man is such a Céladon that, if we do not help him, he will take so much time in overcoming the slightest obstacles that he will not leave us enough to carry out our plan.

You are quite right; it is a pity, and I am as annoyed about it as you, that he should be the hero of this adventure. But what would you have? What is done is done; and it is your fault. I asked to see his reply[1]; it was pitiful. He argued with her interminably to prove that an involuntary sentiment cannot be a crime; as if it did not cease to be involuntary, the moment one ceases to oppose it! It is such a simple idea that it even occurred to the little girl. He complains of his unhappiness in a rather touching way; but his grief is so gentle and seems so great and so sincere that I think it is impossible that a woman who has the opportunity to distress a man to this extent, with so little danger, should not be tempted to gratify the whim. Finally he explains he is not a monk as the little girl believed and undoubtedly this is what he does the best; for, if one went so far as to yield to monastic love, the Knights of Malta would certainly not deserve the preference.

In any case, instead of wasting my time in arguments which would have compromised me and perhaps not have persuaded her, I approved the plan of breaking things off; but I said that in such a case it was more polite to explain one's reasons than to write them, that it was also the custom to return the letters and any other trifles one might have received; and thus appearing to agree with the little creature's ideas I decided her to grant Danceny an interview. We made the arrangements there and then and I undertook to persuade the mother to go out without her daughter; to-morrow afternoon will be the decisive instant. Danceny is already informed; but, for God's sake, if you find an opportunity, persuade this fair shepherd to be a little less languishing; and inform him, since he must be told everything, that the real way of vanquishing scruples is to leave those who have them nothing to lose.

In addition, to prevent a repetition of this ridiculous scene, I did not omit to raise some doubts in the little girl's mind concerning the discretion of confessors; and I can assure you she is now atoning for the fright she gave me by her own fright lest her confessor should tell her mother everything. I hope that when I have talked with her once or twice she will cease telling all her follies to the first person who comes along.[2]

Farewell, Vicomte; gain possession of Danceny and direct him. It would be shameful if we could not do what we chose with two children. If we find it more difficult than we had thought at first, you must stimulate your zeal by remembering that she is de Volanges's daughter; and I, that she is to become ..ourt's wife. Good-bye.

From . . . , 2nd of September, 17—.

LETTER LII

The Vicomte de Valmont to Madame de Tourvel

You forbid me, Madame, to speak to you of my love; but where shall I find the courage requisite to obey you? Occupied solely by a sentiment which should be so soft and which you render so cruel; languishing in the exile to which you have condemned me; living only upon privations and regrets; a prey to torments, the more painful in that they continually remind me of your indifference; must I be compelled to lose the one consolation that is left me, and can I have any other than sometimes to open to you a soul which you fill with distress and bitterness? Will you avert your gaze, in order not to see the tears which are shed owing to you? Will you refuse even the homage of the sacrifices you exact? Would it not be more worthy of you, more worthy of your virtuous and gentle soul, to pity a wretch who is only such because of you, than to wish to aggravate his sufferings still more by a prohibition which is both unjust and harsh?

You pretend to fear love and you will not see that you alone cause the evils of which you accuse it. Ah! No doubt, this is a painful sentiment when the person who inspires it does not share it; but where shall happiness be found if a reciprocal love does not procure it? Tender friendship, a soft confidence, the only confidence which is without reserve, softened griefs, augmented pleasures, enchanted hope, delicious memories—where shall they be found save in love? You calumniate love, you who need but to cease to withstand it in order to enjoy all the good it offers you; and I forget the grief I feel, to undertake its defence.

You force me to defend myself as well; for while I consecrate my life to adoring you, you spend yours in finding fault with me; you already suppose me fickle and deceitful; and by a misuse of certain of my faults—which I myself confessed to you—you take pleasure in confounding what I once was with what I now am. Not content with having abandoned me to the torment of living far from you, you add to it a cruel persiflage about pleasures to which you know how completely you have rendered me insensible. You believe neither my promises nor my vows; well, I have one guarantee to offer you which you at least will not suspect; it is yourself. I only ask you to question yourself in good faith; if you do not believe in my love, if you doubt one moment that you alone reign over my soul, if you are not certain that you have attached this heart, which indeed has hitherto been but too fickle, I consent to bear the pain of this error; I shall groan for it, but I shall not appeal; but if on the contrary, doing justice to us both, you are forced to admit to yourself that you have not, that you will never, have a rival, do not force me, I beg you, to struggle with idle fancies, and at least leave me the consolation of seeing you no longer doubt a sentiment which will, which can only, end with my life. Allow me, Madame, to beg you to reply positively to this part of my letter.

Yet if I abandon that period of my life which seems to harm me so cruelly in your eyes, it is not because I lack reasons to defend it, if needed.

After all, what have I done except not to resist the whirlpool into which I was thrown? Entering the world young and without experience; passed, as it were, from hand to hand by a crowd of women who all hastened to anticipate by their facility a reflection which they felt must be unfavourable to them; was it for me to show the example of a resistance which was not opposed to me? Or should I have punished myself for a momentary error which was often provoked, by a useless constancy which would have only appeared ridiculous? Ah! What save a quick breaking off can justify a shameful choice!

But, I can assert, that intoxication of the senses, perhaps even that delirium of vanity, never reached my heart. Born for love, it might be distracted by intrigue but could not be occupied by intrigue; surrounded with seductive but contemptible objects, my soul was reached by none of them; pleasures were offered me, I sought virtues; and at last I myself believed I was inconstant only because I was delicate and sensitive.

When I saw you I was enlightened; I soon recognised that the charm of love was derived from the qualities of the soul; that they alone could cause its excess and justify it. I felt that it was equally impossible for me not to love you and to love any other but you.

Such, Madame, is the heart to which you fear to confide yourself, upon whose fate you have to pass sentence; but whatever may be the destiny you reserve for it, you will change nothing in the sentiments which attach it to you; they are as unalterable as the virtues which gave birth to them.

From . . . , 3rd of September, 17—.

LETTER LIII

The Vicomte de Valmont to the Marquise de Merteuil

I have seen Danceny, but I have only obtained a half-confidence; he was especially obstinate in concealing the name of the Volanges girl, whom he only spoke of as a very modest woman who was rather pious; except for that he told me his story and especially the last episode fairly accurately. I inflamed him as much as I could and bantered him on his delicacy and scruples; but he appears to set a value upon them and I cannot answer for him; but I shall be able to tell you more after to-morrow. To-morrow I am taking him to Versailles and on the way I shall occupy myself by investigating him.

The meeting which was to take place to-day also gave me some hope; it may be that everything went off as we should desire; and perhaps all we have to do now is to wrench an admission of it and to assemble proofs. This will be an easier task for you than for me; for the little girl is more confiding or, which comes to the same thing, more of a chatterer, than her discreet lover. However, I will do my best.

Good-bye, my fair friend, I am in a great hurry; I shall not see you to-night nor to-morrow; if you learn anything on your part write me a note for my return. I intend to sleep in Paris.

From . . . , 3rd of September, 17—. in the evening.

LETTER LIV

The Marquise de Merteuil to the Vicomte de Valmont

O yes! Indeed it is from Danceny that something is to be learned! If he told you anything, he was boasting. I never knew such a fool in matters of love and more and more I regret our kindness to him. Do you know I was afraid I should be compromised on account of him! And that it was in pure waste! Oh! I shall be revenged, I promise him.

When I arrived yesterday for Madame de Volanges she did not want to go out; she felt unwell; it needed all my eloquence to persuade her and I saw Danceny arriving before we could get away, which would have been the more awkward because Madame de Volanges had told him the night before she would not be at home. Her daughter and I were on thorns. At last we went; and the little girl pressed my hand so affectionately as we said good-bye that, in spite of her determination to break off—which she really believed she meant to do—I augured wonders from the evening.

I was not at the end of my apprehensions. We had scarcely been half an hour with Madame . . . when Madame de Volanges felt ill, seriously ill; and reasonably enough she wanted to go home, which I quite as much did not want, because I was afraid if we surprised the young people (as you might have wagered we should) that my entreaties to the mother to come out might appear suspicious. I adopted the plan of frightening her about her health, which fortunately is not difficult; and I kept her an hour and a half without consenting to take her home, from the fear I pretended to have that the motion of the carriage would be dangerous. We did not get back finally until the time agreed upon. From the shamefaced air I observed on arriving, I must admit I hoped that at least my trouble had not been wasted.

My desire to learn what had happened made me remain with Madame de Volanges, who went to bed immediately; and after having taken supper at her bedside, we left her very soon, under the pretext that she needed rest; and we went into her daughter's room. She on her part had done everything I expected of her; banished scruples, new vows to love forever, etc., etc., in short she carried things out well; but the fool Danceny did not pass one inch beyond the point he was at before. Oh! He is a person one can quarrel with; reconciliations are not dangerous with him!

However, the little girl asserts that he wanted more, but that she was able to defend herself. I would wager that she is boasting or making an excuse; I practically made sure of it. In fact, the fancy came to me to find out what kind of a defence she was capable of. And I, a mere woman, from one remark to another, excited her to a degree. In short, you may believe me, never was a person so liable to an attack on the senses. She is really delightful, dear little thing! She deserves another lover! At least she will have a good woman friend, for I am sincerely attached to her. I have promised her I will train her, and I believe I shall keep my word. I have often felt the need of having a woman in my confidence, and I should prefer her to any other; but I can do nothing with her until she is . . . what she must be; and that is one more reason for being angry with Danceny.

Good-bye, Vicomte; do not call on me to-morrow, unless in

the morning. I have yielded to the entreaties of the Chevalier for an evening in my little house.[1]

From . . . , 4th of September, 17—.

LETTER LV

Cécile Volanges to Sophie Carnay

You are right, my dear Sophie, your prophecies are more successful than your advice. As you predicted, Danceny has been stronger than the confessor, than you, than myself; we are now exactly as we were. Ah! I do not regret it; and if you scold me, it will be because you do not know the pleasure there is in loving Danceny. It is very easy for you to say what I ought to do, there is nothing to prevent you; but if you had felt how much it hurts to see the grief of a person one loves, how his joy becomes yours, and how difficult it is to say *No* when you want to say *Yes*, you would not be surprised at anything; I felt it myself, I felt it very keenly, I do not yet understand it. Do you suppose I can see Danceny cry without crying myself? I assure you it is impossible for me; and when he is pleased, I am as happy as he is. You may talk as you like; what we say does not change what is, and I am quite sure that this is how it is.

I should like to see you in my place . . . No, that is not what I mean, for I certainly would not yield my place to anyone, but I should like you to love someone as well; it is not only because you would understand me better and would scold me less, but because you would be happier or, to speak more accurately, you would only then begin to be happy.

Our amusements, our laughter, all of it, you see were only child's-play; when they are gone they leave nothing behind. But love, ah! love . . . A word, a look, only to know that he is there, that is happiness. When I see Danceny I want nothing else; when I do not see him I want nothing but him. I do not know how it happens; but it is as if everything that pleases me is like him. When he is not with me I think about him; and when I can think about him completely, without being distracted, when I am quite alone for instance, I am still happy; I close my eyes and immediately I think I see him; I remember what he said; and I think I hear him; that makes me sigh; and then I feel a fire, an agitation . . . I cannot keep still. It is like torture and the torture is an inexpressible pleasure.

I even think that when one is in love once it extends to friendship as well. Yet my friendship for you is not changed; it is just as it was at the convent; but the friendship I am speaking about I feel for Madame de Merteuil. It seems to me that I love her more like Danceny than like you, and sometimes I wish she were he. Perhaps the reason for it is that this is not a childish friendship like ours; or perhaps because I see them so often together, which makes me mistake one for the other. In any case, the truth is that between them they make me very happy; and after all I do not think there is anything very wrong in what I am doing. So I ask nothing more than to go on as I am; and the only thing that troubles me is the idea of my marriage; for if Monsieur de Gercourt is like what I am told he is—and I have no doubt of it—I do not know what will become of me. Good-bye, dear Sophie; I still love you very tenderly.

From . . . , 4th of September, 17—.

LETTER LVI

Madame de Tourvel to the Vicomte de Valmont

How could the reply you ask from me, Monsieur, be of service to you? If I believed in your sentiments would that not be still another reason to fear them? And without attacking or defending their sincerity, does it not suffice me, and should it not suffice you, to know that I neither wish nor ought to reply to them?

Suppose you really loved me (and it is only to have done with the matter, that I entertain the supposition) would the obstacles which separate us be any the more surmountable? And should I have anything to do save to wish that you might speedily conquer this love, and above all to aid you to do so to the best of my power by hastening to deprive you of all hope? You admit yourself that "this is a painful sentiment when the person who inspires it does not share it." Well, you are quite aware that it is impossible for me to share it; and even if this misfortune happened to me I should be the more to be pitied without your being the happier. I hope you respect me enough not to doubt it for an instant. Desist then, I beg you, desist from attempting to trouble a heart to which tranquillity is so necessary; do not compel me to regret having known you.

Cherished and respected by a husband whom I love and respect, my duty and my pleasure meet in the same person; I am happy, I ought to be so. If keener pleasures exist, I do not desire them; I do not wish to experience them. Is there anything better than to be at peace with oneself, to spend none but tranquil days, to fall asleep without uneasiness and to wake without remorse? What you call happiness is a mere tumult of the senses, a storm of passion, the sight of which is terrifying even looked at from the shore. Ah! How can one dare these tempests? How dare to embark upon a sea covered with the remains of thousands and thousands of shipwrecks? And with whom? No, Monsieur, I shall remain on land; I cherish the bonds which attach me to it. Even if I could break them, I should not wish to do so; if I did not possess them, I should hasten to acquire them.

Why do you attach yourself to my steps? Why do you persist in following me? Your letters, which should be occasional, follow each other with rapidity. They ought to be sober, and you speak to me of nothing but your mad love. You surround me with the idea of you, more than you did in person. Separated from me in one shape, you reappear in another way. You enjoy embarrassing me by specious reasoning; you elude my own. I do not wish to reply to you again, I shall not reply again . . . How you treat the women you have seduced! With what contempt you speak of them! I am willing to believe that some deserve it; but are all so contemptible? Ah! Doubtless they are, since they betrayed their duty to yield to a criminal love. And that moment they lost everything, even to the respect of him for whom they sacrificed everything. It is a just punishment, but the very idea makes me shudder. But after all what does it matter to me? Why should I concern myself with them or with you? By what right do you trouble my tranquillity? Leave me, do not see me again; write to me no more, I beg you; I insist on it. This letter is the last you will receive from me.

From . . . , 5th of September, 17—.

LETTER LVII

The Vicomte de Valmont to the Marquise de Merteuil

I found your letter yesterday on my arrival. Your anger altogether delighted me. You could not have felt Danceny's mistakes more keenly if they had been directed towards yourself. No doubt it is from vengeance that you are accustoming his mistress to make little infidelities; indeed you are incorrigible! Yes, you are charming and I am not surprised she could resist you less easily than Danceny.

At last I know that fine novel-hero by heart! He has no more secrets! I have told him so often that virtuous love was the supreme good, that one sentiment was worth ten intrigues, that I am myself at the moment both in love and timid—in short he found my way of thinking so conformable to his own that in his enchantment at my candour he told me everything and vowed to me an unreserved friendship. This does not help us on with our plan. First of all, it appeared to me his idea was that an unmarried girl deserves much more deference than a woman, since she has more to lose. He thinks especially that a man cannot be justified in placing a girl in the necessity of marrying him or living dishonoured, when the girl is infinitely richer than the man, as she is in his case. The mother's confidence, the girl's candour, everything intimidates and restrains him. The difficulty would not be to combat his reasonings, however true they may be. With a little skill and the help of passion, they could soon be destroyed; the more so, since they lend themselves to ridicule and the authority of custom is on our side. But what prevents my having any influence over him, is that he is happy as he is. Indeed, if first loves appear in general more virtuous and, as they say, more chaste; if they are at least slower in their progress; it is not, as people think, from delicacy or timidity, but because the heart, surprised by an unknown sentiment, hesitates as it were at every step to enjoy the charm it feels, and because this charm is so powerful upon a fresh heart that it forgets every other pleasure. This is so true, that a libertine in love—if a libertine can be in love—becomes at that very moment less eager to enjoy. In short there is only the difference of more and less between Danceny's conduct with the Volanges girl and mine with the chaste Madame de Tourvel.

To inflame our young man he needs more obstacles than he has met with. Above all he needs more mystery, for mystery leads to boldness. I am not far from thinking that you have harmed us by serving him so well; your conduct would have been excellent with *an accustomed* man who would merely have had desires; but you might have forseen that to a young man, virtuous and in love, the great value of favours is that they are the proof of love; and consequently that the more certain he was of being loved the less enterprising he would be. What is to be done now? I do not know; but I have ceased to hope that the little girl will be taken before her marriage; we shall have had our trouble for nothing; I am sorry, but I see no help for it.

While I am discoursing here, you are doing something better with your Chevalier. This reminds me that you have promised an infidelity in my favour; I have your promise in writing and I do not mean to make it a *la Châtre*.[1] I admit that it has not yet fallen due; but it would be generous on your part not to wait until then; and for my part I will account to you for the interest. What do you say to this, my fair friend? Are you not wearied by your constancy? Is this Chevalier so very wonderful? Oh! Let me do what I want; I hope to compel you to admit that if you have found some merit in him, it is because you have forgetten me.

Good-bye, my fair friend; I embrace you as I desire you; I defy all the Chevalier's kisses to have as much ardour.

From . . . , 5th of September, 17—.

LETTER LVIII

The Vicomte de Valmonte to Madame de Tourvel

How have I deserved, Madame, the reproaches you make me and the anger you display towards me? The liveliest but yet the most respectful attachment, the most complete submission to your slightest wishes; there in two phrases is the history of my sentiments and of my conduct. Overwhelmed by the distress of an unhappy love, I had no consolation but that of seeing you; you ordered me to deny myself that; I obeyed without a murmur. As a reward for this sacrifice you gave me permission to write to you, and now you wish to deprive me of this one pleasure. Shall I let it be taken from me without attempting to defend it? No, certainly not. Ah! how could it fail to be dear to my heart? It is the one thing which is left me, and I hold it from you.

My letters, you say, are too frequent! Remember, I beg you, that during the ten days my exile has lasted I have not passed a moment without thinking of you, and yet you have only received two letters from me. "I speak to you of nothing but my love!"

Ah! What am I to say, if not what I think? All I could do was to weaken the expression of it; and you can believe me, I only allowed you to see what it has been impossible for me to hide. You end up by threatening not to answer me again. And so you are not content with treating harshly a man who prefers you to all else and who respects you even more than he loves you; you wish to add scorn to it! And why these threats and this anger? Why should you need them? Are you not certain to be obeyed, even when your orders are unjust? Is it possible for me to thwart any of your desires—have I not already given you a proof? Will you abuse the power you have over me? After you have rendered me unhappy, after you have become unjust, will it be so easy for you to enjoy that tranquillity which you assert is so necessary to you? Will you never say to yourself: "He made me the mistress of his fate and I made him miserable; he implored my aid and I watched him without pity." Do you know how far my despair may go? No.

To calculate my misery, you would have to know how much I love you and you do not know my heart.

You are sacrificing me to what? To imaginary fears. And who creates them in you? A man who adores you, a trustworthy man over whom you will never cease to have complete power. What do you fear, what can you fear, from a sentiment which you will always be able to control as you please, but your imagination creates monsters for itself and you attribute the terror they cause you to love. A little confidence and these phantoms will disappear.

A wise man has said that fears can almost always be dissipated by discovering their cause.[1] This truth is especially applicable to love. Love, and your fears will vanish. Instead of the objects which terrify you, you will find a delicious sentiment, a tender and submissive lover; and all your days, impressed by happiness, will leave you no regret except that you have wasted some in indifference. Since I have recognised my errors, and exist only for love, I myself regret a time which I thought was passed in pleasures;

and I feel that it is for you alone to make me happy. But, I beg you, do not let the pleasure I have in writing to you be disturbed by the fear of displeasing you. I do not wish to disobey you; but I am at your knees, I implore the happiness you would deprive me of, the only happiness you have left me; I cry to you, hear my prayers and see my tears; ah! Madame, will you refuse me?

From . . . , 7th of September, 17—.

LETTER LIX

The Vicomte de Valmont to the Marquise de Merteuil

Tell me, if you know, what is meant by this raving of Danceny's? What has happened and what has he lost? Perhaps his fair one is annoyed by his eternal respect? One must be just; anybody would be annoyed by less. What shall I say to him this evening, at the meeting he asks of me, which I have given him at hazard? Assuredly I shall not waste my time in listening to his lamentations, if this is not to lead us anywhere. Amorous complaints are not good to listen to except in recitatives or in ariettas. Tell me what has happened and what I am to do; otherwise I shall desert, to avoid the boredom I foresee. Can I have a talk with you this morning? If you are *engaged*, at least write me a word and give me the cues for my part.

Where were you yesterday? I never succeed in seeing you now. Really, it was not worth keeping me in Paris in September. Make up your mind, for I have just received a very pressing invitation from the Comtesse de B . . . to go and see her in the country; and, as she writes me amusingly enough, "her husband has a splendid wood which he carefully preserves for the pleasures of his friends." Well, you know I have some rights over that wood; and I shall go and re-visit it if I am not useful to you. Good-bye, remember Danceny will be with me at four o'clock.

From . . . , 8th of September, 17—.

LETTER LX

The Chevalier Danceny to the Vicomte de Valmont

(Enclosed with the preceding)

Ah! Monsieur, I am in despair, I have lost everything. I dare not confide to paper the secret of my distress; but I need to pour it out into the bosom of a faithful and trustworthy friend. At what hour can I see you and ask consolation and advice? I was so happy the day I opened my soul to you! And now what a difference! Everything has changed for me. What I suffer on my own account is the least part of my torture; my anxiety on behalf of a far dearer object is what I cannot endure. More fortunate than I, you can see her, and I expect of your friendship that you will not refuse me this step; but I must see you, I must tell you about it. You will pity me, you will help me; I have no hope but in you. You are sensitive, you know what love is, and you are the only person in whom I can confide; do not refuse me your assistance.

Farewell, Monsieur, the only consolation I feel in my grief is to remember I have a friend like you. Let me know, I beg you, at what hour I shall find you in. If it is not this morning, I should wish it to be early in the afternoon.

From . . . , 7th of September, 17—.

LETTER LXI

Cécile Volanges to Sophie Carnay

My dear Sophie, pity your Cécile, your poor Cécile; she is very miserable. Mamma knows everything; I cannot conceive how she can have suspected anything and yet she has discovered everything. Yesterday evening, Mamma indeed seemed to me a little out of humour but I did not pay much attention to it; and while we waited for the end of her game, I even chatted very gaily with Madame de Merteuil who had supped here, and we talked much of Danceny. Yet I do not think we could have been overheard. She went away and I retired to my apartment.

I was undressing, when Mamma came in and sent my waiting-woman away; she asked me for the key of my writing-table, and the tone in which she made this demand caused me to shake so much that I could scarcely stand up. I pretended not to be able to find it; but at last I had to obey. The first drawer she opened was precisely that which contained the Chevalier Danceny's letters. I was so upset that when she asked me what they were, I could make no answer except that they were nothing; but when I saw her begin to read the first letter which came to hand I had only time to reach an armchair before I felt so ill that I lost consciousness. As soon as I came to myself, my mother, who had called my waiting-woman, told me to go to bed and left the room. She took all Danceny's letters with her. I shudder every time I think that I must appear before her again. I have done nothing but cry all night.

I am writing to you at dawn, in the hope that Josephine will come. If I can speak to her alone, I shall beg her to deliver a little note which I am going to write to Madame de Merteuil. If not, I shall put it into your letter and you will be kind enough to send it as if it were from you. She is the only person from whom I can receive any consolation. At least we will speak of him for I do not hope ever to see him again. I am very unhappy! Perhaps she will be kind enough to take charge of a letter for Danceny. I dare not confide in Josephine for this purpose, and still less in my waiting-woman; for perhaps it was she who told my mother I had letters in my writing-desk.

I shall not write to you at more length, because I want to have time to write to Madame de Merteuil, and also to Danceny in order to have my letter quite ready if she will take charge of it. After that I shall go to bed again so that they will find me there when they come to my room. I shall say I am ill to avoid going to see Mamma. It will not be much of a lie; indeed I suffer more than if I were feverish. My eyes burn from having cried so long, and I have a weight on my stomach which prevents me from breathing. When I think that I shall never see Danceny again, I wish I were dead. Good-bye, my dear Sophie. I cannot talk to you any further; I am suffocated by tears.

From . . . , 9th of September, 17—.

[Note: Cécile Volange's letter to the Marquise has been suppressed, because it only contained the same facts as the preceding letter and with less details. The Chevalier Danceny's letter could not be found. The reason for this will be seen in *Letter LXIII*, from Madame to the Vicomte (C. de L.)]

LETTER LXII

Madame de Volanges to the Chevalier Danceny

Monsieur, after having abused a mother's confidence and the innocence of a child, you will doubtless not be surprised if you are received no longer in a house where you have repaid the proofs of a most sincere friendship by a complete forgetfulness of all good behaviour. I prefer to ask you not to come to my house again, instead of giving orders at my door, which would compromise us all through the remarks the footman could not fail to make. I have a right to hope that you will not force me to make use of this method. I warn you also that if in the future you make the slightest attempt to continue my daughter in the aberration you have plunged her, an austere and eternal retirement will remove her from your solicitation. It is for you to see, Monsieur, whether you will have as little scruple in causing her misfortune as you have had in attempting her dishonour. For my part, I have made my choice and she knows it.

You will find enclosed a packet of your letters. I expect you to send me in exchange all my daughter's letters; and that you will assist in leaving no trace of an event whose memory cannot be recalled by me without indignation, by her without shame, and by you without remorse. I have the honour to be, etc.

From . . . , 7th of September, 17—.

LETTER LXIII

The Marquise de Merteuil to the Vicomte de Valmont

Yes, indeed, I will explain Danceny's note to you. The event which made him write it, is my work and is, I think, my masterpiece. I have not wasted my time since your last letter and I said like the Athenian architect: "What he has said I will do."

This fine novel-hero needs obstacles, he is drowsing in his felicity! Oh! Let him apply to me, I will find work for him; and if I am not mistaken his sleep will not be so tranquil. He had to be taught the value of time and I flatter myself that he now regrets the time he has wasted. He needed more mystery, you said too; well, he will not lack that necessity. I have one good point; I need only to be showed my mistakes and I never rest until I have retrieved them. Learn then what I have done.

When I returned home two mornings ago, I read your letter and thought it enlightening. Convinced that you had very plainly pointed out the cause of the trouble, I employed myself entirely in thinking how to remove it. However, I began by going to bed; for the indefatigable Chevalier had not allowed me a moment's sleep and I thought I was drowsy; but not at all; I was entirely occupied with Danceny, and my desire to drag him out of his indolence or to punish him for it prevented me from shutting my eyes, and it was not until I had thoroughly worked out my plans that I obtained a couple of hours' rest.

The same evening, I called on Madame de Volanges and, in accordance with my plan, I told her in confidence that I was sure there existed a dangerous acquaintance between her daughter and Danceny. This woman, so clear-sighted against you, was so blinded that she replied to me at first that I must be wrong; that her daughter was a child, etc., etc. I could not tell her all I knew; but mentioned glances, remarks, "which had alarmed my virtue and my friendship." Finally I talked almost as well as a devotee would have done; and, to strike the decisive blow, I went so far as to say that I thought I had seen a letter given and received. That reminds me, I added, that one day she opened a drawer of her writing-table in my presence, and I saw in it a lot of papers, which no doubt she is keeping. Do you know whether she has any frequent correspondence? Here Madame de Volanges's face changed and I saw a few tears swim in her eyes. "I thank you, my real friend," she said, pressing my hand, "I shall look into this."

After this conversation, which was too short to be suspicious, I went up to the young person. I left her soon after, to ask the mother not to compromise me with her daughter; which she promised the more willingly, since I pointed out to her how convenient it would be if the child should attain enough confidence in me to open her heart to me and put me in a position to give her "my careful advice." What convinces me that she will keep her promise, is that I have no doubt but that she wishes to impress her daughter with her own acuteness. In this way I shall be able to keep up my tone of friendship with the girl, without appearing deceitful in Madame de Volanges's eyes; which I desired to avoid. Moreover, in the future I shall have the advantage of remaining as long and as secretly with the girl as I wish, without the mother ever taking offence.

I profited by it that very evening; and when my card-game was over I took the child aside into a corner and started her on the subject of Danceny, about whom she is inexhaustible. I amused myself by exciting her with the pleasure she would have in seeing him the next day; there was no kind of folly I did not make her express. I had to give her back in hope what I took from her in reality; and then all this will make her feel the blow more acutely and I am convinced that the more she suffers the more eager she will be to console herself for it on the first occasion. Moreover it is good to accustom to such events someone whom one destines to great adventures.

After all, can she not pay with a few tears for the pleasure of having her Danceny? She is wild about him! Well, I promise her she shall have him and that but for this storm she would never have had him. It is a bad dream, the awakening from which will be delicious; and, looking at it all round, it seems to me she ought to be grateful to me; and if I have mixed a little malice with it, one must be amused:

> Fools are here below for our minor pleasures.[1]

I left at last, very pleased with myself. Either Danceny, I said to myself, animated by obstacles, will redouble his love, and then I shall have served him to the best of my power; or if he is a mere fool, as I have sometimes been tempted to think, he will be in despair and will consider himself beaten; in this case, at least I shall be avenged on him, as far as it was in my power, and so doing I shall have increased the mother's esteem, the daughter's friendship, and the confidence of both. As to Gercourt, who is the primary object of my attention, I shall be very unlucky if, mistress of his wife's mind as I am and shall be even more, I do not find a thousand ways of making him what I wish him to be. I went to bed with these pleasant ideas; I therefore slept well and woke up very late.

At my waking, I found two notes, one from the mother and one from the daughter; I could not keep from laughing when I found literally the same phrase in both: "it is from you alone I expect any consolation." Is it not indeed amusing to console for and against, and to be the only agent of two directly opposite interests? Here I am like the Divinity, receiving the contrary prayers of blind mortals and changing nothing in my immutable decrees.

However I quitted this august role to take on that of the consoling angel; and in accordance with the precept, I have visited my friends in their affliction.

I began by the mother; I found her in a state of grief which already partly avenges me for the annoyances she made you undergo with your fair prude. Everything has succeeded perfectly; my one fear was lest Madame de Volanges might profit by the moment to gain her daughter's confidence, which would have been very easy if the language of gentleness and kindness had been employed towards her, and if the advice of reason had been given with the air and tone of indulgent tenderness. Fortunately, she armed herself with severity; she arranged it all so badly indeed that I had nothing to do but to applaud. It is true she nearly spoiled all our plans by her determination to send her daughter back to the convent, but I parried this thrust; and I have persuaded her merely to threaten it in case Danceny should continue his attentions; this was to force them both to a circumspection I think necessary for success.

I then went to the daughter. You will not believe how grief has improved her! If she becomes even a little coquettish I warrant you she will often cry; but this time she was crying without malice aforethought. Struck by this new charm which I had not observed in her before and which I was very glad to notice, I gave her only clumsy consolations at first which increased rather than allayed her grief; and in this way I brought her to a state where she was really choking. She ceased crying and for a moment I was afraid of convulsions. I advised her to go to bed which she agreed to; I acted as her waiting-woman; she had not made her toilet and soon her scattered hair fell on her shoulders and breasts which were entirely uncovered. I kissed her; she let me take her in my arms and her tears began to flow again without effort. Heavens! How beautiful she was! Ah! if Magdalene was like this she must have been far more dangerous as a penitent than as a sinner.

When the disconsolate fair one was in bed I began to console her in earnest. First of all I reassured her about her fear of the convent. I awoke the hope in her of seeing Danceny secretly; and sitting on the bed I said to her: "If he were here"; and then embroidering on that theme I led her from one diversion to another until she forgot she was afflicted. We should have separated perfectly pleased with each other if she had not wished to entrust me with a letter for Danceny, which I consistently refused. Here are my reasons, which no doubt you will approve.

First of all it was compromising me with Danceny; and though this was the only reason I could give the little girl, there are nevertheless many other reasons between ourselves. Would it not be risking the fruit of my labours if I gave our young people so soon such an easy way of soothing their distress? And then, I should not be sorry to compel them to entangle a few servants in this adventure; for if it turns out well, as I hope it will, it must be known immediately after the marriage; and there are few more certain methods of spreading it about; or, if by some miracle they did not talk, we would talk, and it will be more convenient to lay the indiscretion upon them.

You must put this idea into Danceny's head to-day; and since I am not certain of the Volanges's waiting-woman (whom she seems not to trust) tell him of mine, my faithful Victoire. I shall take care that the application succeeds. This idea pleases me all the more because the confidence will only be useful to us and not to them; for I have not reached the end of my story.

While I was refusing to take charge of the little girl's letter, I feared every moment that she would ask me to put it in the post-box; which I could not have refused. Happily, either from agitation, or from ignorance, or because she thinks less of the letter than of the reply (which she could not receive in this way) she did not speak of it; but to prevent this idea coming to her or at least to prevent her making use of it, I made up my mind immediately; and returning to her mother I persuaded her to take her daughter away for some time, to take her to the country . . . and where? Does not your heart beat with joy? . . . To your aunt, to the old Rosemonde woman. She is to let her know about it to-day; and thus you have a pretext to return to your devotee who will not be able to reproach you with the scandal of a *tête à tête*: and thanks to my care, Madame de Volanges herself will repair the harm she has done you.

But listen to me, and do not busy yourself so eagerly with your own affairs that you lose sight of this; remember I am interested in it.

I want you to become the correspondent and adviser of the two young people. Inform Danceny of this journey and offer him your services. Do not allow any difficulty except that of delivering your letter of credit to the fair one's hands; and immediately overcome the obstacle by telling him that my waiting-woman is the way. There can be no doubt that he will accept; and as a reward for your trouble you will have the confidence of an unspoiled heart, which is always an interesting thing. Poor little girl! How she will blush when she hands you her first letter! Really, the part of confident, against which prejudices exist, seems to me a very pleasant relaxation when one is occupied elsewhere; which will be the case with you.

The result of this intrigue will depend upon your care. You must decide on the time when the actors must be brought together. The country offers a thousand means; and Danceny will assuredly be ready to go there at your first signal. A night, a disguise, a window . . . anything. But if the little girl returns here in the same state she goes away I shall blame you. If you think she needs any encouragement from me, let me know. I think I have given her a good enough lesson on the danger of keeping letters to dare to write to her now; and I still mean to make her my pupil.

I think I have forgotten to tell you that her suspicions about the betrayal of her correspondence fell first of all upon her waiting-woman, and that I directed them against the confessor. It is killing two birds with one stone.

Good-bye, Vicomte; I have been writing to you a long time and it has delayed my dinner; but my letter was dictated by self-esteem and friendship; both are chatterers. However, it will reach you at three o'clock and that is all you need.

Complain of me now if you dare; and if you are tempted, go and have another look at the Comte de B . . .'s wood. You say he keeps it for his friends' pleasure! And is the man everybody's friend? But good-bye, I am hungry.

From . . . , 9th of September, 17—.

LETTER LXIV

The Chevalier Danceny to Madame de Volanges
(Sent with *Letter LXVI* from the Vicomte to the Marquise)

Without seeking to justify my conduct, Madame, and without complaining of yours, I cannot but be afflicted by an event which makes three persons miserable, all three of whom are worthy of a happier fate. I am more grieved that I am its cause than that I am its victim, and ever since yesterday I have been trying to reply to you without finding strength to do so. Yet I have so many things to tell you that I must make an effort; and if this letter lacks order and continuity you will be sufficiently conscious of the painfulness of my situation to grant me some indulgence.

Permit me first to protest against the first phrase in your letter. I dare to assert that I abused neither the confidence nor the innocence of Mademoiselle de Volanges; I respected both in my actions. My actions alone were within my control; and if you should make me responsible for an involuntary sentiment I am not afraid to add that the feelings inspired in me by your daughter were such as might displease you but could not offend you.

On this matter, which moves me more than I can say I ask none but yourself as judge and my letters as witnesses.

You forbid me to enter your house in the future and I shall certainly submit to everything you are pleased to order in this matter; but will not this sudden and total absence give as much occasion to the observations you wish to avoid as the order which, for that very reason, you did not wish to give at your door? I insist the more on this point, since it is far more important for Mademoiselle de Volanges than for me. I beg you then to weigh everything carefully and not to allow your severity to warp your prudence. I am convinced that nothing but your daughter's interests will determine your resolve and shall await new orders on your part.

But in case you should permit me to call upon you sometimes, I promise you, Madame (and you can count upon my promise), not to abuse these occasions by attempting to speak privately to Mademoiselle de Volanges or to convey any letter to her. The fear of what might compromise her reputation binds me to this sacrifice; and the happiness of seeing her sometimes will console me. This part of my letter is also the sole reply I can make to what you say about the fate to which you destine Mademoiselle de Volanges, a fate which you wish to make dependent upon my conduct. To promise you more would be to deceive you. A vile seductor can suit his plans to circumstances and calculate according to events; but the love which animates me permits me only two sentiments—courage and constancy.

What! Shall I consent to be forgotten by Mademoiselle de Volanges, to forget her myself? No, no, never. I shall be faithful to her; she has received my oath and I renew it to-day. Pardon me, Madame, I digress; I must return.

There remains one other matter for me to discuss with you; that of the letters you ask from me. I am truly pained to add a refusal to the harm you think I have committed; but I beg you will listen to my reasons and will deign to remember, in order to appreciate them, that the sole consolation for the misfortune of having lost your friendship is the hope of retaining your esteem.

Mademoiselle de Volanges's letters which were always so precious have become much more so at this moment. They are the one treasure left to me; they alone still recall to me a sentiment which creates the whole charm of my life; yet, you may believe me, I should not hesitate a moment to sacrifice them, and my regret at being deprived of them would yield to the desire to prove my respectful deference to you; but powerful considerations restrain me and I am sure that you yourself will not be able to condemn them.

It is true you have Mademoiselle de Volanges's secret. But allow me to tell you I have every reason to believe that this was the result of a surprisal and not of confidence. I do not presume to blame a proceeding which was authorised, perhaps, by maternal solicitude. I respect your rights but they do not go so far as to relieve me of my duties. The most sacred of all is never to betray the confidence placed in us. I should be failing in my duty if I exposed to the eyes of another the secrets of a heart which desired only to unveil them to mine. If your daughter consents to confide them to you, let her speak; her letters are useless to you. If on the contrary she wishes to keep her secret to herself, doubtless you will not expect that I should make them known to you.

As to the silence in which you desire this event to remain buried, be at rest, Madame; in all that concerns Mademoiselle de Volanges, I can defy even the heart of a mother. To relieve you completely of all uneasiness, I have foreseen everything. The precious packet which hitherto has been inscribed "Papers to be burned"; is now inscribed: "Papers belonging to Madame de Volanges." The course I am taking will also prove to you that my refusal is not based on the fear that you would find in these letters a single sentiment of which you personally might have reason to complain.

This is a very long letter, Madame. Yet it will not be sufficiently long if it leaves you the least doubt of the honesty of my sentiments, of my very sincere regret at having displeased you and of the profound respect with which I have the honour to be, etc.

From . . . , 9th, of September. 17—.

LETTER LXV

The Chevalier Danceny to Cécile Volanges

(Sent open to the Marquise de Merteuil
in *Letter LXVI* from the Vicomte)

O, my Cécile, what will become of us? What god will save us from the misfortunes which threaten us? May love at least give us courage to endure them! How can I describe to you my astonishment, my despair, at the sight of my letters, at reading Madame de Volanges's note? Who can have betrayed us? Whom do your suspicions fall upon? Did you do something imprudent? What are you doing now? What has been said to you? I want to know all about it and I know nothing. But perhaps you yourself know nothing more than I do.

I send you your Mamma's note and a copy of my reply. I hope that you will approve of what I say to her. I want you to approve also the steps I have taken since this fatal event; the purpose of them all is to have news from you and to give you news of me; and, who knows? perhaps to see you again, more freely than before.

Can you imagine, my Cécile, what pleasure it would be to be together again, to be able to swear once more an eternal love, and to see in our eyes, to feel in our souls, that this vow will never be broken? What griefs would not be forgotten in so sweet a moment? Well, I hope to see it come, and I owe it to the very steps I beg you to approve. What do I say? I owe it to the consoling solicitude of the tenderest friend and my one request is that you will permit this friend to be yours also.

Perhaps I ought not to have given your confidence without your permission? But I have unhappiness and necessity for my excuse. Love it was that guided me; it is love which claims your indulgence, which begs you to forgive a necessary confidence without which we should remain perhaps separated forever.[1] You know the friend of whom I speak; he is the friend of the woman you love most. It is the Vicomte de Valmont.

My design, in going to him, was first of all to ask him to persuade Madame de Merteuil to take charge of a letter to you. He did not think this method would succeed; but in default of the mistress he answers for the waiting-woman, who has obligations to him. She will hand you this letter and you can give her your reply.

This assistance will be of no use to us if, as Monsieur de Valmont believes, you are going to the country immediately. But in that case he himself will help us. The woman to whose house you are going is his relative. He will take advantage of this pretext to go there at the same time as you do; and our correspondence will pass through his hands. He even promises that if you will allow yourself to be guided he will arrange for us to see each other without any risk of compromising you at all.

And now, my Cécile, if you love me, if you pity my misfortune, if, as I hope, you share my regrets, will you refuse your confidence to a man who will be our guardian angel? But for him I should be reduced to the despair of being unable even to soften the sorrows I cause you. They will end, I hope; but, my dear one, promise me not to abandon yourself to sorrow too much, do not let yourself be prostrated. The idea of your pain is unendurable torture to me. I would give my life to make you happy! You know it. May the certainty that you are adored bring some consolation to your soul! Mine needs your assurance that you forgive Love the ills he makes you suffer.

Good-bye, my Cécile; good-bye, my dear one.

From . . . , 9th of September, 17—.

LETTER LXVI

Vicomte de Valmont to the Marquise de Merteuil

You will see, my fair friend, when you read the two enclosed letters that I have carried out your plan well. Although they are both dated to-day, they were written yesterday in my house and under my eyes; that to the little girl says everything we could wish. I can only bow before the profundity of your insight, if I may judge of it by the success of your proceedings. Danceny is all on fire; and assuredly at the first occasion you will have nothing more to reproach him with. If his fair *ingénue* is docile all will be over soon after his arrival in the country; I have a hundred methods ready. Thanks to your care I am now positively *Danceny's friend*; the only thing he needs now is to be a *Prince.*[1]

Danceny is still very young! Would you believe I could not persuade him to promise the mother that he would give up his love; as if it were very difficult to make a promise when you have decided not to keep it! It would be deceitful, he kept repeating; an edifying scruple, is it not, especially as he wants to seduce the daughter; but that is what men are! They are all equally base in their designs and their weakness in carrying them out they call probity.

It will be your business to prevent Madame de Volanges's being scared at the little pranks our young man has allowed himself in his letter; preserve us from the convent; and try to make her give up her request for the little girl's letters. First of all he will not give them up, he does not want to, and I agree with him; here love and reason tally. I have read these letters, I have devoured their boredom. They may become useful. I will explain.

In spite of all our prudence, there might be a scandal; it would prevent the marriage, would it not, and destroy all our plans for Gercourt. But since I want to avenge myself on the mother, in this case I intend to dishonour the girl. By choosing carefully from this correspondence and only producing part of it, the little

Volanges girl could be made to appear as if she had taken all the first steps and absolutely thrown herself at his head. Some of the letters might even compromise her mother, and would at least *stain* her with an unpardonable negligence. I quite realise that the scrupulous Danceny would refuse at first but since he would be personally attacked, I think he could be persuaded. It is a thousand to one against the luck turning this way; but we must foresee everything.

Farewell, my fair friend; it would be very agreeable of you to come and sup to-morrow with the Maréchale de . . .; I was unable to refuse. I suppose there is no need to recommend secrecy towards Madame de Volanges about my country projects; she would soon decide to stay in town; while, once she arrives there, she will not leave the next day; and if she gives us only a week I will answer for everything.

From . . . , 9th September, 17—.

LETTER LXVII

Madame de Tourvel to the Vicomte de Valmont

I did not desire to answer you again, Monsieur, and perhaps the embarrassment I feel at this moment is itself a proof that indeed I ought not to do so. However I do not wish to leave you any subject for complaint against me; I wish to convince you that I have done everything for you I could do.

I gave you permission to write to me, you say? I admit it; but when you remind me of this permission do you think I forget on what conditions it was given to you? If I had kept to them as much as you have departed from them, would you have received a single reply from me? Yet this is the third; while you are doing everything that must force me to break off this correspondence, it is I who am employed in trying to continue it. There is one way of doing so, but only one; and if you refuse to take it, it will be, whatever you may say, a proof to me of how little value you set upon it.

Abandon a language which I cannot and will not listen to; renounce a sentiment which offends and frightens me, to which perhaps you ought to be less attached by remembering that it is the object which separates us. Is this sentiment the only one you are able to feel, and shall love have one fault the more in my eyes by excluding friendship? And will you yourself commit the fault of not desiring to have as a friend her in whom you have wished for more tender sentiments? I am unwilling to think so; this humiliating idea would revolt me, would estrange me from you forever.

In offering you my friendship, Monsieur, I give you all I have, all that I can bestow. What more can you desire? To yield to this gentle sentiment, so well suited to my heart, I await only your consent and the promise I exact from you that this friendship shall suffice for your happiness. I shall forget everything which may have been said to me; I shall rely on you to justify my choice.

You see my frankness; it should prove to you my confidence; it depends entirely on you to increase it further; but I warn you that the first word of love destroys it forever and brings back all my fears, above all that it will become for me the signal for an eternal silence towards you.

If, as you say, you have abandoned your errors, would you not prefer to be the object of a virtuous woman's friendship than of a guilty woman's remorse? Good-bye, Monsieur; you will realise that after having spoken thus I can say nothing more until you have replied to me.

From . . . , 9th September, 17—.

LETTER LXVIII

The Vicomte de Valmont to Madame de Tourvel

How can I reply to your last letter, Madame? How can I dare to be outspoken, when my sincerity may ruin me with you? No matter, it must be so; I shall have the courage to do it. I tell myself, I repeat to myself, that it is better to deserve you than to obtain you; and though you should refuse me forever a happiness I shall ceaselessly desire, I must at least prove to you that my heart is worthy of it.

What a pity it is that, as you put it, I have abandoned my errors. With what transports of joy I should then have read this very letter to which I fear to reply to-day! In it you speak to me "frankly," you express your "confidence" in me, you finally offer me your "friendship." What treasures, Madame, and how I regret I cannot profit by them! Why am I no longer what I was?

If I were so; if I felt only an ordinary inclination for you, that light desire, the child of seduction and pleasure, which is called love nowadays, I should hasten to take advantage of all I could obtain. With little delicacy concerning means so long as they procured the success, I should encourage your frankness in order to discover your thoughts; I should want your confidence with the purpose of betraying it; I should accept your friendship in the hope of deluding it . . . What! This picture frightens you, Madame? . . . Well! It would be my portrait if I told you that I consent to be nothing but your friend . . .

Shall I consent to share with anyone else a sentiment that comes from your soul? If ever I should say so, do not believe me. From that moment I should be seeking to deceive you; I might still desire you, but I should certainly have ceased to love you.

It is not that amiable frankness, soft confidence, tender friendship are valueless in my eyes . . . But love, true love, the love you inspire, uniting all these feelings, giving them more energy, could not lend itself like them to that tranquillity, to that coldness of soul which allows comparisons, which even endures preferences. No, Madame, I shall not be your friend; I shall love you with the tenderest, the most ardent and yet the most respectful love; you may bring it to despair, but you cannot destroy it.

By what right do you presume to dispose of a heart whose homage you refuse? From what refinement of cruelty is it that you envy me even the happiness of loving you? That is mine, it is independent of you. I shall take care to defend it. Though it is the source of my woes, it is also their remedy.

No, once again, no. Persist in your cruel denials, but leave me my love. You delight in rendering me unhappy! Well, so be it; try to weary my courage; I shall at least be able to compel you to decide my fate and perhaps some day you will do me more justice. It is not that I hope ever to make you feel this emotionally: but without being persuaded, you will be convinced, you will say to yourself: I misjudged him.

Let us put it more plainly, you do yourself an injustice. To know you without loving you, to love you without being constant to you, are both equally impossible; and in spite of the modesty which embellishes you, it must be easier for you to complain of, than to be surprised by, the feelings you create. As for me, whose sole merit is that of having appreciated you, I desire not to lose you; and far from agreeing to your insidious offers, I renew at your feet my oath to love you forever.

From . . . , 10th of September, 17—.

LETTER LXIX

Cécile Volanges to the Chevalier Danceny

(Note written in pencil and copied out by Danceny)

You ask me what I am doing; I love you, and I weep. My mother does not speak to me; she has taken away my paper, pens and ink; I am using a pencil which luckily was left me and I write to you on a piece of your letter. I must approve all you have done; I love you too much not to take any means of getting news of you and giving you my news. Although I did not like Monsieur de Valmont and did not believe him to be your friend, I shall try to accustom myself to him and I shall love him for your sake. I do not know who betrayed us; it can only have been my waiting-woman or my confessor. I am very miserable; to-morrow we are going to the country; I do not know for how long. Great Heaven! Never to see you again! I have no more space. Good-bye; try to read this. These pencilled words will perhaps be effaced, but never the sentiment engraved in my heart.

From . . . , 10th of September, 17—.

LETTER LXX

The Vicomte de Valmont to the Marquise de Merteuil

I have something important to tell you, my dear friend. Yesterday, as you know, I supped with the Maréchale de . . .; you were mentioned and I expressed, not all the good I think of you, but all the good I do not think. Everybody appeared to be of my opinion and the conversation was languishing as always happens, when we say nothing but good of our neighbour, when a contradictor arose; it was Prévan.

"Heaven forbid," said he as he got up, "that I should doubt Madame de Merteuil's virtue; but I dare to think she owes more to her levity than to her principles. It is perhaps more difficult to pursue her than to please her; since in running after a woman one never fails to meet others on the way, since, all things considered, these other women may be as good or better than she is, some men are distracted by a new inclination, others desist from lassitude; and perhaps she has had to defend herself less than any woman in Paris. For my part," he added, (encouraged by the smiles of several women), "I shall only believe in Madame de

Merteuil's virtue after I have killed six horses in paying court to her."

This mockery succeeded like all those which arise from malice; and during the laughter it excited, Prévan sat down and the general conversation changed. But the two Comtesses de B . . ., who were beside our unbeliever, continued the subject with him in their private conversation, which happily I was in a position to over-hear.

The challenge to move your tenderness was accepted. The promise to tell everything was given; and of all the promises which might be given in this adventure that would assuredly be the most religiously kept. But you are forewarned, and you know the proverb . . .

It remains for me to tell you this Prévan, whom you do not know, is infinitely agreeable and still more dextrous. If you have sometimes heard me say the contrary, it is only because I do not like him, that I enjoy thwarting his successes and that I know what power my opinion has with some thirty of our most fashionable women.

Indeed, by this means I prevented him for a long time from appearing on what we call the theatre of fashion; and he performed prodigies without acquiring any reputation. But the noise of his triple adventure, by turning all eyes upon him, has given him the confidence he lacked hitherto and has made him really formidable. In short he is perhaps the only man I should now be afraid of meeting on my path; and apart from your own interests, you would do me a real service by making him a little ridiculous on the way. I leave him in good hands and I hope that when I return he will be a lost man.

In return I promise you to conduct successfully your pupil's adventure and to pay as much attention to her as to my fair prude.

The latter has just sent me a plan of capitulation. Her whole letter shows her desire to be deluded. It is impossible to offer a more convenient and more usual method. She wants me to be her friend. But I, who like new and difficult methods, do not intend to let her off so easily; and I shall certainly not have taken so much trouble with her in order to end up with an ordinary seduction.

On the contrary, my plan is that she shall feel, and feel thoroughly, the value and extent of each sacrifice she makes me; not to lead her so fast that remorse cannot follow her; to make her virtue expire in a slow agony; to hold her attention continually upon this painful spectacle; and not to grant her the happiness of having me in her arms until I have forced her to admit her desire for it. After all, I am not worth much if I am not worth the trouble of being asked for. And can I take a less vengeance on a lofty woman who appears to blush to admit that she adores?

I have therefore refused this precious friendship and have held to my title of lover. Since I am aware that this title, which at first appears merely a dispute about words, is really an important thing to obtain, I took great care with my letter and tried to put into it that disorder which alone can render sentiment. In short I talked as much nonsense as I could; for without talking nonsense, there is no tenderness; and I think that is the reason why women are so much our superiors in love-letters.

I ended up mine with a piece of flattery and that is another result of my profound observation. When a woman's heart has been exercised for some time, it needs rest; and I have noticed that flattery is the softest pillow one can offer any of them. Good-bye, my fair friend; I leave to-morrow. If you have any orders to give me for the Comtesse de . . ., I shall stay at her house at least for dinner. I am sorry to go without seeing you. Send me your sublime instructions and help me with your wise counsels at this decisive moment.

Above all things resist Prévan; and may I one day compensate you for this sacrifice! Good-bye.

From . . . , 11th of September, 17—.

LETTER LXXI

The Vicomte de Valmont to the Marquise de Merteuil

My rattle-headed servant has left my portfolio in Paris! My fair one's letters, those of Danceny to the Volanges girl, were all left behind, and I need them all. He is just leaving to make amends for his stupidity; and while he is saddling his horse, I will give you an account of last night; for I beg you to believe I do not waste my time.

The adventure of itself is a very small thing; it was only a revival with the Vicomtesse de M . . . But it interested me in its details. Moreover I am glad to let you see that if I have a talent for misleading women I have none the less, when I wish, that of excusing them. I always take the most difficult or the most amusing course; and I do not blame myself for a good action, so long as it gives me exertion or amusement.

I found the Vicomtesse here, and when she added her requests to the entreaties of the others that I should spend the night in the Chateau: "Well! I consent," I said to her, "on condition that I spend it with you." "Impossible," she replied, "Vressac is here." Till that moment I had merely meant to say something polite; but that word "impossible" roused me, as it always does. I felt humiliated at being sacrificed to Vressac and I resolved not to endure it; so I insisted.

The circumstances were not favourable. Vressac has been clumsy enough to arouse the Vicomte's suspicions; so that the Vicomtesse cannot receive him in her own house; and this visit to the good Comtesse had been arranged between them with the purpose of snatching a few nights. At first the Vicomte even showed his annoyance at meeting Vressac there; but since he is even more a sportsman than a jealous husband he stayed none the less; and the Comtesse, who is still the same as she was when you knew her, after putting the wife in the main corridor, lodged the husband on one side and the lover on the other and left them to arrange it between them. The evil genius of both of them caused me to be lodged opposite.

That very day (that is to say yesterday), Vressac who, as you may suppose, is flattering the Vicomte, went shooting with him in spite of his dislike for sport, and intended to console himself at night in the wife's arms for the boredom the husband caused him all day; but I considered he would need a rest and devoted my attention to the means of persuading his mistress to give him time for it.

I succeeded and made her agree to quarrel with him over this day's shooting, to which obviously he had only consented for her sake. A worse pretext could not have been found; but no woman possesses to a higher degree than the Vicomtesse that talent common to them all of putting caprice in the place of common sense and of being never so difficult to sooth as when she is in the wrong. Moreover it was not a convenient time for explanation; and, as I only wanted one night I agreed that they should patch matters up the next day.

Vressac then was sulked with when he came back. He tried to ask the reason; she started a quarrel with him. He attempted to justify himself; the husband, who was present, served as a pretext for breaking off the conversation; finally he tried to profit by a moment when the husband was out of the room to request that he might be given a hearing at night. It was then that the Vicomtesse was sublime. She grew indignant with the audacity of men who, because they have received favours from a woman, think they have the right to abuse her even when she has reason to complain of them; and having changed the topic thus skilfully, she talked delicacy and sentiment so well that Vressac was rendered mute and confused and I myself was tempted to think she was right; for you must know that as a friend of them both I was present at the conversation.

She finally declared positively that she would not add the fatigues of love to those of shooting and that she would blame herself for disturbing such soft pleasures. The husband came back; the wretched Vressac, who thereby lost the opportunity of replying, addressed himself to me; and after he had told me his grievances at great length (which I knew as well as he did) he begged me to speak to the Vicomtesse and I promised him to do so. I did indeed speak to her; but it was to thank her and to arrange with her the hour and means of our meeting.

She told me that since she was lodged between her husband and her lover, she had thought it more prudent to go to Vressac than to receive him in her apartment; and that since I was opposite her she thought it safer also to come to me; that she would arrive as soon as her waiting-woman had left her alone; and that I had only to leave my door ajar and to wait for her.

Everything was carried out as we had agreed; she came to my room about one o'clock in the morning.

> . . . in the simple attire
> Of a beauty just snatched from sleep.[1]

Since I am without vanity, I will not dwell upon the details of the night; but you know me, and I was satisfied with myself.

At dawn we had to separate. And here the interest begins. The scatter-brain thought she had left her door ajar; we found it closed with the key inside; you can have no idea of the expression of despair with which the Vicomtesse said to me at once: "Ah! I am ruined." It must be admitted that it would have been amusing to leave her in this situation, but could I allow a woman to be ruined *for* without being ruined *by* me? And ought I, like the majority of men, to let myself be dominated by events? A way out had to be found. What would you have done, my fair friend? This is what I did, and it succeeded.

I soon found out that the door in question could be broken in by making a great deal of noise. I therefore persuaded the Vicomtesse, not without difficulty, to give piercing cries of terror, such as "Thieves!" "Murder!" etc., etc. And we agreed that at the first cry I should break down the door while she rushed into bed. You cannot think how much time was needed to bring her to the point even after she had agreed to do it. However it was the only thing to do and at the first kick the door gave way.

It was fortunate for the Vicomtesse that she did not waste any time. For at the same moment the Vicomte and Vressac were in the corridor and the waiting-woman rushed to her mistress's room.

I alone kept my head and I made use of it to extinguish a night-light which was still burning and to throw it on the ground; for you will realise how ridiculous it would have been to feign this panic terror with a light in the room. I then abused the husband and the lover for their heavy sleeping, assuring them that the cries, at which I had run out and my efforts to break down the door had lasted at least five minutes.

The Vicomtesse, who had recovered her courage in her bed, seconded me well enough and vowed by all the gods that there was a thief in her apartment. She protested with more sincerity that she had never been so afraid in her life We looked everywhere and found nothing, when I pointed out the overturned night-light and concluded that no doubt a rat had caused the damage and the terror; my opinion was accepted unanimously and after a few hackneyed pleasantries about rats the Vicomte was the first to return to his room and his bed, begging his wife to have quieter rats in future.

Vressac remained alone with us and went up to the Vicomtesse to tell her that this was a vengeance of Love; upon which she looked at me and replied: "He must have been angry for he has avenged himself amply; but," she added, "I am quite exhausted, I must go to sleep."

I was in an expansive mood; consequently, before we separated I pleaded Vressac's cause and brought about a reconciliation. The two lovers embraced and I in my turn was embraced by them both. I had no more interest in the Vicomtess's kisses; but I admit that Vressac's gave me pleasure. We went out together, and after I had received his lengthy thanks we each went back to bed.

If you think this story amusing, I do not ask you to keep it secret. Now that I have had my amusement it is just that the public should have its turn. At the moment I am only speaking of the story; perhaps we shall very soon say as much of the heroine.

Good-bye; my servant has been waiting an hour; I delay only long enough to embrace you and to warn you above everything to beware of Prévan.

From the Château of . . . , 13th of September, 17—.

LETTER LXXII

The Chevalier Danceny to Cécile Volanges

(Not delivered till the fourteenth)

O my Cécile! How I envy Valmont's lot! To-morrow he will see you. He will deliver this letter to you and I, languishing far from you, shall drag out my weary existence between regrets and misery. My love, my tender love, pity me for my misfortune; above all pity me for yours; it is against them that courage abandons me.

How dreadful it is for me to be the cause of your misery! But for me, you would be happy and tranquil. Can you forgive me? Tell me, ah! tell me that you forgive me; tell me also that you love me, that you will love me forever. I must have you repeat it to me. It is not that I doubt it; but it seems to me that the more certain I am of it the sweeter it is to hear it said. You love me, do you not? Yes, you love me with all your soul. I do not forget that this was the last word I heard you speak! How I gathered it to my heart! How profoundly it is engraved there! And with what delight mine responded to it!

Alas! In that moment of happiness I was far from foreseeing the dreadful fate which awaited us. Let us think, my Cécile, of how to mitigate it. If I am to believe my friend, you have only to give him the confidence he deserves and we shall achieve it.

I admit I was pained by the unfavourable idea you seem to have of him. In that I recognised your mother's prejudices; it was from deference to them that for some time I neglected this truly amiable man who to-day is doing everything for me, who, in short, is working to re-unite us when your Mamma has separated us. I beg you, my dear, to look upon him with a more favourable eye. Remember he is my friend, that he wishes to be yours, that he can procure me the happiness of seeing you. If these reasons do not convince you, my Cécile, you do not love me as much as I love you, you do not love me as much as you did love me. Ah! If ever you should love me less . . . But no, my Cécile's heart is mine; it is mine for life; and if I have to dread the grief of an unhappy love, her constancy at least will save me from the tortures of a betrayed love.

Good-bye, my charming love; do not forget that I am suffering and that it only depends upon you to render me happy, perfectly happy. Listen to my heart's prayer, and receive the tenderest kisses of love.

Paris, 11th of September, 17—.

LETTER LXXIII

The Vicomte de Valmont to Cécile Volanges

(Enclosed with the preceding letter)

The friend who assists you has learned that you have no means of writing and has already provided for it. In the ante-chamber of the apartment you occupy, under the large wardrobe on the left, you will find a stock of paper, pens and ink, which he will renew when wished, and which he thinks you can leave in the same place if you do not find a safer one.

He asks you not to be offended if he seems to pay no attention to you in company and only to look upon you as a child. This behaviour seems to him necessary in order to create the confidence he needs to be able to work more efficaciously for his friend's happiness and your own. He will try to create opportunities of speaking to you when he has something to tell you or to hand to you; and he hopes to succeed if you are zealous in assisting him.

He also advises you to return to him one by one the letters you receive, so that there may be less risk of compromising you.

He ends by assuring you that if you will give him your confidence he will use all his energy to soften the persecution inflicted by a too cruel mother upon two persons, one of whom is already his best friend while the other appears to him to deserve the tenderest interest.

From the Château de . . . , 14th of September, 17—.

LETTER LXXIV

The Marquise de Merteuil to the Vicomte de Valmont

How long is it, my friend, that you have been so easily terrified? Is this Prévan so very formidable? See how simple and modest I am! I have often met this proud conqueror; and I have scarcely looked at him! It needed no less an event than your letter to make me pay attention to him. Yesterday I repaired my injustice. He was at the Opera, almost opposite me, and I took note of him. He is certainly handsome, very handsome; fine and delicate features! He must improve when seen at close quarters! And you say that he wants to have me! Assuredly, he will do me both honour and pleasure! Seriously, I have a caprice for him, and I here confide to you that I have taken the first steps.

As we left the Opera, he was two steps from me, and in a loud voice I arranged with the Marquise de . . . to sup on Thursday with the Maréchale. I think this is the only house where I can meet him. I have no doubt he heard me . . . Suppose the ungrateful wretch does not come? Tell me, do you think he will come? Do you know that if he does not come I shall be out of humour all the evening? You see, he will not find so much difficulty in following me, and, which will surprise you more, he will find still less in pleasing me. He says he wishes to kill six horses in paying court to me! Oh! I will save the lives of those horses. I shall never have the patience to wait so long as that. You know it is not one of my principles to make a man languish when once I have made up my mind, and I have made it up for him.

Admit now that it is a pleasure to talk reason to me. Has not your important advice had a great success? But what do you expect; I have been vegetating so long; it is more than six weeks since I allowed myself a frolic. This one comes along; can I deny myself this? Is not the motive of it worth the trouble? Is there anyone more acceptable, in whatever sense you take the word?

You yourself are compelled to do him justice; you do more than praise him, you are jealous of him. Well! I make myself judge between you; but first of all I must have information and that is what I want to obtain. I shall be a just judge and you shall both be weighed in the same balance. I have your records, and your affair is completely investigated. Is it not just that I should now busy myself with your adversary? Come, submit yourself with a good grace, and to begin with, tell me, I beg you, what is the triple adventure of which he is the hero? You speak of it as if I had never known anything else and this is the first I have heard of it. Apparently it happened while I was away at Geneva and your jealousy must have prevented you from writing it to me. Atone for this fault at once; remember that nothing which concerns him is indifferent to me. It seems to me that when I returned people were still talking about it; but I was busy with other matters and I rarely listen to anything of this kind which is not a topic of the day.

Even if what I ask of you should annoy you a little, is it not the least return you can make for the trouble I have taken on your behalf? Was it not I who brought you in touch with your Madame de Tourvel again, when your stupidities had separated you from her? Was it not I again who put into your hands the means of avenging yourself on the intense zeal of Madame de Volanges? You have complained so often of the time you wasted in seeking your adventures! Now you have them under your hand. Love, hatred, you have only to choose; they all sleep under the same roof; you can double your existence, caress with one hand and strike with the other.

Again it is to me that you owe the adventure of the Vicomtesse. I am very pleased with it; but, as you say, it must be talked about; for if the occasion led you, as I can well understand, to prefer mystery to scandal for the moment, it must yet be admitted that the woman did not deserve such good treatment.

Moreover, I have reason to complain of her. The Chevalier de Belleroche thinks her prettier than I like; and for many reasons I should be glad to have a pretext for breaking with her; well, there is no more convenient pretext than to be able to say "It is impossible to meet that woman."

Good-bye, Vicomte; remember that in your position, time is precious; I shall spend mine in procuring Prévan's happiness.

Paris, 15th of September, 17—.

LETTER LXXV

Cécile Volanges to Sophie Carnay

> [Note: In this letter Cecile Volanges relates in the fullest detail everything relative to her in the events which the reader has seen at the end of the first part, *Letter XLI*, and those following. This repetition has been suppressed. She speaks at last of the Vicomte de Valmont and expresses herself as follows. (C. de L.)]

. . . I assure you he is a very extraordinary man. Mamma speaks very ill of him; but the Chevalier Danceny speaks very well of him and I think he is in the right. I have never seen such an able man. When he gave me Danceny's letter, it was before everybody, and nobody saw anything; it is true I was very frightened because I had not been warned of anything; but after this I shall be in readiness. I already understand what he wants me to do to convey him my reply. It is very easy to arrange with him, for his looks say anything he wants. I do not know how he does it; he told me in the note I mentioned to you that he would appear not to pay any attention to me in Mamma's presence; and indeed you would say he never thinks of me; and yet every time I seek his eyes, I am certain to meet them at once.

There is a close friend of Mamma's here whom I did not know, who also seems not to like Monsieur de Valmont at all, although he is very attentive to her. I am afraid he will soon grow weary of the life we lead here and that he will return to Paris; it would be a great pity. He must have a good heart to have come here on purpose to help his friend and me! I should like to show him my gratitude, but I do not know what to do to get an opportunity to speak to him; and if I did find one, I should be so bashful that perhaps I should not know what to say to him.

I can only speak freely to Madame de Merteuil, when I speak of my love; perhaps I should be embarrassed even with you, whom I tell everything, if it was in conversation. I have often felt with Danceny himself, as if in spite of myself, a certain fear which prevented me from saying everything I was thinking. Now I blame myself for it, and I would give everything in the world to find a moment to tell him once, only once, how much I love him. Monsieur de Valmont has promised him that if I allow myself to be guided, he will procure us an opportunity of seeing each other. I shall do whatever he wishes; but I cannot imagine that it is possible.

Good-bye, my dear, I have no more space.[1]

From the Château de . . . , 14th of September, 17—.

LETTER LXXVI

The Vicomte de Valmont to the Marquise de Merteuil

Either your letter is a joke which I have not understood; or when you wrote to me you were in a very dangerous delirum. If I knew you less, my fair friend, I should really be very frightened; and whatever you may say, I am not easily frightened.

However much I read and re-read you I get no further; for your letter cannot be taken in the obvious sense it presents. What did you mean?

Was it only that it is unnecessary to take such precautions against an enemy so little dangerous? But in this case you may be wrong. Prévan is really amiable; he is more so than you think; above all he has the very useful talent of interesting a woman in his love by his skill of speaking of it in company and before everyone, by making use of the first conversation he finds. There are few women who avoid the snare of replying to him because they all have pretentions to subtlety and none of them likes to miss an opportunity of showing it. Now, you know well enough that a woman who consents to speak of love very soon ends by accepting it, or at least by acting as if she had done so. By this method, which he has really perfected, he has the further advantage of often calling women themselves as witness to their own defeat; and what I tell you I have seen myself.

I was only in the affair at second hand; I have never been intimate with Prévan; but there were six of us; and the Comtesse de P . . ., while she thought herself very sly, and indeed appeared, to everybody who was not in the secret, to be carrying on a general conversation, told us everything in the greatest detail, both how she had yielded to Prévan and all that had passed between them. She related this so confidently that she was not even disturbed by a smile which came to all six of us at the same time; and I shall always remember that when one of us tried to excuse himself by feigning to doubt what she was saying, or rather what she appeared to be saying, she replied gravely that assuredly none of us knew so much about it as she; and she was not afraid even to address Prévan and to ask him if she had been mistaken in a single word.

I had reason to believe then that this man was dangerous for everybody; but is it not enough for you, Marquise, that he is handsome, very handsome, as you say yourself? Or that he should make upon you "One of those attacks which you are pleased sometimes to reward for no reason except that you think them well made?" Or that you should have thought it amusing to yield for any other reason? Or . . . How can I tell? How can I divine the thousands and thousands of caprices which govern a woman's mind, by which alone you are still related to your sex? Now that you are warned of the danger I have no doubt you will easily escape it; but still you had to be warned of it; I return then to my text—what did you mean?

If it is only a joke about Prévan, in addition to its being a very long one, it was not of any use with me; it is in society that you must fasten some ridicule upon him, and I repeat my request to you in this matter.

Ah! I think I have the key to the enigma! Your letter is a prophecy, not of what you will do, but of what he will believe you ready to do at the moment of the fall you are preparing for him. I quite approve of this plan, but it demands very great caution. You know as well as I that for public effect, to have a man or to receive his attentions is absolutely the same thing, unless the man is a fool; and Prévan is far from being that. If he obtains nothing but an appearance, he will boast of it and that will be enough. The fools will believe in it, the malicious will appear to believe in it; what resource will you have? I am really afraid. It is not that I doubt your skill; but it is the best swimmers who get drowned.

I do not think I am more stupid than others. I have found a hundred, I have found a thousand ways of dishonouring a woman; but when I have tried to find a way to save her from dishonour, I have never found the possibility of it. Even with you, my fair friend, whose conduct is a masterpiece, I have thought a hundred times that you had more luck than skill.

But after all perhaps I am seeking a reason where there is none. I wonder that for an hour I have been seriously considering what is assuredly only a jest on your part. You will banter me! Well! So be it; but hurry up and let us speak of something else. Of something else! I am wrong, it is always the same thing; always women to have or to ruin, and often both together.

Here, as you have very well pointed out, I have the means of exercising myself in both sorts, but not with the same facility. I foresee that vengeance will be swifter than love. The Volanges girl has surrendered, I will answer for it; she is dependent simply on the opportunity, and I shall take care to provide one. But it is not the same with Madame de Tourvel: she is an annoying woman, I do not understand her; I have a hundred proofs of her love, but a thousand of her resistance; and indeed I am afraid she will escape me.

The first effect produced by my return made me hope for something better. You may suppose that I wanted to judge of it for myself; and to be sure of seeing her first impulses, I sent no one on ahead of me and calculated my journey to arrive while they were at table. In fact, I dropped from the sky like a Divinity at the Opera, who comes to wind up the plot.

Having made enough noise on entering to turn their looks upon me, I could see at one glance the joy of my old aunt, the annoyance of Madame de Volanges and the disconcerted pleasure of her daughter. My fair one was in a seat where her back was turned to the door; at that moment she was engaged in cutting something and did not even turn her head; I spoke to Madame de Rosemonde and at the first word, the tender devotee recognised my voice and gave a cry in which I thought I noticed more love than surprise and fear. By this time I was far enough into the room to see her face; the tumult of her soul, the combat of her thoughts and sentiments, were painted on it in twenty different ways. I sat down at table beside her; she did not know what she was doing or what she was saying. She tried to continue eating; and could not; at last, less than a quarter of an hour afterwards, her embarrassment and her pleasure became stronger than she, and she could think of nothing better than to ask permission to leave the table; she escaped into the park, under the pretext that she needed the air. Madame de Volanges wished to go with her; the tender prude would not allow it; no doubt she was but too happy to find a pretext for being alone and yielding without constraint to the soft emotions of her heart!

I cut the dinner as short as I could. Desert had scarcely been served when that infernal Volanges woman, urged apparently by the necessity of injuring me, rose from her place to go and find my charming invalid; but I had foreseen this plan and I thwarted it. I feigned to take this one rising for the general rising. And as I stood up at the same time the Volanges girl and the local *curé* allowed themselves to be led by this double example; so that Madame de Rosemonde found herself sitting alone at table with the old Knight-Commander T . . . and they both decided to leave too. We all went then to rejoin my fair one whom we found in a grove near the *Château*; and since she wanted solitude and not a walk she was just as glad to return with us as to make us stay with her.

As soon as I was certain that Madame de Volanges would have no opportunity of speaking to her alone, I thought of carrying out your orders, and busied myself with your pupil's interests. Immediately after the coffee, I went up to my room and entered the others', to reconnoitre the ground; I made my dispositions to provide for the little girl's correspondence; and after this first kindness, I wrote a note to inform her of it and to ask for her confidence; I put my note in with Danceny's letter. I returned to the drawing-room. There I found my fair one stretched out on a sofa in a delicious lassitude.

This spectacle by awakening my desires animated my glances; I felt they must be tender and pressing, and I placed myself in such a way that I could make use of them. Their first effect was to make my heavenly prude lower her large, modest eyes. For some time I looked at this angelic face; then wandering over all her person I amused myself by guessing at the curves and forms through a light but always troublesome garment. After I had descended from the head to her feet, I returned from the feet to her head. My fair friend, that soft gaze was fixed upon me; immediately it was lowered again; but, desirous of favouring its return, I turned away my eyes. Then there was established between us that tacit convention, the first treaty of timid love, which to satisfy the mutual need for looking at each other allows glances to follow each other until it is time for them to mingle.

I was so sure that my fair one was completely absorbed in this new pleasure that I took it upon myself to watch over our common safety; but after I had assured myself that a fairly brisk conversation prevented our being noticed by the company, I tried to persuade her eyes to speak their language frankly. With that purpose I first of all surprised a few glances but with so much reserve that her modesty could not be alarmed by it; and to put the timid creature more at her ease I appeared myself as embarrassed as she. Little by little our eyes grew accustomed to meeting each other and remained steady for a longer time; finally they ceased to relinquish each other and I perceived in hers that soft langour which is the happy signal of love and desire; but it was only for a moment; she soon recovered herself and, not without some shamefacedness, changed her attitude and her gaze.

Not wishing to leave her any doubt that I had noticed her different movements, I got up quickly and asked her with an air of anxiety if she felt ill. Immediately everyone surrounded her. I let them all pass in front of me; and as the little Volanges girl who was working at her embroidery near the window, needed a little time to leave her frame, I seized this opportunity to give her Danceny's letter.

I was a little way from her; I threw the epistle on her knee. She did not at all know what to do with it. You would have laughed at her look of surprise and embarrassment; however I did not laugh, for I was afraid so much clumsiness would betray us. But a strongly marked glance and gesture at length made her understand that she had to put the letter in her pocket.

The rest of the day brought nothing interesting. What has happened since may perhaps lead to events which will please you, at least in the matter of your pupil; but it is better to spend one's time in carrying out plans than in relating them. Besides this is the eighth page I have written and I am tired; so, good-bye.

You will realise without my telling you that the little girl has replied to Danceny.[1] I have also had a reply from my fair one, to whom I wrote the day after my arrival. I send you the two letters. You will read them or you will not read them; for this perpetual repetition, which already is ceasing to amuse me, must be very insipid for any one not directly concerned.

Once more, good-bye. I still love you very much; but I beg you, if you speak to me of Prévan again, do it in such a way that I understand you.

From the Château de . . . , 17th of September, 17—.

LETTER LXXVII

The Vicomte de Valmont to Madame de Tourvel

What can be the reason, Madame, for your cruel pertinacity in avoiding me? How can it be that the most tender eagerness on my part should only obtain from you a behaviour which would scarcely be legitimate towards a man of whom you had the greatest reason to complain? What! I am brought back to your feet by love; and when a fortunate chance places me beside you, you prefer to feign an indisposition and to alarm your friends rather than consent to remain near me! How often yesterday did you not turn away your eyes to deprive me of the favour of a look? And if for one instant I was able to see less severity in it, the moment was so short that it seemed as if you desired less to let me enjoy it than to make me feel what I lost by being deprived of it.

That, I dare to say, is not the treatment love deserves nor the treatment friendship would tolerate; and yet you know I am animated by one of these sentiments and I think I have reason to believe that you did not decline the other. What have I done to lose that precious friendship, of which no doubt you thought me worthy since you offered it to me? Have I harmed myself by confiding in you, and will you punish me for my frankness? Are you not afraid at least of abusing them both? Indeed was it not in your bosom, my friend, that I laid the secret of my heart? Was it not with you alone that I felt obliged to refuse conditions which I had only to accept to procure myself the means of not keeping and perhaps of exploiting to my advantage? Would you, by so undeserved a harshness, force me to believe that I had only to deceive you in order to obtain more indulgence?

I do not regret a line of conduct which I owed you, which I owed myself; but by what fatality is it that each praiseworthy action becomes the signal of a new misfortune for me! It was after I had given rise to the only praise you have yet deigned to give my conduct, that for the first time I had to deplore the misfortune of having displeased you. It was after I had proved to you my complete submission by depriving myself of the happiness of seeing you, solely to reassure your delicacy, that you wished to break off all correspondence with me, to take from me that poor compensation for the sacrifice you had exacted and to deprive me even of the love which alone could have given you a right to it. Finally it is after I spoke to you with a sincerity which even the interests of my love could not weaken that you now fly me as if I were a dangerous seducer whose perfidy you had detected.

Will you never grow weary of being unjust? At least tell me what new faults can have brought you to such severity and do not refuse to give me the orders you wish me to follow; when I promise to carry them out is it claiming too much to ask to know what they are?

From . . . , 15th of September, 17—.

LETTER LXXVIII

From Madame de Tourvel to the Vicomte de Valmont

You appear surprised by my conduct, Monsieur, and you are not far from demanding an explanation of it, as if you had the right to blame it. I confess I should have thought I had more reason than you to be surprised and to complain; but since the refusal contained in your last reply I have adopted the course of enclosing myself in an indifference which leaves no occasion for observations or for reproaches. However, since you ask for explanations, and since, thanks be to Heaven, I feel nothing in myself which can prevent me from giving them, I am willing once more to come to an explanation with you.

Anyone who read your letters would think me unjust or capricious. I think I deserve that no one should think that of me; it seems to me that you especially are least of all in a position to think so. No doubt you felt that by compelling my justification you would force me to recall all that has passed between us. Apparently you thought you had only to gain by this examination; since, for my part, I do not think I have anything to lose by it, at least in your eyes, I am not afraid to undertake it. Indeed it is perhaps the only means of knowing which of us two has the right to complain of the other.

From the first day of your arrival at this Château, Monsieur, I think you will admit that at least your reputation gave me the right to be reserved with you; and without being taxed with an excess of prudery I think I might have limited myself merely to expressions of the coldest politeness. You yourself would have treated me with indulgence and would have thought it quite simple for so inexperienced a woman to have not even the merit necessary to appreciate your own. Assuredly that would have been the prudent course; and it would have been the less difficult for me to follow because (I will not conceal it from you) when Madame de Rosemonde came to inform me of your arrival I was compelled to recollect my friendship for her, and her friendship for you, in order not to let her see how much I was annoyed by this news.

I willingly admit that you showed yourself at first under a more favourable aspect than I had imagined; but you will admit in your turn that it did not last long and that you soon grew weary of a constraint for which you apparently did not think yourself sufficiently recompensed by the favourable opinion it gave me of you.

It was then that you took an unfair advantage of my good faith and confidence; you did not shrink from speaking to me about a sentiment which you must have known would offend me; and, while you did nothing save aggravate your faults by multiplying them, I was seeking an opportunity to forget them by offering you a chance to atone for them, at least in part. My request was so just that you yourself did not feel you could refuse it; but you made a right out of my indulgence and profited by it to ask me for a permission which, no doubt, I ought not to have granted, yet which you obtained. Certain conditions were attached to it, none of which you have observed; and your correspondence has been such that every one of your letters made it a duty for me not to reply. It was at the moment when your obstinacy compelled me to send you away from me that, perhaps from a blameworthy condescension, I tried the one means which could permit me to bring you back; but of what value is a virtuous sentiment in your eyes? You scorn friendship; and in your mad intoxication you count miseries and shame as nothing, you seek only pleasures and victims.

You are as inconsiderate in your proceedings as you are inconsequental in your reproaches; you forget your promises, or rather you make it an amusement to break them; after having consented to go away from me, you return here without being recalled; without any regard for my requests and reasons, without having even the politeness to inform me beforehand, you did not shrink from exposing me to a surprise whose effect (although it was assuredly a very simple matter) might have been interpreted in a sense unfavourable to me by the other persons present. You did not try to distract attention from, or to dissipate, the embarrassment you had caused; you appeared to devote all your attention to increasing it still more. At table you chose a place exactly beside mine; a slight indisposition forced me to leave before the others and instead of respecting my solitude you persuaded everyone to come and disturb it. Returned to the drawing-room, I found you beside me if I took a step; if I said a word it was always you who replied. The most insignificant word served you as a pretext for returning to a conversation I did not wish to hear, which might even have compromised me; for indeed, Monsieur, however skilful you may be, I think others are able to understand what I understand.

Thus I am forced by you into immobility and silence, and none the less you continue to pursue me; I cannot raise my eyes without meeting yours. I am continually obliged to look away; and from a most incomprehensible thoughtlessness you fixed the eyes of the company upon me at a moment when I should have liked to be able even to avoid my own.

And you complain of my conduct! And you are surprised at my eagerness to avoid you! Ah! Rather blame my indulgence, rather be surprised that I did not leave at the moment you arrived. I ought perhaps to have done so; and you will compel me to this violent but necessary course if you do not cease your offensive pursuit. No, I do not forget, I shall never forget what I owe to myself, what I owe to the bonds I have formed, bonds I respect and cherish; and I beg you to believe that if ever I found myself reduced to the unhappy dilemma of having to sacrifice them or sacrifice myself, I should not hesitate for an instant. Good-bye, Monsieur.

From . . . , 16th of September, 17—.

LETTER LXXIX

The Vicomte de Valmont to the Marquise de Merteuil

I meant to go shooting this morning; but the weather is detestable. I have nothing to read but a new novel, which would bore even a school-girl. It will be at least two hours until luncheon; so in spite of my long letter yesterday I shall chat to you again. I am quite sure not to weary you, for I shall talk to you about the "very handsome Prévan." How is it you have not heard of his famous adventure which separated "the inseparables?" I wager the first word will bring it back to your mind. However, here it is, since you want it.

You remember that all Paris was surprised that three women, all three of them pretty, all three of them with the same talents, and able to advance the same claims, should remain in intimate friendship from the moment of their entry into society. At first the reason was thought to be their extreme timidity; but soon they were surrounded by a numerous group whose attentions they shared, and were made aware of their worth by the eagerness and regards of which they were the objects; yet their union became all the closer; and you would have said that the triumph of one was always the triumph of the two others. It was hoped that the moment of love at least would create some rivalry. Our charming ladies vied with each other for the honour of being the apple of discord; I should myself have entered the ranks if the great vogue of the Comtesse de . . . at that time had allowed me to be unfaithful to her before I obtained the favour I sought.

However our three beauties made their choice at the same carnival, as if in concert; and far from exciting the storms which had been expected it only rendered their friendship more interesting through the charm of mutual confidences.

The crowd of rejected suitors then joined the crowd of jealous women and this scandalous constancy was submitted to public censure. Some asserted that in this society of "inseparables" (so they were called then) the fundamental law was community of property and that it applied even to love; others maintained that if the three lovers were exempt from male rivals, they had female rivals; some went so far as to say that the lovers had only been received from decency and had merely obtained a title with no functions.

These rumours, true or false, did not have the expected effect. On the contrary the three couples felt they would be lost if they separated at this moment; they adopted the course of weathering out the storm; the public, which wearies of everything, soon wearied of a fruitless satire. Carried away by its natural frivolity, it turned to other things; and then, coming back to this with its usual inconsistency, it changed censure into praise. Since everything here is a matter of fashion, the enthusiasm gained ground; it was becoming a positive delirium when Prévan undertook to verify these prodigies and to settle public opinion and his own about them.

He therefore sought out these models of perfection. He was easily admitted into their society and considered this a favourable omen. He knew very well that happy people are not so easily accessible. He very soon saw, in fact, that this boasted happiness was like the happiness of kings—more envied than desirable.

He noticed that these supposed inseparables began to look for outside pleasures and even to seek distractions; and from this he concluded that the bonds of love or friendship were already slackened or broken and that only the ties of vanity and habit retained some strength.

However, the women, united by the emergency, preserved among themselves the appearance of the same intimacy; but the men, who were freer in their proceedings, discovered they had duties to fulfil or other affairs to follow; they still complained of them but had ceased to avoid them and their evenings were rarely complete.

This behaviour on their part was useful to the assiduous Prévan who, naturally placed beside the neglected lady of the day, took occasion to offer alternately, as circumstance permitted, the same attentions to each of the three friends. He easily perceived that to make a choice among them would be his ruin; that the false shame of finding herself the first to be unfaithful would frighten whichever woman he preferred; that the wounded vanity of the two others would make them enemies of the new lover and that they would not fail to oppose him with all the severity of high principles; and finally that jealousy would certainly revive the attentions of a rival who might still be formidable. Everything would have become an obstacle; but in his triple project everything became easy; each woman was indulgent because she was concerned in it and each man because he thought he was not.

Prévan, who at that time had only one woman to victimise, was so fortunate as to have her become conspicuous. Her position as a foreigner and her adroit refusal of a great prince's attentions had fixed upon her the observation of the court and the town; her lover shared the honour of it and made use of it with his new mistresses. The one difficulty was to carry on the three intrigues at the same time, for their progress was necessarily regulated by the slowest; indeed I had it from one of his confidents that his greatest difficulty was to delay one of them who was ready to begin nearly a fortnight before the others.

At last the great day arrived. Prévan, who had obtained the consent of all three, was now master of the proceedings and arranged them as you will see. One of the three husbands was away, the other was going at dawn the next day, the third was in Paris. The inseparable women were to sup with the future widow; but the new Master had not allowed the old Servants to be invited. The morning of the same day he made three packets from the letters of his mistress; he accompanied one with a portrait he had received from her, the second with a true lover's knot she had painted herself, the third with a lock of her hair; each one took this third of a sacrifice as a complete sacrifice and consented in exchange to send a letter, breaking off relations, to the disgraced lover.

That was a good deal; but it was not enough. She whose husband was in Paris had only the daytime at her disposal; it was agreed that a feigned indisposition should excuse her from supping with her friend and that the whole evening should be Prévan's; she whose husband was away granted the night; and dawn, the moment of the third husband's departure, was marked down by the last for the rendezvous.[1]

Prévan, who neglected nothing, then ran to the fair foreigner, carried and created there the ill-humour he needed and only left after having made a quarrel which assured him twenty-four hours of liberty. Having thus made his dispositions, he went home with the purpose of taking some rest; but other affairs were waiting for him.

The letters breaking off relations had been a flash of light to the disgraced lovers; each one of them could have no doubt but that he was sacrificed to Prévan; and their annoyance at being trifled with was added to the ill-humour which is almost always created by the little humiliation of being left; all three, without communicating with each other, but as if in concert, resolved to have satisfaction and took the course of demanding it from their favoured rival.

He therefore found three challenges waiting for him; he accepted them honourably; but not wishing to lose either the pleasures or the fame of this adventure he fixed the meetings for the morning of the next day and assigned them all three to the same place and the same hour. This was at one of the gates of the Bois de Boulogne.

When evening came he ran his triple career with equal success, at least he afterwards boasted that each of his new mistresses thrice received the gage and oath of his love. Here, as you may suppose, the story lacks proof; all the impartial historian can do is to point out to the incredulous reader that excited vanity and imagination can achieve prodigies, and moreover that the morning that was to follow so brilliant a night might be held to excuse him from consideration of the future. In any case the following facts are more certain.

Prévan arrived punctually at his rendezvous; there he found his three rivals, a little surprised at meeting each other; and perhaps each of them already partly consoled by finding he had companions in misfortune. He addressed them in an affable and sprightly way and made them the following speech which has been faithfully reported to me:

"Gentlemen," said he, "by finding yourselves met together here you have doubtless guessed that all three of you have the same reason for complaint against me. I am ready to give you satisfaction. Decide by lot among yourselves which of the three shall first attempt a vengeance to which you have all an equal right. I have brought no second and no witnesses. I had none for the offence; I ask none for the reparation." Then, giving way to his gambling propensities, he added: "I know one rarely wins 'seven and the go'[2]; but whatever be the fate which is in store for me, a man has always lived long enough when he has had time to acquire the love of women and the esteem of men!"

While his astonished adversaries gazed at each other in silence and their delicacy was perhaps calculating that this triple combat was not an equal match, Prévan went on: "I will not conceal from you that the night I have just passed has fatigued me cruelly. It would be generous on your part to allow me to recruit my strength. I have given orders for a breakfast to be prepared near here. Do me the honour of accepting it. Let us breakfast together and above all let us breakfast gaily. We may fight for such trifles but I think they should not be allowed to spoil our temper."

The breakfast was accepted. Prévan, they say, had never been so amiable. He had the skill to humiliate none of his rivals; to persuade them that they would all easily have had the same successes, and above all to make them admit they would not have missed the opportunity any more than he had. Once these facts were admitted the rest arranged itself. And so before the breakfast was over they had already repeated ten times that such women did not deserve that honest men should fight for them This idea brought cordiality with it, wine fortified the cordiality; so that a few moments afterwards, it was not sufficient to bear no ill will, and they swore an unreserved friendship.

Prévan no doubt was as glad of this result as of the other, but he was unwilling to lose any of his celebrity. Consequently he adroitly altered his plans to fit the circumstances. "After all," said he to the three offended lovers, "you should avenge yourselves not upon me but upon your faithless mistresses. I offer you the opportunity. Like yourselves, I feel already an injury which I shall soon share; for if each of you has been unable to hold one of them, can I hope to hold all three of them? Your quarrel becomes mine. Come and sup this evening in my house, and I hope to delay your vengeance no longer." They wanted him to explain, but he replied in the tone of superiority which the circumstances authorised him to take: "Gentlemen, I think I have proved to you that I have some ability in managing; rely on me."

They all consented; and after having embraced their new friend, separated until the evening to await the result of his promises.

Without wasting any time Prévan returned to Paris, and, in accordance with custom, visited his new conquests. He persuaded all three of them to come that evening and sup privately with him at his house. Two of them indeed made some difficulties, but what is there left to refuse on the morning after? He arranged the meetings with an hour between each, the time needed for his plan. After these preparations, he went away, informed the three other conspirators and all four went off gaily to wait for their victims.

They heard the first one arrive. Prévan appeared alone, received her with an air of alacrity, and led her to the sanctuary of which she thought herself the divinity; then, disappearing on some slight pretext, he immediately sent the outraged lover in his stead.

As you may suppose, the confusion of a woman who is not yet accustomed to adventures rendered the triumph very easy, at this moment; every reproach which was not made was counted as a grace; and the fugitive slave, delivered up again to her former master, was only too glad to hope for pardon by returning to her first chains. The treaty of peace was ratified in a more solitary place, and the empty stage was filled in turn by the other actors practically in the same way, and assuredly with the same ending.

However, each of the women still thought herself the only one concerned. Their astonishment and embarrassment increased when the three couples met for supper; but their confusion was at its height when Prévan reappeared among them and had the cruelty to make excuses to the three faithless ladies—excuses which gave away their secret and showed them plainly to what extent they had been duped.

However they sat down to table and soon regained countenance; the men accepted the situation, the women submitted to it.

All had hatred in their hearts; but their words were none the less tender; gaiety awakened desire which in turn lent new charms to their merriment. This astounding orgy lasted until morning; and when they separated the women had every reason to think themselves forgiven but the men had kept their resentment and the next day broke matters off without the possibility of a reconciliation; and not content with leaving their light mistresses they completed their vengeance by making the adventure public. Since that time one of the women is in a convent and the other two are languishing in exile in their country estates.[3]

This is the story of Prévan; it is for you to see whether you wish to add to his fame and harness yourself to his triumphal chariot. Your letter really made me uneasy and I am impatiently waiting a wiser and clearer answer to the last letter I wrote you.

Good-bye, my fair friend; beware of the amusing or capricious ideas which always seduce you too easily. Remember that in the career you are following, intelligence is not enough and that a single imprudence may become an irreparable misfortune. And finally sometimes allow prudent friendship to guide your pleasures.

Good-bye, I still love you as much as if you were reasonable.

From . . . , 18th of September, 17—.

LETTER LXXX

The Chevalier Danceny to Cécile Volanges

Cécile, my dearest Cécile, when will the day come for us to meet again? Who will teach me to live so far from you? What will give me the needful strength and courage? Never, no, never, can I endure this fatal absence. Every day adds to my misery and I see no end to it! Valmont who promised me help and consolation, Valmont neglects and perhaps has forgotten me. He is near her whom I love; he knows not what one suffers when far from her.

He did not write to me when sending me your last letter. Yet it is he who should inform me when and how I am to see you. Has he then nothing to tell me? You yourself do not speak to me of it; can it be that you have ceased to share my longing for it? Ah! Cécile, Cécile, I am very unhappy. I love you more than ever; but this love which is the charm of my life is becoming its torment.

No, I cannot continue to live thus, I must see you, I must, if only for a moment. When I get up I say to myself: I shall not see her. I go to bed saying: I have not seen her. These long, long days bring not a moment of happiness. All is privation, all is regret, all is despair; and all these misfortunes come to me from her whence I expected all my pleasures! Add to these killing griefs my uneasiness about yours and you will have an idea of my situation. I think of you continually, yet never think of you without uneasiness. If I see you grieved and unhappy, I suffer for all your griefs; if I see you tranquil and consoled, my own griefs are redoubled. Everywhere I find unhappiness.

Ah! It was not thus when you dwelt in the same place as I! Then all was pleasure. The certainty of seeing you embellished even the moments of absence; the time I was compelled to pass far from you brought me nearer to you as it glided away. The use I made of it was always concerned with you. If I carried out my duties, they rendered me more worthy of you; if I cultivated some talent, I hoped I should please you more. Even when the distractions of society took me away from you I was not separated from you. At the theatre I tried to guess what would have pleased you; a concert reminded me of your talents and our sweet occupation. In company and on walks I seized upon the slightest resemblance to you. I compared you with everyone; you had the advantage everywhere. Each moment of the day was marked by a new homage and each evening I brought them as a tribute to your feet.

And now, what have I left? Painful regrets, eternal privations, and a faint hope which is diminished by Valmont's silence and changed into uneasiness by yours! We are separated only by ten leagues and this distance, so easily traversed, becomes an insurmountable obstacle! And when I implore my friend and my mistress to help me overcome it, both remain cold and calm! Far from helping me, they do not even answer me.

What has happened to Valmont's active friendship? Above all what has happened to those tender feelings of yours which rendered you so ingenious in discovering means for us to see each other every day? I remember that sometimes, though I still desired it, I was forced to sacrifice it to prudence or to duties; what did you not say to me then? By how many pretexts did you combat my reasons! And recollect, my Cécile, my reasons always yielded to your wishes. I do not make that a merit; I had not even the merit of sacrifice. What you desired to obtain, I burned to grant. But now I ask in my turn; and what is my request? To see you for a moment, to give and receive once more the vow of an eternal love. Is it no longer your happiness as it is mine? I repel this disheartening idea which would put the finishing touch to my misfortunes. You love me, you will always love me; I believe it, I am sure of it, I will never doubt it; but my situation is terrible and I cannot endure it much longer. Good-bye, Cécile.

Paris, 18th of September, 17—.

LETTER LXXXI

The Marquise de Merteuil to the Vicomte de Valmont

How much I pity you for your fears! How they prove my superiority over you! And you think to teach me, to lead me? Ah! My poor Valmont, what a distance there still is from you to me! No, not all the pride of your sex will suffice to fill up the interval which separates us. Because you could not carry out my plans, you think them impossible! Proud, weak creature! It well befits you to try to gauge my methods and to judge of my resources! Really, Vicomte, your advice put me out of humour and I cannot conceal it from you. That you should mask your incredible clumsiness towards your Madame de Tourvel by displaying to me as a triumph the fact that for a moment you disconcerted this timid woman—that I allow. When you boast of obtaining a look, a single look, I smile and grant it you. And when, feeling in spite of yourself the unimportance of your action, you try to hide it from my notice by flattering me with your sublime effort in bringing together two children, who are both of them burning to see each other, and who, let me observe in parenthesis, owe the ardour of that desire to me alone—that too I will admit. And when finally you make use of these brilliant actions to tell me in a doctoral tone that "it is better to spend one's time in carrying out plans than in talking about them"—well, such vanity does not harm me and I forgive it. But that you should think I need your prudence, that I should go astray if I did not defer to your opinion, that I ought to sacrifice to it a pleasure, a fantasy—really, Vicomte, you are becoming too proud of the confidence I have in you!

And what then have you done that I have not surpassed a thousand times? You have seduced, ruined even, a number of women; but what difficulties did you have to overcome? What obstacles to surmount? Where is there in that any merit which is really yours? A handsome face, the result of mere chance; grace, which is almost always given by experience; wit indeed, but mere chatter would take its place if needed; a quite praiseworthy impudence, but probably due solely to the facility of your first successes; if I am not wrong, those are all your methods; for, as for the celebrity you have been able to acquire, I think you will not expect me to rank very high the art of creating or seizing the opportunity of a scandal.

As to prudence and shrewdness, I do not mention myself—but where is the woman who would not possess more than you? Why! Your Madame de Tourvel leads you like a child.

Believe me, Vicomte, people rarely acquire the qualities they can dispense with. You fought without risk and necessarily acted without wariness. For you men defeats are simply so many victories the less. In this unequal struggle our fortune is not to lose and your misfortune not to win. If I granted you as many talents as we have, still how much we should surpass you from the continual necessity we have of using them!

Let us suppose, if you like, that you display as much skill in conquering us as we do in defending ourselves or in yielding; yet you will admit this skill becomes useless to you after success. You are wholly occupied with your new inclination and you give yourselves up to it without fear and without reserve; its duration does not matter to you.

Indeed, these ties reciprocally given and received (I speak the jargon of love) can be drawn tighter or broken by you alone as you choose; we are but too happy if in your inconsistences you prefer mystery to publicity and content yourselves with a humiliating abandonment without making the idol of one day a victim of the next!

But if an unfortunate woman is the first to feel the weight of her chain, what risks she runs if she tries to escape it, if she even dares to raise it. It is with trembling that she attempts to send away the man whom her heart repels with an effort. If he persists in staying, she has to yield from fear what she granted to love:

> Her arms still open though her heart is closed.

She must unloose with prudent skill the very ties you would have broken. She is at the mercy of her enemy and is without resource if he is without generosity; and how can she hope it of him, when, though he is sometimes praised for having it, he is never blamed for lacking it?

Doubtless you will not deny these truths which are so obvious they are commonplaces. But if you have seen me directing events and opinions and making these formidable men the toy of my caprices or my fantasies; depriving some of will, others of the power of harming me; if in accordance with my changing tastes I have turn by turn attached to my train or cast far from me

> Those unthroned tyrants now become my slaves[1];

if through these frequent revolutions I have kept my reputation intact; ought you not to have concluded that, since I was born to avenge my sex and to dominate yours, I must have created methods unknown to anybody but myself?

Ah! Keep your advice and your fears for those unbalanced women who rave of their "sentiment," whose excited imagination would make one think nature had placed their senses in their heads; who, having never reflected, always confuse love with the lover; who, in their foolish illusion, think that the one man with whom they have sought pleasure is the sole depository of it; who, being truly superstitious, give the priest the respect and faith which is due only to the divinity.

Fear also for those who are more vain than prudent and who will not allow themselves to be abandoned when necessary.

Tremble above all for those women, active in their idleness, whom you call "tender," of whom love takes possession so easily and with such power; women who feel the need to occupy themselves with it even when they do not enjoy it and who abandoning themselves unreservedly to the ebullition of their ideas, give birth through them to those sweet letters which are so dangerous to write; women who are not afraid to confide these proofs of their weakness to the person who causes them, imprudent women, who cannot see their future enemy in their present lover.

But what have I in common with these incautious women? When have you seen me depart from the rules I have prescribed for myself or lose my principles? I say my principles and I say it advisedly; for they are not, like other women's principles, given at hazard, accepted without reflection and followed from habit; they are the result of my profound meditations; I created them and I can say that I am my own work.

I entered society, an unmarried girl, at a time when my condition compelled me to silence and inaction; I made use of it to observe and to reflect. While I was thought light-headed or inattentive because I paid little attention to the conversations they were careful to hold with me, I carefully gathered those they tried to hide from me.

That useful curiosity which served to instruct me, also taught me to dissimulate; I was often forced to conceal the objects of my attention from the eyes of those about me and I tried to direct my own eyes at will; from this I gained that ability to simulate when I chose that preoccupied gaze you have praised so often. Encouraged by this first success, I tried to govern the different expressions of my face in the same way. Did I experience some grief, I studied to show an air of serenity, even one of joy; I carried my zeal so far as to cause myself voluntary pain and to seek for an expression of pleasure at the same time. I worked over myself with the same care and more trouble to repress the symptoms of an unexpected joy. In this way I acquired that power over my features by which I have sometimes seen you astonished.

I was still very young and practically without any interest; but I only possessed my thoughts and I felt indignant that they might be snatched from me or surprised in me against my will. Furnished with these first arms, I made trial of them. I was not content with not allowing people to see through me, I amused myself by showing myself under various aspects; certain of my movements, I now watched over what I said; I regulated both according to circumstances or even only according to my fancy. From that moment on, my way of thinking was for myself alone and I only showed what it was useful for me to render visible.

This labour upon myself fixed my attention on the expression and character of faces; from that I obtained a penetrating glance which experience however, has taught me not to rely upon entirely, but which in general has rarely deceived me.

I was not yet fifteen and already I possessed the talents to which the greater part of our politicians owe their reputation; and I considered myself still in the first elements of the science I wished to acquire.

As you may suppose, I was like all young girls and tried to guess at love and its pleasures. But I had never been to a convent, I had no intimate friend, and I was watched by a vigilant mother; I had only vague ideas which I could not settle; even nature which I have subsequently had every reason to be grateful to, had not given me any indication. It was as if she was silently working to perfect her labour. My head alone was in a ferment; I did not desire to enjoy, I wanted to know; my desire to learn suggested the means to me.

I realised that the only man with whom I could discuss this topic without compromising myself was my confessor. I made up my mind at once; I overcame my little bashfulness; and boasting of a fault I had not committed, I accused myself of having done "everything that women do." That was my expression; but when I spoke in this way I did not know at all what idea I was expressing. My hope was not entirely disappointed or entirely fulfilled; the fear of betraying myself prevented me from obtaining information; but the good father made it out to be so wicked that I concluded the pleasure must be extreme; and the desire to enjoy it succeeded the desire to know what it was.

I do not know where this desire might have led me and as I was then quite without experience a single occasion might perhaps have ruined me; fortunately for me my mother informed me a few days later that I was to be married and immediately the certainty of knowing extinguished my curiosity and I reached Monsieur de Merteuil's arms a virgin.

I awaited with confidence the moment which would enlighten me and I needed premeditation to show embarrassment and fear. This first night, which is generally thought to be so cruel or so sweet, was for me simply an opportunity for experience; I noticed very carefully both pain and pleasure and in these different sensations I saw nothing but facts to collect and to meditate upon.

This kind of study very soon began to please me; but faithful to my principles, and feeling perhaps instinctively that no one should be further from my confidence than my husband, I resolved from the mere fact that I was moved by it to appear unmoved to my husband. This apparent coldness was afterwards the unshakable foundation of his blind confidence; on further reflection I added to this the air of heedlessness which my years sanctioned; and he never thought me so childish as in the moments when I praised him with most boldness. However, I must confess I allowed myself at first to be carried away in the whirlpool of society and I gave myself up entirely to its futile distractions. But when after some months Monsieur de Merteuil took me to his gloomy country estate, the fear of boredom brought back the taste for study; and finding myself surrounded by people whose difference of rank placed me above all suspicion, I made use of this to give my experiments a wider field. It was there that I made certain that love which they tell us is the cause of our pleasures, is at most only the pretext for them.

Monsieur de Merteuil's illness interrupted these soft occupations; I had to follow him to town whither he went for medical aid. He died, as you know, shortly afterwards; and although, taking it all round, I had no reason to complain of him, I felt none the less keenly the value of the liberty my widowhood would give me and I promised myself to make good use of it.

My mother expected I should go into a convent or return to live with her. I refused both courses; all I granted to decency was to return to the country again, where I still had a few observations to make.

I supported them with the assistance of reading; but do not think that it was all of the kind you suppose. I studied our manners in novels, our opinions in philosophers; I even sought in the severest moralists what it was they exacted of us and in this way I learned what one could do, what one ought to think and what one ought to appear. Once I was certain on these three points, the last alone presented some difficulties in practice; I hoped to conquer them and I meditated the means.

I began to grow weary of my rustic pleasures which were too monotonous for my active head; I felt a need for coquetry to reconcile me with love, not to feel it veritably but to inspire and to feign it. In vain I had been told and had read that this sentiment could not be feigned; I saw that to do so successfully one had only to join the talent of a comedian to the mind of an author. I practised myself in both arts and perhaps with some success; but instead of seeking the vain applause of the theatre, I resolved to employ for my happiness what others sacrifice to vanity.

A year passed away in these different occupations. My period of mourning allowed me to reappear and I returned to town with my plans; I had not foreseen the first obstacle I met with.

My long solitude, my austere retirement had given me a veneer of prudery which frightened all our most charming men; they kept away and left me at the mercy of a crowd of wearisome creatures who all wanted to marry me. The difficulty was not in refusing them; but several of these refusals displeased my family and I wasted on these private annoyances a time which I had promised myself to employ so charmingly. I was therefore obliged, in order to recall the one party and to get rid of the other, to display a few imprudences and to employ the care I had meant to use for the preservation of my reputation in harming it. I succeeded easily, as you may believe. But not being carried away by any passion, I only did what I considered necessary and measured the doses of my heedlessness with prudence.

As soon as I obtained the objective I wished to reach, I returned on my steps and gave the honour of my amendment to those women who claim merit and virtue since they can have no pretention to charm. This was a master stroke which was of more value to me than I had hoped. These grateful Duennas constituted themselves my defenders; and their blind zeal for what they called their work was carried to such an extent that at the slightest remark any one dared to make about me, the whole band of prudes exclaimed that it was a scandal and an insult. The same method secured me also the support of our women with pretentions, who, persuaded that I had given up the idea of following the same course as themselves, chose me as the object of their praises every time they wished to prove that they did not talk scandal about everybody.

However my preceding conduct had brought back the lovers; and to steer cautiously between them and my faithless female protectors I gave myself out to be a tender-hearted but fastidious woman, whose excessive delicacy supplied her with weapons against love.

Then I began to display on the great stage the talents I had procured myself. My first care was to acquire the reputation of being invincible. To attain this I always pretended to accept the attentions of those men only who did not please me. I employed them usefully in gaining me the honours of resistance, while I yielded myself fearlessly to the accepted lover. But my feigned timidity never allowed him to accompany me into society; and thus the gaze of the company was always fixed upon the rejected lover.

You know how quickly I make up my mind; that is because I have observed a woman's secret is almost always betrayed by the preliminary attentions. Whatever one may do the tone is never the same after as before the success. This difference does not escape an attentive observer; and I have found it less dangerous to be mistaken in my choice than to allow myself to be fathomed. In this way I have the further advantage of removing the appearances by which alone we are judged.

These precautions and the further ones of never writing and never giving up any proof of my defeat may appear excessive, but have never seemed to me sufficient. I had descended into my own heart and I studied in it the heart of others. There I saw that everybody keeps a secret in it which he must not allow to be revealed—a truth which antiquity appears to have known better than we and of which the story of Samson may be only an ingenious parable. Like a new Delilah I always used my power as she did to surprise this important secret. Ah! How many of our modern Samsons are there whose hair I keep under my scissors! I have ceased to be afraid of them; they are the only men I have sometimes let myself humiliate. With the others I was more pliant; the art of making them unfaithful to avoid appearing fickle, a feigned friendship, an apparent confidence, a few generous actions, the flattering idea each of them had that he was my only lover, these procured me their discretion. Finally, when these methods failed me I have foreseen the time of breaking things off and I have smothered with ridicule or calumny any credence these dangerous men might have obtained.

You see me constantly practise what I tell you here; and you doubt my prudence! Well! Think of the time when you paid your first attentions to me; none ever flattered me so much; I desired you before I had seen you. Seduced by your reputation, I felt you were lacking to my glory; I burned to wrestle with you hand to hand. This was the only one of my inclinations which ever had a moment's power over me. And yet if you had wished to ruin me, what means could you have found? Vain talk which leaves no trace behind it, which your very reputation would have helped to make suspicious, and a train of improbable facts, the accurate relation of which would have sounded like a badly written novel.

Since then I have indeed given up to you all my secrets; but you know the interests which unite us and whether, of us two, it is I who should be charged with imprudence.[2]

Since I am in the way of giving you information, I will do so exactly. I can hear you telling me that at least I am at the mercy of my waiting-woman; indeed, although she has not the secret of my feelings, she has the secret of my actions. When you spoke to me of her before, I simply answered that I was sure of her; and the proof that this reply was sufficient for your tranquility is that you have since on your own account confided dangerous secrets to her. But now that you are offended about Prévan and have lost your head I suppose you will not believe me any more on my bare word. You must therefore be instructed.

First of all, this girl is my foster-sister, and that tie which to us does not appear to be one is not without strength for people of her class; moreover I have her secret and better still; she was the victim of a love-folly which would have ruined her if I had not saved her. Her parents, bristling up with honour, wanted nothing less than to have her shut up. They applied to me. I saw at a glance how useful their anger might be to me. I supported them and asked for the order, which I obtained. Then, suddenly going over to the side of clemency to which I brought her parents also, and making use of my credit with the old Minister, I made them all consent to leave the order in my hands, and to give me the power of witholding it or demanding its execution according as I should determine from the girl's future conduct. She knows therefore that her fate is in my hands; and if, which is impossible, these powerful methods should not stop her, is it not obvious that her conduct made public and her authentic punishment would soon deprive her words of all credence? To these precautions which I call fundamental, are added a thousand others, either local or casual, which are suggested to me by reflection and habit; to give them in detail would be circumstantial but the use of them is important and you must take the trouble to collect them from my whole conduct if you wish to understand them.

But to suppose that I have taken so much trouble and shall not enjoy the fruits of it; that after I have raised myself above other women by painful labour, I should consent to crawl like them between impudence and timidity; that above all I should fear a man to the extent of seeing my safety nowhere but in flight—no, Vicomte, never. I must conquer or perish. As to Prévan, I want to have him and I will have him; he wants to tell it and he shall not tell it; that is our romance in a phrase. Good-bye.

From . . . , 20th of September, 17—.

LETTER LXXXII

Cécile Volanges to the Chevalier Danceny

Heavens, how your letter distressed me! I did well to be so impatient to receive it! I hoped to find consolation in it and now I am more upset than before I received it. I cried a lot in reading it; I do not blame you for that; I have already cried many times on account of you without that distressing me. But this time it is not the same thing.

What do you mean when you say your love becomes a torment to you, that you cannot live any longer in this way, nor endure your situation any more? Are you going to stop loving me because it is not so agreeable as it was before? It seems to me that I am no happier than you are, on the contrary; and yet I only love you the more. If Monsieur de Valmont has not written to you, it is not my fault; I could not ask him to do so, because I have not been alone with him and we have arranged never to speak to each other in other peoples' presence; and that is on your account too, so that he can sooner do what you desire. I do not say that I do not desire it also, and you ought to be certain of this; but what do you expect me to do? If you think it is so easy, find the way, I ask nothing better.

Do you think it is very pleasant for me to be scolded every day by Mamma, who never said anything harsh to me before; but on the contrary? It is worse now than if I were at the convent.

But I consoled myself by remembering that it was for your sake; there were even moments when I felt I was glad of it; but when I see that you are angry as well and that without it being my fault at all, I am more upset than by anything that has happened to me up to now.

It is difficult even to receive your letters and if Monsieur de Valmont were not so kind and so clever I should not know what to do, and it is still more difficult to write to you. All the morning I dare not, because Mamma is close at hand and keeps coming into my room. Sometimes I can in the afternoon, under the pretext of singing or playing on my harp; and even then I must break off at every line so that they can hear I am working. Luckily my waiting-woman sometimes goes to sleep in the evening and I tell her I can easily go to bed myself, to make her go away and leave me the light. And then I must get under my bed-curtains so that no light can be seen, and then I must listen for the slightest noise so that I can hide everything in my bed if anyone comes. I wish you were here to see! You would see that one has to be very much in love to do all this. Indeed it is true I do all I can and I wish I could do more.

Certainly I do not refuse to tell you that I love you and and that I shall always love you; I never said it with a better heart; and you are angry! Yet you assured me before I had said it to you that it was enough to make you happy. You cannot deny it; it is in your letters. Although I have them no longer, I remember it as well as when I read them every day. And because we are absent from each other, you do not think in the same way! But this absence will not last forever, perhaps? Heavens! How unhappy I am! And you are the cause of it!

About your letters—I hope you have kept those that Mamma took from me and sent back to you; there must come a time when I shall not be so oppressed as I am now, and you will give them all back to me. How happy I shall be when I can keep them all without anybody being able to object! Now I give them all back to Monsieur de Valmont, because it would be too risky; in spite of that I never give them back to him without feeling very distressed.

Good-bye, my dear. I love you with all my heart. I shall love you all my life. I hope you are not grieved any more now; if I were sure of it I should cease to be grieved myself. Write me as soon as you can, for I feel I shall be sad until then

From the Château de . . . , 23rd of September, 17—.

LETTER LXXXIII

The Vicomte de Valmont to Madame de Tourvel

Pray, Madame, let us continue the intercourse so unhappily interrupted! How can I finally prove to you how much I differ from the odious portrait of me which has been made to you; above all how can I enjoy again that charming confidence you began to show me! What charms you lent to virtue! How you embellished all virtuous sentiments and forced one to cherish them! Ah! That is where you seduce; that is the most powerful seduction; it is the only one which is both powerful and respectable.

Doubtless it is enough to see you, to have the desire of pleasing you; to listen to you in company, to have this desire increased; but he who has the happiness to know you better, who can sometimes read in your soul, soon yields to a more noble enthusiasm and, touched by veneration as well as by love, adores in you the image of all virtues. I was perhaps created to love them and follow them more than others, and, after being carried away from them by a few errors, I was brought back to them by you; through you I felt all their charm again. Will you make this new love a crime in me? Will you condemn your own work? Would you reproach yourself even for the interest you might take in it? What evil can you fear from so pure a sentiment, what sweetness might you not taste in it?

My love frightens you, you think it violent, extreme? Temper it by a gentler love; do not refuse the power I offer you, which I swear never to escape from, which I dare to think would not be entirely a loss for virtue. What sacrifice could appear difficult to me were I sure your heart would reward me for it! What man is so unhappy as to be unable to enjoy the privations he imposes on himself; as not to prefer a word or a look that is granted to all the enjoyments he might ravish or surprise! And you thought I was that man! And you feared me! Ah! Why does your happiness not depend upon me! How I would avenge myself upon you, by making you happy! But that soft power is not produced by sterile friendship; it is granted only to love.

That word intimidates you! And why? A more tender attachment, a closer union, a single thought, the same happiness and the same griefs—what in these is foreign to your soul? Yet such is love! Such at least is the love you inspire and I feel! Above all it is that love which calculates without interest and can appreciate actions on their merits, not from their value; the inexhaustible treasure of tender minds, everything becomes precious when it is done by love or for love.

What is there frightening in these truths which are so easy to grasp, so sweet to practise? What fears can be caused you by a tender man whom love allows no happiness but yours? That is to-day the only wish I have; I would sacrifice everything to carry it out except the sentiment which inspires it; and, if you but consent to share that sentiment itself, you shall govern it yourself. But do not let us allow it to divide us any longer when it ought to unite us. If the friendship you have offered me is not a vain word, if, as you said to me yesterday, it is the softest feeling known to your soul; I do not refuse to allow friendship to bargain between us; but if friendship becomes the judge of love she must consent to listen to it, refusal to hear it would become an injustice and friendship is not unjust. A second interview will have no more disadvantages then the first; chance may again furnish the opportunity; you yourself can appoint the time. I am willing to believe that I am wrong; would you not rather amend me than combat me, and have you any doubts of my docility? If that annoying third person had not interrupted us, perhaps I should already be entirely of your opinion; who knows how far your power might go?

Shall I confess it? Sometimes I am afraid of this invincible power to which I yield myself without daring to calculate, of that irresistible charm which renders you the sovereign of my thoughts as of my actions. Alas! Perhaps it is I who should dread this interview I ask for! When it is over perhaps I shall find myself changed by my promises and doomed to burn with a love which I feel can never be extinguished, without daring even to implore your aid! Ah! Madame, for pity's sake, do not abuse your power! Ah! But if they would make you happier, if by them I should appear more worthy of you—what pains would not be softened by these consoling ideas! Yes, I feel it; to speak to you again is to give you more powerful weapons against me, is to yield myself more completely to your will. It is easier to defend oneself against your letters; they are indeed your very words, but you are not there to lend them strength. Nevertheless, the pleasure of listening to you makes me brave the danger; at least I shall have the happiness of having done everything for you even against myself; and my sacrifices will become a homage. I am but too happy to prove to you in a thousand ways, as I feel it in a thousand shapes, that, not even excepting myself, you are, you always will be the dearest object of my heart.

From the Château de . . . , 23rd September, 17—.

LETTER LXXXIV

The Vicomte de Valmont to Cécile Volanges

You saw how we were thwarted yesterday. All day long I was unable to hand you the letter I had for you; I do not know if I shall find more opportunity to-day. I am afraid of compromising you by showing more zeal than skill; and I should not forgive myself an imprudence which would be so fatal to you and would cause my friend's despair by making you eternally unhappy. However I know love's impatiences. I feel how irksome it must be for you in your position to experience any delay in the only consolation you can enjoy at this time. I have been considering means of putting aside these obstacles and I have discovered one which will be easy to carry out, if you take some trouble.

I think I have noticed that the key to your bedroom door which opens on to the corridor, is always on your Mamma's mantlepiece. With that key everything would be easy, as you will realise; but in default of it I will procure you one like it, which will do instead. To carry this out it will be enough if I have it for one or two hours. You ought easily to find an opportunity to take it, and in order that it shall not be missed I enclose one of my own, which is rather like it, so that the difference cannot be seen unless it is tried; which they will not do. You must merely be careful to put a faded blue ribbon on it like the ribbon on yours.

You must try to have this key at breakfast-time to-morrow or the day after; because it will be easier to give it me then and can be returned to its place for evening, when your Mamma is more likely to look closely at it. I could return it to you at dinner time if we have a proper understanding.

You know that when we go from the drawing-room to the dining-room, Madame de Rosemonde always comes last. I will give her my arm. You have only to leave your tapestry frame slowly or to let something drop, so that you stay behind; you will then be able to take the key which I shall be careful to hold behind me. Immediately after you have taken it you must not neglect to rejoin my old aunt and to caress her. If by chance you let the key drop, do not be disconcerted; I will pretend I did it and I answer for everything.

Your Mamma's lack of confidence in you and her harsh treatment of you justify this little deceit. Moreover it is the only way for you to continue to receive Danceny's letters and to transmit yours to him; any other is really too dangerous and might completely ruin you both; my prudent friendship would blame me if I continued to employ that method.

Once master of the key we shall still have to take some precautions against the noise of the door and the lock; but they are very easy. Under the same wardrobe where I hid your paper you will find oil and a feather. You sometimes go to your room when you are alone; you must make use of this time to oil the lock and the hinges. The only thing to beware of is to take care of oil-stains which would be evidence against you. You must also wait until it is night-time because if this is done with the intelligence of which you are capable, no trace of it will remain the next morning.

However, if it is noticed do not hesitate to say that it was the floor-polisher of the *Château*. In this case you must specify the time and even what he said to you; as, for example, that he takes this precaution against rust for all locks which are not used. For you will realise it is not probable that you should be a witness of this without asking the reason. It is the small details which give probability, and probability renders lies without consequence by taking away the desire to verify it.

After you have read this letter, I beg you to re-read it and to pay attention to it; first, because you must know well anything you wish to do well; and then in order to assure yourself that I have omitted nothing. I am unaccustomed to employ artifice on my own account and have no great practice in it; nothing less than my keen friendship for Danceny and the interest I have in you could have persuaded me to make use of these methods, however innocent they may be. I hate anything which looks like deceit; that is my character. But I am so moved by your misfortunes that I would attempt anything to soften them.

You will realise that when this communication is once established between us it will be much easier for me to procure you the interview with Danceny which he desires. However do not speak to him yet about all this; you would only increase his impatience, and the moment to satisfy it has not yet come. I think you ought to calm his impatience rather than to irritate it. I rely on your delicacy in this. Good-bye, my fair pupil; for you are my pupil. Love your tutor a little and above all be docile with him; you will find it to your advantage. I am busying myself with your happiness and you may rest assured I shall find my own in it.

From . . . , 24th of September, 17—.

LETTER LXXXV

The Marquise de Merteuil to the Vicomte de Valmont

At last you will be tranquillised, and above all you will do me justice. Listen, and never again confound me with other women. My adventure with Prévan is ended; ended! Do you understand what that means? You shall now judge whether he or I should boast. The relation will not be so amusing as the action; but it would not be just that you who have done nothing but argue well or ill about this affair should receive as much pleasure from it as I, who gave my time and my trouble.

However, if you have some grand stroke to make, if you have some enterprise to attempt in which this dangerous rival seems to you to be feared, come here. He leaves the field free to you, at least for some time; perhaps he will never rise again after the blow I have dealt him.

How fortunate you are to have me as your friend! I am your good fairy. You languish far from the beauty who attracts you; I say a word and you find yourself near her. You wish to avenge yourself on a woman who does you harm; I show you the place where you can strike her and hand her over to your discretion. Finally it is I whom you invoke to remove a dangerous competitor from the lists, and I grant your prayer. Indeed if you do not pass your life in thanking me you are an ingrate. I return to my adventure and go back to the beginning.

The rendezvous made in such a loud voice as I left the Opera was overheard as I had hoped. Prévan was there; and when the Maréchale told him politely that she was happy to see him twice in succession at her days, he was careful to reply that since Monday evening he had altered a thousand arrangements to be able to come that evening. "A word to to the wise!" However as I wished to know with more certainly whether or not I was the real object of this flattering eagerness, I wanted to compel the new wooer to chose between me and his ruling passion. I declared I should not gamble; and he on his part discovered a thousand pretexts for not gambling; so my first triumph was over *lansquenet*.

I took possession of the Bishop of . . . for my conversation; I chose him because of his acquaintance with the hero of the day, to whom I wished to give every opportunity of addressing me. I was also very glad to have a respectable witness who, if needed, could depose to my conduct and to my speech. This arrangement succeeded.

After the usual vague remarks, Prévan soon made himself master of the conversation and assumed different tones in turn to try which would please me. I refused that of sentiment, as not believing in it; by being serious I stopped his gaiety which seemed to me too airy for a beginning; he fell back on delicate friendship and it was under that commonplace flag that we began our mutual attack.

The Bishop did not go down to supper; Prévan therefore gave me his arm and was naturally placed beside me at table. I must be just and admit he showed great skill in keeping up our private conversation while appearing only to concern himself with the general conversation, the whole weight of which seemed to fall on him. At dessert he spoke of a new play to be given on Monday next at the Français. I expressed some regret at not having my box; he offered me his which I first refused, as is customary; to which he replied wittily enough that I did not understand him, that he would certainly not sacrifice his box to someone he did not know, but that he simply informed me the Maréchale had the use of it. She lent herself to the jest and I accepted.

When we had returned to the drawing-room, he asked for a place in the box, as you may suppose, and when the Maréchale, who treats him very kindly, promised it to him "if he were good," he made it the occasion for one of those conversations with a double meaning, in which you boasted his talents. He knelt down like an obedient child (as he said) under the pretext of asking her advice and imploring her judgment, and said many flattering and quite tender things which it was easy for me to apply to myself. After supper several people gave up gambling and the conversation was more general and less interesting, but our eyes were very eloquent. I say our eyes; I ought to have said his eyes, for mine had only one expression, that of surprise. He must have thought I was astonished and excessively preoccupied by the prodigious effect he was making on me. I think I left him very well satisfied; I was just as pleased.

The following Monday I went to the Français as we had agreed. In spite of your literary curiosity I can tell you nothing of the play except that Prévan has a marvellous talent for flattery and that the play was a failure; that is all I learnt. It was with regret I saw the evening ending, for really it gave me great pleasure; and to prolong it I invited the Maréchale to come and sup with me; this gave me a pretext to invite the amiable Flatterer who only asked for time to run and postpone an engagement with the Comtesses de P . . .[1] This name brought back all my anger; I saw clearly that he was going to begin his confidences; I recollected your wise advice and promised myself . . . to pursue the adventure, feeling sure I should cure him of this dangerous indiscretion.

As he was a stranger among my guests, who were not numerous that evening, he owed me the customary attentions; and so when we went to supper he offered me his hand. When I took it I was malicious enough to make mine tremble slightly and to walk with lowered eyes and more rapid breathing. I appeared to foresee my defeat and to fear my conqueror. He noticed it perfectly; and the traitor immediately changed his tone and behaviour. He had been gallant, he became tender. Circumstances compelled him to much the same remarks, but his gaze became less sprightly and more caressing, the inflection of his voice softer and his smile was no longer that of artifice but of contentment. In his talk he gradually subdued the fire of his sallies, and delicacy replaced wit. I ask you, would you have done better?

On my part I became absent-minded, to such an extent that the company was forced to notice it; and when I was reproached with it, I was skilful enough to defend myself awkwardly and to cast at Prévan a quick but timid and disconcerted glance, such as would make him think that my one fear was lest he should guess the cause of my confusion.

After supper I took advantage of the time when the good Maréchale was telling one of those stories she always tells, to place myself on my ottoman in the languid attitude caused by a tender reverie. I was not sorry that Prévan should see me thus, and indeed he honoured me with a very particular attention. As you may suppose, my timid glances did not dare to seek my conqueror's eyes; but when directed towards him in a more humble way they soon informed me that I had obtained the effect I wished to produce. It was then necessary to convince him that I shared it; and so, when the Maréchale announced that she was leaving, I exclaimed in a soft and tender voice: "Ah! I was so comfortable there!" However, I got up; but before taking leave of her, I asked her plans, to have a pretext to tell mine and to let it be known I should be at home on the day after the morrow. After which everyone left.

I then began to reflect. I had no doubt that Prévan would make use of the sort of rendezvous I had given him; that he would come early enough to find me alone and that the attack would be sharp; but I was also very sure that with my reputation he would not treat me with that brusqueness which men of any experience only employ with women who have had intrigues or with those who have had no experience; and I saw my success was certain if he uttered the word "love"; above all if he tried to obtain it from me.

How convenient it is to have to deal with you "men of principles!" Sometimes a bungling lover disconcerts one by his timidity or embarrasses one by his passionate raptures; it is a fever which, like others, has its cold shiverings and its burning, and sometimes varies in its symptoms. But it is so easy to guess your prearranged advance! The arrival, the bearing, the tone, the remarks—I knew what they would all be the evening before. I shall therefore not tell you our conversation which you will easily supply. You will only observe that in my feigned defence I gave him all the help in my power; embarrassment, to give him time to speak; poor arguments, for him to combat; fear and suspicion, to encourage his protestations; and the perpetual refrain on his part, "I only ask one word of you"; and the silence on my part, which only seemed to let him wait in order to make him desire it more; and through all this, a hand taken a hundred times, always withdrawn and never quite refused. One could spend all day in this way; we spent a mortal hour; we should perhaps be there still if we had not heard a carriage come into my court-yard. This fortunate occurrence naturally made his entreaties more pressing; and I, seeing that the moment had come when I was safe from any sudden attack on his part, prepared myself by a long sigh and granted the precious word. Someone was announced and soon after there was a numerous company.

Prévan asked if he might come the next morning and I consented; but, careful of my defence, I ordered my waiting-woman to remain in my bedroom during the whole of this visit, and you know that from there you can see everything that is going on in my dressing-room, which was where I received him. Free in our conversation, and both having the same desire, we were soon in agreement; but it was necessary to get rid of this importunate spectator, which was what I expected.

Then, I gave him my own version of my private life and easily convinced him that we should never find a moment of liberty; that he must consider as a miracle the opportunity we had enjoyed the day before and that even this was far too dangerous for me to risk, since at any moment someone might come into my drawing-room. I did not fail to add that all these customs had grown up because hitherto they had never been an impediment to me; and at the same time I insisted upon the impossibility of altering them without compromising myself in the eyes of my servants. He tried to look sad, to be out of humour, to tell me my love was slight; and you can guess how much that touched me! But to strike the decisive blow, I called in tears to my help. It was precisely the situation of "Zaïre, you are weeping."[2] The power he thought he had over me, and the hope it gave him that he could take me as he wished, stood him in stead of all the love of Orosmane.

After this theatrical incident, we returned to our arrangements. As the daytime was impossible, we considered the night; but my porter became an insurmountable obstacle and I would not allow Prévan to try to bribe him. He then proposed the small door into my garden; but I had foreseen that and I invented a dog which was tranquil and silent by day but a real demon at night. The facility with which I entered into all these details naturally emboldened him; and so at last he proposed the most ridiculous expedient, which was the one I accepted.

First of all, his servant was as trustworthy as he was; in this he did not deceive me, for one was as little trustworthy as the other. I was to give a big supper in my house; he would be there and would arrange to leave alone. The skilful confidant would call the carriage, open the door, and Prévan, instead of getting in, would nimbly slip away. His coachman would not be able to perceive it; and so everyone would think he had gone and yet he would be in my house; it remained to be seen if he could get to my room. I admit that at first my difficulty was to allege against this project reasons sufficiently weak for him to be able to destroy them; he replied by instances. According to him, nothing was more usual than this way; he had often made use of it himself; it was even the method he usually adopted, as being the least dangerous.

Subjugated by this irrefutable authority, I candidly admitted that there was a concealed stairway which led up to my boudoir; that I could leave the key in it and he could lock himself in and wait without much risk until my women had left; and then, to make my consent more probable, a moment afterwards I would not allow it, I was only brought to consent on condition of his perfect submission, his restraint . . . Ah! What restraint! In short I was willing to prove my love to him, but not to satisfy his.

I forgot to tell you that he was to leave by the small garden door; he had only to wait until dawn; the Cerberus would then say not a word. Not a soul passes at that hour and the servants are in their deepest sleep. If you are surprised at this heap of silly arguments, it is because you have forgotten our mutual situation. Where was the necessity for better ones? He asked nothing better than that everything should be known and I was quite sure it would not be known. The appointment was fixed for two days afterwards.

Notice that the affair had been arranged and that nobody had yet seen Prévan alone with me. I meet him at supper at a friend's house; he offers her his box for a new play and I accept a seat in it. I invite this woman to supper during the play and in Prévan's presence. It is almost impossible for me not to ask him too. He accepts and two days afterwards pays me the visit demanded by custom. It is true he comes to see me the next morning; but, apart from the fact that morning visits no longer count, it was in my power to consider this improper; and in fact I put him in the class of my less intimate friends by a written invitation to a formal supper. I can say like Anette: "But that is all, after all!"

When the fatal day came, the day on which I was to lose my virtue and my reputation, I gave my instructions to my faithful Victoire and she carried them out as you will soon see.

The evening came. There were already a number of people with me when Prévan was announced. I received him with a marked politeness, which showed how little intimate I was with him; and I put him in the Maréchale's party, as being that through which I had made his acquaintance. The evening produced nothing but a very short note, which the discreet lover found a way to hand me and which I burned in accordance with my custom. He announced in it that I could count upon him; and this essential word was surrounded by all the parasitic words about love, happiness, etc., which are never found missing on such an occasion.

At midnight when the card parties were finished, I proposed a short *macédoine*.[3] I had the double intention of giving Prévan his opportunity to leave and at the same time rendering it noticeable; which could not fail to happen from his reputation as a gambler. I was also glad that, if necessary, people could remember I had not been in a hurry to remain alone.

The play lasted longer than I had thought. The devil tempted me and I yielded to the desire to go and tempt the impatient prisoner. I was proceeding towards my ruin when I reflected that once I had surrendered to him it would no longer be in my power to keep him in the costume of decency necessary to my plans. I had the strength to resist. I turned back and, not without ill humour, returned to my place at the eternal game. At last it ended and everyone went away. I rang for my women, undressed very quickly and sent them away at once.

Can you see me, Vicomte, in my light toilette, walking timidly and carefully, and opening the door to my conqueror with a trembling hand? He saw me—a flash is not quicker. How shall I tell you? I was overcome, completely overcome, before I could say a word to stop him or to defend myself. He then wished to put himself into a more comfortable situation, more suitable to the circumstances. He cursed his clothes which, he said, kept him at a distance from me; he wished to combat me with equal weapons, but my extreme timidity opposed this plan and my tender caresses did not leave him time for it. He busied himself with other matters.

His privileges were doubled and his pretentions returned; and then: "Listen to me," I said, "Up till now you will have a very agreeable story to tell the two Comtesses de P . . . and a thousand others; but I am curious to know how you will relate the end of the adventure." So saying, I rang the bell with all my strength. This time it was my turn, and my action was quicker than his speech. He had only begun to stammer when I heard Victoire running and calling the men-servants whom she had kept in her room as I had ordered. Then taking a lofty tone and raising my voice, I continued: "Go, Monsieur, and never appear before me again." Thereupon a crowd of my servants rushed in.

Poor Prévan lost his head and thought he saw an ambush in what at bottom was only a pleasantry, and rushed for his sword. It turned out badly for him; for my footman who is brave and strong, seized him round the body and threw him on the ground. I admit I was in mortal terror. I called to them to stop and ordered them to let him go free, merely making sure that he left my house. My servants obeyed me; but there was a great clamour among them and they were indignant that anyone should have dared to insult "their virtuous mistress." They all went out with the unhappy Chevalier, with noise and scandal as I had wished. Victoire alone remained and during this time we occupied ourselves in repairing the disorder of my bed.

My servants came back still in a tumult; and I, "still thoroughly upset," asked them by what good fortune it was that they were still up; and Victoire told me that she had given a supper to two of her friends, that they had sat up in her room; in short, everything we had agreed on together. I thanked them all and sent them away, but ordered one of them to go for my doctor. It appeared to me that I had a right to fear the effects of "my profound shock"; and this was a certain means of giving circulation and celebrity to the news.

The doctor came, pitied me very much and only ordered me rest. I ordered Victoire to go out early in the morning and gossip about it in the neighbourhood.

Everything succeeded so well that before midday and as soon as it was daylight with me, my devout neighbour was at my bedside to know the truth and the details of this horrible adventure. For an hour I was forced to lament with her over the corruption of the age. Shortly afterwards, I received the enclosed note from the Maréchale. Finally just before five to my great astonishment M . . .[4] arrived. He said he came to apologise to me for the fact that an officer in his corps could have insulted me to such an extent. He had only learned it when dining with the Maréchale and had immediately sent orders to Prévan to consider himself under arrest. I asked pardon for him and it was refused. I then thought that as an accomplice I ought to comply with the order myself and at least keep myself under close arrest. I ordered my door to be closed and said I was unwell.

You owe this long letter to my solitude. I shall write one to Madame de Volanges, who will surely read it in public, and you will see how the story ought to be told.

I forgot to tell you that Belleroche is in a rage and is determined to call Prévan out. Poor boy! Fortunately I shall have time to calm him down. Meanwhile I shall rest myself for my head is weary with writing. Good-bye, Vicomte.

From the Château de . . . , 25th of September, 17—.
(In the evening.)

LETTER LXXXVI

The Maréchale de . . . to the Marquise de Merteuil
(Note enclosed in the preceding letter)

Heavens! What is this I hear, my dear Madame? Is it possible that little Prévan should attempt such abominations? And upon you too! What one is liable to! And so we shall not be safe in our own houses! Really, such happenings are a consolation for being old. But what I shall never feel consoled for, is that I was partly the cause of your having received such a monster in your house. I promise you that if what I am told is true, he shall never set foot in my house again; all decent people will adopt the same course with him, if they act as they ought.

I am told that it made you ill, and I am anxious about your health. Pray, give me news of your dear self; or let me hear by one of your women, if you cannot do it yourself. I should have hastened to you this morning, but for the baths which the doctor will not allow me to interrupt; and this afternoon I must go to Versailles about that affair of my nephew's.

Farewell, dear Madame; count upon my sincere and lifelong friendship.

Paris, 25th of September, 17—.

LETTER LXXXVII

The Marquise de Merteuil to Madame de Volanges

I am writing to you from my bed, my good, dear friend. A most disagreeable event, the most impossible to foresee has made me ill with shock and grief. Assuredly I have nothing to reproach myself with; but for a virtuous woman who preserves the modesty befitting her sex, it is always so painful to have public attention directed upon her, and I would give anything in the world to have been able to avoid this unfortunate adventure; I am not yet certain whether I shall not go to the country to wait for it to be forgotten. This is what happened.

At the Maréchale de . . .'s house I met a Monsieur de Prévan whom you will certainly know by name and whom I did not know otherwise. But as I met him in this house I think I had every reason to believe him an honourable man. In his person he is quite handsome, and he seemed to me not to be lacking in intelligence. Chance and the tedium of gambling left me the only woman between him and the Bishop of . . ., while every one else was occupied with *lansquenet*. We all three conversed until supper time. At table a new play was spoken of, which gave him the opportunity to offer his box to the Maréchale, who accepted it; and it was agreed that I should have a place in it. It was for last Monday, at the Français. As the Maréchale was coming to sup with me after the play, I invited this gentleman to accompany her, and he came. Two days later he paid me a call which was spent with the usual talk and without anything noticeable occurring. The next day he came to see me in the morning, which appeared to me a little improper; but I thought that instead of letting him feel it by my manner of receiving him, it would be better to warn him by some act of politeness that we were not as intimately known to each other as he seemed to think. For that reason I sent him the same day a very dry and very ceremonious invitation for a supper I gave the day before yesterday. I did not speak four times to him the whole evening; and he, on his part, left as soon as he had finished his game of cards. You will admit that hitherto nothing could appear less likely to lead to an adventure; after the card parties there was a "*macédoine*" which kept us up until nearly two o'clock; and then I went to bed.

At least a mortal half hour after my women had gone, I heard a noise in my room. I opened my curtains in great terror and saw a man come in by the door which leads to my boudoir. I gave a piercing cry; and by the glimmer of my night-light I recognised this Monsieur de Prévan who, with incredible effrontery, told me not to be alarmed, that he would enlighten me upon the mystery of his behaviour and that he begged me to make no noise. So saying, he lighted a candle; I was so startled I could not speak. His tranquil and easy air petrified me I think even more. But he had not spoken two words when I saw what this pretended mystery was; and, as you may suppose, my only answer was to ring as hard as I could.

By an incredible good fortune all the servants on duty had remained up with one of my women and had not yet gone to bed. As my waiting-woman was coming up to me she heard me speaking with great warmth, was frightened, and called them all up. You may imagine what a scene it was! My servants were furious; at one moment I thought my footman would kill Prévan. I confess that at that time I was very glad to be in force; but when reflecting on it today I should have preferred that only my waiting-woman had come. She would have sufficed and perhaps I should have avoided the scandal which so distresses me.

Instead of that, the noise awoke the neighbours, the servants gossiped about it, and ever since yesterday all Paris is talking about it. Monsieur de Prévan is under arrest by the orders of his commanding officer who had the politeness to call upon me; to apologise, he said. This imprisonment will increase the talk; but I was unable to get it changed. The Town and the Court have inquired at my door which I have closed to everybody. The few persons I have seen tell me that people do me justice, and that the public indignation against Monsieur de Prévan is intense; certainly he deserves it, but that does not remove the unpleasantness of this adventure.

Moreover this man must surely have some friends and his friends must be evil; who knows, who can know what they will invent to harm me? Heavens, how unfortunate a young woman is! She has done nothing when she has preserved herself from scandal, she must overcome even calumny.

Pray tell me what you would have done and what you would do in my place; in brief, all you think. I have always received from you the kindest consolation and the wisest advice; and it is from you I most like to receive them.

Farewell, my dear and good friend; you know the sentiments which will forever attach me to you. I kiss your amiable daughter.

Paris, 26th of September, 17—.

END OF PART TWO

PART III

LETTER LXXXVIII

Cécile Volanges to the Vicomte de Valmont

In spite of all my pleasure, Monsieur, in receiving letters from M. le Chevalier Danceny, and although I desire no less than he does that we should see each other again without our being prevented, yet I dare not do what you propose. First of all, it is too dangerous; the key you wish me to put in place of the other does indeed resemble it; but still there is a difference between them, and Mamma looks at everything and notices everything. Moreover, although it has not been used since we have been here, by some misfortune it might be; and if it were noticed, I should be ruined for ever. And then it seems to me it would be very wrong; to make a double key in that way—it is too much! It is true that you would be the person who would be kind enough to take charge of it; but in spite of that, if it were known, the blame and the fault would none the less be laid on me, because you would have done it for me. Lastly, I twice tried to take it and certainly it would be very easy if it were anything else; but I do not know why, I always began to tremble and had not the courage to do it. I think it is better to remain as we are.

If you still have the kindness to be as complaisant as you have been, you will always find means of giving me a letter. Even with the last one, but for the accident which made you turn round quickly at the wrong moment, we should have done it easily. I know that you cannot always be thinking about this, as I do; but I prefer to have more patience and not to take such risks. I am sure Monsieur Danceny would say as I do; for every time he wanted something which was too hard for me to do, he always agreed that it should not be done. I shall hand you at the same time as this letter, Monsieur, your own letter, Monsieur Danceny's and your key. I am none the less grateful to you for all your kindnesses and I beg you will continue them. It is true I am very unhappy, and but for you I should be much more so; but after all, it is my mother; we must be patient. And so long as Monsieur Danceny loves me and you do not abandon me, perhaps there will come a happier time.

I have the honour to be, Monsieur, with much gratitude, your most humble and most obedient servant.

From . . . , 26th of September, 17—.

LETTER LXXXIX

The Vicomte de Valmont to the Chevalier Danceny

If your affairs do not always progress as fast as you would like, my friend, you must not blame me alone. I have more than one obstacle to overcome here. Madame de Volanges's vigilance and severity are not the only ones; your young friend herself hinders me with some. Either from coldness or timidity, she does not always do what I advise her and yet I think I know what ought to be done better than she.

I had found a simple, convenient and certain way of delivering your letters to her and even of facilitating later the interviews you desire but I cannot persuade her to make use of it. I am the more distressed by this since I see no other way to bring you together; and even with your correspondence I am continually in fear of compromising all three of us. Well, you will realise that I desire neither to run that risk myself nor to expose either of you to it.

Yet I should be really pained if your little friend's lack of confidence should prevent me from being useful to you; perhaps you would do well to write to her about it. Consider what you wish to do, it is for you alone to decide; for it is not enough to serve one's friends, they must be served in the manner they wish. This might also be a further means of making sure of her feelings for you; for the woman who keeps a will of her own does not love as much as she says.

It is not that I suspect your mistress of inconstancy; but she is very young; she is very much afraid of her Mamma who, as you know, seeks only to harm you; and perhaps it would be dangerous to remain too long without occupying her with you. But do not grow too uneasy about what I tell you. I have really no reason for suspicion; it is merely the solicitude of friendship. I do not write you at more length because I too have some affairs on my own account. I am not so far forward as you; but I love as much and that is consoling; and even if I do not succeed on my own account, I shall think I have employed my time well if I can be useful to you. Good-bye, my friend.

From the Château de . . . , 26th of September, 17—.

LETTER XC

Madame de Tourvel to the Vicomte de Valmont

I very much hope, Monsieur, that this letter will give you no pain; or, if it must give you pain, that at least it may be softened by that I feel in writing to you. You ought to know me enough now to be quite sure I would not willingly distress you; but doubtless you yourself would be just as unwilling to plunge me into an eternal despair. I call upon you then, in the name of the tender friendship I promised you, in the name even of those sentiments, perhaps keener but certainly not more sincere, which you have for me, not to let us see each other again; leave me; and, until then, let us especially avoid those private and too dangerous interviews in which through some inconceivable power I pass my time in listening to what I ought not to hear, without being able to tell you what I wish to say.

Even yesterday, when you came and joined me in the park, my sole object was to tell you what I am writing to you to-day; and yet what did I do except occupy myself with your love . . . with your love, to which I ought never to reply! Ah! I beseech you, leave me.

Do not fear that my absence should ever alter my feelings for you; how should I succeed in conquering them, when I have no longer the courage to combat them? You can see it, I confess everything; I am less afraid of admitting my weakness than of yielding to it; but the power I have lost over my feelings, I have kept over my actions; yes, I shall keep it, I am resolved upon it, were it at the cost of my life.

Alas! It was not long ago that I felt certain I should never have to engage in such struggles. I felt glad of it and perhaps glorified myself on it too much. Heaven has punished, cruelly punished my pride; but full of pity at the moment even when it strikes us, it warns me again before I fall; and I should be doubly guilty if I continued to be lacking in prudence when I am already made aware that I have no more strength.

You have said a hundred times that you would not want a happiness bought by my tears. Ah! Let us talk no more of happiness, but allow me to regain some tranquillity.

By granting my request, what new rights will you not acquire over my heart? And as they will be founded upon virtue, I shall not need to defend myself from them. What delight I shall take in my gratitude! I shall owe to you the pleasure of enjoying a delicious sentiment without remorse. But now, on the contrary, I am frightened by my feelings, by my thoughts, and dread equally to occupy myself with you or with myself; even the very idea of you terrifies me; when I cannot fly from it, I combat it; I do not drive it away, but repel it.

Is it not better for us both to end this state of uneasiness and anxiety? O you, whose ever tender soul remained the friend of virtue even in the midst of its errors, you will respect my painful situation, you will not reject my prayer! A softer but not less tender interest will take the place of these violent agitations; then, breathing by your favours, I shall cherish my existence, and I shall say in the joy of my heart: "The joy I feel is owing to my friend."

In submitting to a few slight privations, which I do not impose upon you but which I ask of you, shall you think you are buying too dearly the cessation of my torments? Ah! If I could make you happy by only consenting to be unhappy myself, believe me, I should not hesitate a moment . . . but to become guilty! . . . No, my friend, no, rather death a thousand times.

Already I am assailed by shame, I am on the eve of remorse, and I shrink from others and myself; I blush in company and shiver in solitude; I have nothing but a life of grief; I shall have no tranquillity unless you consent. My most praiseworthy resolutions are insufficient to reassure me; I made this resolution yesterday and yet I passed the night in tears.

Behold your friend, she whom you love, confused and supplicating, begging you for rest and innocence. Ah God! But for you would she ever have been reduced to this humiliating request? I do not blame you at all; I feel too much myself how difficult it is to resist an imperious sentiment. A wail is not a murmur. Do from generosity what I am doing from duty, and to all the sentiments you have inspired in me I shall add that of an eternal gratitude. Good-bye, good-bye, Monsieur.

From . . . , 27th of September, 17—

LETTER XCI

The Vicomte de Valmont to Madame de Tourvel

I am in consternation at your letter, Madame, and still do not know how to reply to it. Doubtless if there must be a choice between your unhappiness and mine, it is for me to sacrifice myself, and I do not hesitate; but it seems to me that things of this importance ought above all to be discussed and made clear; and how can that be done, if we are never to speak to each other nor see each other again?

What! When we are united by the softest feelings, shall a vain terror suffice to separate us, perhaps for ever! In vain will tender friendship and ardent love claim their rights; their voices will not be heard. And why? What is this pressing danger which threatens you? Ah! Believe me; such fears, so lightly conceived, are already, it seems to me sufficiently powerful reasons for confidence.

Permit me to tell you that I find in this a trace of the unfavourable impressions you have been given about me. A woman does not tremble in the presence of the man she esteems; above all she does not send away the man she has thought worthy of a certain friendship; it is the dangerous man who is feared and shunned.

And yet who was ever more respectful and more submissive than I? Already, as you see, I am watching what I say; I no longer allow myself those names which are so soft, so dear to my heart, and which my heart does not cease to give you in secret. I am no longer the faithful and unhappy lover receiving advice and consolations from a tender and sensitive friend; I am the prisoner before his judge, the slave before his master. No doubt these new titles imply new duties; I promise to carry them all out. Hear me and if you condemn me, I will accept the sentence and go. I promise more; do you prefer the despotism which judges without hearing? Do you feel in yourself the courage to be unjust? Command and I will still obey.

But let me hear this judgment or this order from your mouth. And why? you will say in your turn. Ah! If you put that question, you little know love and my heart! Is it then nothing to see you once more? Ah! Even if you carry despair into my soul perhaps a consoling look will prevent me from succumbing to it. And then if I must renounce love, renounce friendship, for which alone I exist, at least you will see your work and your pity will be left me; even if I did not deserve that slight favour I think I am ready to pay dearly enough for it to hope to obtain it.

What! You are about to send me away from you! You agree that we shall become strangers to each other! What do I say? You desire it; and while you assure me that my absence will not alter your sentiments, you only hasten my departure to labour more easily at their destruction.

You already speak to me of replacing them by gratitude. And so you offer me the sentiment which a stranger would obtain from you for the slightest service, which even your enemy would obtain by ceasing to do you harm! And you wish my heart to be content with that! Question your own; if your lover, if your friend, should one day speak to you of their gratitude, would you not say to them indignantly: "Leave me, you are ungrateful."

I pause here and call upon your indulgence. Forgive the expression of a grief which you have created; it will do no harm to my complete submission. But in my turn I call upon you, in the name of those soft sentiments, which you call upon yourself, not to refuse to hear me; and at least from pity for the utter distress into which you have plunged me, do not put off the moment. Good-bye, Madame.

From . . . , 27th of September, 17—, in the evening.

LETTER XCII

The Chevalier Danceny to the Vicomte de Valmont

O my friend! Your letter froze me with terror . . . Cécile . . . Oh God! Is it possible? Cécile no longer loves me. Yes, I see that dreadful truth through the veil your friendship throws over it. You wish to prepare me to receive a mortal blow; I thank you for your pains, but can Love be imposed upon? It anticipates what concerns it; it does not hear its fate, but divines it. I have no longer any doubt of mine; speak to me plainly, you can do so, and I beg you to. Tell me everything; what gave birth to your suspicions, what confirmed them? The slightest details are precious. Above all try to recollect her words. One word for another may change a whole phrase; the same word has sometimes two meanings . . . you may have been deceived. "Alas! I am still trying to flatter myself. What did she say to you? Does she blame me for anything? Does she not at least excuse herself for her faults? I ought to have foreseen this change from the difficulties which for some time she has found in everything. Love does not admit so many obstacles.

What ought I to do? What do you advise me? Should I try to see her? Is that impossible? Absence is so cruel, so disastrous . . . and she has refused a way to see me! You do not tell me what it was! If it was indeed too dangerous, she well knows I should not wish her to risk too much. But then I know your prudence; and, to my misfortune, cannot but believe it.

What am I to do now? How shall I write to her? If I let her see my suspicions they may perhaps grieve her; and if they are unjust, should I ever forgive myself for having distressed her? If I hide them from her, it is deceiving her, and I cannot use dissimulation with her. Oh! If she could know what I suffer, my pain would touch her. I know she is tender, her heart is excellent and I have a thousand proofs of her love. Too much timidity, some embarrassment, she is so young! And her mother treats her with such severity! I shall write to her; I shall restrain myself; I shall only ask her to rely implicitly upon me. Even if she refuses again, at least she cannot be angry at your request; and perhaps she will consent.

I send you a thousand apologies, both for her and for myself. I assure you that she feels the value of the pains you are taking and that she is grateful for them. This is not suspicion on her part, it is timidity. Be indulgent; it is the fairest trait of friendship. Yours is very precious to me and I do not know how to acknowledge all you are doing for me. Good-bye, I am going to write to her at once.

I feel all my fears return; who could have believed it would ever be difficult for me to write to her! Alas! But yesterday it was my dearest pleasure.

Good-bye, my friend; continue your exertions and pity me.

Paris, 27th of September, 17—.

LETTER XCIII

The Chevalier Danceny to Cécile Volanges

(Enclosed with the preceding)

I cannot conceal from you how distressed I was to learn from Valmont of the small confidence you continue to have in him. You know he is my friend, that he is the only person who can bring us together again; I thought these claims would have been sufficient for you; I see with pain that I was wrong. Can I hope at least that you will tell me your reasons? Will you still find certain difficulties to prevent your doing so? Yet without your aid I cannot guess the mystery of this conduct. I dare not suspect your love, and doubtless you would not dare to betray mine. Ah! Cécile! . . .

Is it then true that you have refused a way of seeing me? A way which is "simple, convenient and certain?"[1] And this is how you love me! So short an absence has greatly altered your sentiments. But why deceive me? Why tell me that you still love me, that you love me more than ever? Has your Mamma destroyed your candour as well as your love? If at least she has left you some pity you will not hear without pain of the dreadful torments you cause me. Ah! I should suffer less in dying.

Tell me, is your heart closed to me for ever? Have you entirely forgotten me? Thanks to your refusal, I do not know when you will hear my complaints nor when you will reply to them. Valmont's friendship had made our correspondence certain; but you did not want it; you thought it difficult, you preferred that it should be seldom. No, I no longer believe in love, in good faith. Ah! Who is to be believed if Cécile has deceived me?

Answer me. Is it true you love me no more? No, that is impossible; you delude yourself; you calumniate your own heart. A passing fear, a moment of discouragement, which love has soon banished; is it not so, my Cécile? Ah! Doubtless it is, and I am wrong to accuse you. How happy I should be to be wrong! How glad I should be to make you tender apologies, to repair this moment of injustice by an eternity of love!

Cécile, Cécile, have pity on me! Consent to see me, take any way of doing it! See what is the result of absence! Fears, suspicions, perhaps coldness! One look, one word, and we shall be happy. But what! Can I still speak of happiness? Perhaps it is lost for me, lost for ever. Tormented by fear, urged cruelly between unjust suspicions and more cruel truth, I cannot fix upon any thought; I retain only sufficient existence to suffer and to love you. Ah! Cécile! You alone have the right to render existence dear to me; and I expect from the first word you pronounce the return of happiness or the certainty of an eternal despair.

Paris, 27th of September, 17—.

LETTER XCIV

Cécile Volanges to the Chevalier Danceny

I understand nothing of your letter, except the pain it causes me. What has Monsieur de Valmont told you, what can have made you believe I do not love you? Perhaps it would be very fortunate for me, for assuredly I should be less tortured; and it is very hard, when I love you as I do, to see that you always think I am wrong, and that instead of consoling me, it is you who always cause me the pains which grieve me most. You think that I am deceiving you and that I tell you things which do not exist! You have a pretty idea of me! But even if I were as untruthful as you reproach me with being, what interest should I have in being so? Assuredly, if I did not love you any more I should only have to say so and everyone would praise me for it; but unluckily it is stronger than I am; and then it must needs be for someone who has no gratitude at all to me for it!

What have I done to make you so angry? I did not dare take the key because I was afraid Mamma would notice it and that it would cause me more trouble and you too on my account; and then too, because it seemed wrong to me. But only Monsieur de Valmont had spoken to me about it; I could not know whether you wanted it or not since you knew nothing about it. Now I know that you want it, do I refuse to take the key? I shall take it tomorrow; and then we shall see what you will have to say. Monsieur de Valmont may well be your friend; I think I love you quite as much as he can like you, at least; and yet he is always right and I am always wrong. I assure you I am very angry. That is all the same to you, because you know I am pacified at once; but when I have the key I can see you when I want; and I assure you I shall not want to, if you act like this. I would rather suffer the grief which comes from myself than from you; consider what you wish to do.

If you would allow it, we should love each other so much!

And at least we should have no troubles except those inflicted on us by other people! I assure you that if it were in my control you would never have to complain of me; but if you do not believe me, we shall always be very unhappy and it will not be my fault. I hope we shall soon be able to see each other, and that then we shall have no more reasons to hurt each other as we do now.

If I could have foreseen this, I should have taken the key at once; but, truly, I thought I was doing right. Do not be angry with me, I beg you. Do not be sad any more, and always love me as much as I love you; then I shall be quite content. Good-bye, my dear.

From the Château de . . . , 28th of September, 17—.

LETTER XCV

Cécile Volanges to the Vicomte de Valmont

I beg you, Monsieur, to be good enough to return me the key you gave me to put in place of the other; since everybody desires it, I too must agree to it.

I do not know why you told Monsieur Danceny I did not love him any more; I do not think I have ever given you any reason to think so; and it has pained him very much and me too. I know you are his friend; but that is not a reason for grieving him, nor me either. You would do me a great pleasure to tell him the contrary the next time you write to him and that you are sure of it; for he has more confidence in you; and when I have said a thing and I am not believed, I do not know what else to do. You can be at rest about the key; I remember perfectly everything you told me in your letter. However, if you still have it and would give it to me at the same time as the key, I promise you to pay great attention to it. If it could be tomorrow when we are going in to dinner, I could give you the other key at breakfast the next day and you would return it to me in the same way as the first. I should prefer it not to be a long time, because there would be less risk of Mamma noticing it.

And then, once you have this key, will you have the kindness to make use of it to get my letters; in this way Monsieur Danceny will have news of me more often. It is true that this will be more convenient than it is now; but at first it frightened me too much; I beg you to excuse me and I hope you will continue to be as complaisant as in the past. I shall always be very grateful for it.

I have the honour to be, Monsieur, your most humble and most obedient servant.

From . . . , 28th of September, 17—.

LETTER XCVI

The Vicomte de Valmont to the Marquise de Merteuil

I wager that ever since your adventure you have every day been expecting my compliments and praise; I have no doubt but that my long silence has put you a little out of temper; but what do you expect? I have always thought that when one had nothing but praise to give a woman one can be at rest about her and occupy oneself with something else. However, I thank you on my own account and congratulate you on yours. To make you perfectly happy, I am even willing to admit that this time you have surpassed my expectations. After that, let us see if on my side I have at least partly fulfilled yours.

I do not want to talk to you about Madame de Tourvel; her slow advance displeases you. You only like completed affairs. Drawn-out scenes weary you; but I have never tasted the pleasure I now enjoy in this supposed tardiness.

Yes, I like to see, to watch this prudent woman impelled, without her perceiving it, upon a path which allows no return, and whose steep and dangerous incline carries her on in spite of herself, and forces her to follow me. There, terrified by the peril she runs, she would like to halt and cannot check herself. Her exertions and her skill may render her steps shorter; but they must follow one upon the other. Sometimes, not daring to look the danger in the face, she shuts her eyes, lets herself go, and abandons herself to my charge. More often her efforts are revived by a new fear; in her mortal terror she would like to try to turn back once again; she exhausts her strength in painfully climbing back a short distance and very soon a magic power replaces her nearer the danger from which she had vainly tried to fly. Then having no one but me for guide and for support, without thinking of reproaching me for the inevitable fall, she implores me to retard it. Fervent prayers, humble supplications, all that mortals in their fear offer to the divinity, I receive from her; and you expect me to be deaf to her prayers, to destroy myself the worship she gives me, and to use in casting her down that power she invokes for her support! Ah! At least leave me the time to watch these touching struggles between love and virtue.

What! Do you think that very spectacle which makes you rush eagerly to the theatre, which you applaud there wildly, is less interesting in reality? You listen with enthusiasm to the sentiments of a pure and tender soul, which dreads the happiness it desires and does not cease to defend itself even when it ceases to resist; should they only be valueless for him who gives birth to them? But these, these are the delicious enjoyments this heavenly woman offers me every day; and you reproach me for lingering over their sweetness! Ah! The time will come only too soon when, degraded by her fall, she will be nothing but an ordinary woman to me.

But in speaking to you of her I forget that I did not wish to speak of her. I do not know what power attaches me to her, ceaselessly brings me back to her, even when I insult it. Let me put aside the dangerous thought of her; let me become myself again to deal with a more amusing subject. It concerns your pupil, now become mine, and I hope you will recognise me here.

For some days I had been better treated by my ter
and in consequence, being less preoccupied with her, I
that the Volanges girl is indeed very pretty; and that if it w
silly to be in love with her like Danceny, perhaps it was none the less silly on my part not to seek with her a distraction rendered necessary by my solitude. I also thought it just to pay myself for the trouble I am giving myself for her; I remembered too that you had offered it to me before Danceny had any claims to it; and I thought myself authorised to claim some rights in a property he only possessed through my refusal and rejection. The little person's pretty look, her fresh mouth, her childish air, even her awkwardness fortified these sage reflections; I resolved to act upon them and the enterprise was crowned by success.

You are wondering already by what means I have so soon supplanted the cherished lover; what seduction is suitable to this age, this inexperience? Spare yourself the trouble, I used none. While you skilfully handled the arms of your sex and triumphed by subtlety; I returned to man his imprescriptible rights and overcame by authority. I was sure of seizing my prey if I could come at it; I needed no ruse except to approach her and that I made use of hardly deserves the name.

I profited by the first letter I received from Danceny for his fair one, and after having notified her of it by the signal agreed upon between us, instead of using my skill to give it her I used it to find means not to give it; I feigned to share the impatience I had created, and after having caused the difficulty I pointed out the remedy for it.

The young person's room has a door opening on the corridor; but naturally the mother had taken the key. It was only a question of getting possession of it. Nothing could have been easier to carry out; I only asked to have it at my disposition for a couple of hours and I guaranteed to have one like it. Then correspondence, interviews, nocturnal rendez-vous, all became convenient and certain. However, would you believe it? the timid child was afraid and refused. Another would have been nonplussed by this; but I only saw in it an opportunity for a more piquant pleasure. I wrote to Danceny to complain of this refusal and I acted so well that our scatter-brain had no rest until he had obtained, exacted even, from his timorous mistress her consent to grant my request and to yield wholly to my discretion.

I confess I was very pleased to have changed parts in this way and to have the young man do for me what he thought I should do for him. This idea doubled the value of the adventure in my eyes; therefore as soon as I had the precious key, I hastened to make use of it; this was last night.

After making sure that everything was quiet in the *Château*, I armed myself with a dark lantern, made the toilet which suited the hour and was demanded by the situation, and paid my first visit to your pupil. I had caused everything to be prepared (and by herself) to be able to enter noiselessly. She was in her first sleep and in the sleep of her age; so that I came up to her bedside without awakening her. At first I was tempted to go further and to try to pass as a dream; but fearing the effect of surprise and the noise it brings with it, I preferred to arouse the pretty sleeper cautiously, and in fact I succeeded in preventing the cry I feared.

After having calmed her first fears, I risked a few liberties since I had not come there to talk. Doubtless she had not been well informed at her convent as to how many different perils timid innocence is exposed, and all it has to guard to avoid a surprisal; for giving all her attention and all her strength to defending herself from a kiss, which was only a false attack, she left all the rest without defence. How could I not profit by it! I therefore changed my movement and immediately took post. Here we were both very nearly lost; the little girl was terrified and tried to scream in good faith; fortunately her voice was quenched in tears. She also threw herself towards her bell-rope, but my skill restrained her arm in time.

"What are you doing," I then said to her, "You will ruin

yourself forever. Suppose someone comes; what does it matter to me? Whom will you convince that I am not here by your wish? Who but you could have given me the means of coming in? And will you be able to explain the use of this key I hold from you, which I could only have had through you?"

This short harangue calmed neither her grief nor her anger: but it brought about submission. I do not know if I achieved the tone of eloquence; it is at least true that I did not have its gestures. With one hand occupied by force and the other by love what orator could pretend to grace in such a situation? If you imagine it correctly, you will at least admit it was favourable to attack; but I do not understand anything and, as you say, the simplest woman, a mere school-girl, leads me like a child.

For all her distress she felt she had to adopt some course and come to terms. Since prayers found me inexorable she had to come to offers. You will suppose that I sold this important post very dearly; no, I promised everything for a kiss. It is true that, having taken the kiss, I did not keep my promise; but I had good reasons. Had we agreed that it should be taken or given? After much bargaining, we agreed on a second and it was said that this should be received. Then I guided the timid arms around my body, I held her more amorously in one of mine, and the soft kiss was indeed received, well received, perfectly received, in short, love himself could not have done better.

Such good faith deserved a reward; and so I immediately granted the request. The hand was withdrawn; but I do not know by what chance I found myself in its place. You suppose that there I was very eager, very active, do you not? Not at all. I have begun to like slow methods, I tell you. Once certain of arriving, why hurry on the journey so fast?

Seriously, I was very glad to observe for once the power of opportunity, and here I found it divested of all other aid. Yet she had to combat love, and love supported by modesty or shame, fortified above all by the annoyance I had given, which was considerable. Opportunity was alone; but it was there, all was offered, all was present, and love was absent.

To make certain of my observations I was cunning enough to use no more force than she could combat. Only if my charming enemy abused my facility, and was ready to escape me, I restrained her by the same fear whose happy effects I had already made proof of. Well, without my taking any other exertion, the tender mistress, forgetting her oaths, yielded at first and ended up by consenting; not but that after the first moment reproaches and tears returned in concert; I do not know whether they were true or feigned; but, as always happens, they ceased as soon as I busied myself with giving reason for others. In short, from frailty to reproaches and from reproaches to frailty, we did not separate until we were satisfied with each other and had both agreed on the rendez-vous for this evening.

I did not go back to my room until dawn. And I was worn out with fatigue and lack of sleep; however I sacrificed both to the desire of being at breakfast in the morning; I have a passion for observing behaviours the morning after. You can have no idea of what this was. There was embarrassment in her countenance! Difficulty in walking! Eyes continually lowered, and so large and so tired! That round face had grown so much longer! Nothing could be more amusing. And for the first time, her mother, (alarmed by this extreme change) showed quite a tender interest in her! And Madame de Tourvel too was very attentive about her! Oh! Her attentions are only lent; a day will come when they can be returned to her and that day is not far off. Good-bye, my fair friend.

From the Château, 1st of October, 17—.

LETTER XCVII

Cécile Volanges to the Marquise de Merteuil

Ah Heaven! Madame, how distressed I am! How miserable I am! Who will console me in my grief? Who will advise me in my present state of embarrassment? This Monsieur de Valmont . . . and Danceny! No, the idea of Danceny fills me with despair . . . How can I relate it to you? How can I tell you? . . . I do not know what to do. Yet my heart is full . . . I must speak to someone and you are the only person in whom I can, in whom I dare confide. You are so kind to me! But do not be so now; I am not worthy of it; how shall I put it? I do not desire your kindness. Everyone here has showed an interest in me to-day; they all increased my pain. I felt so much that I did not deserve it! Scold me, rather; scold me well, for I deserve it; but afterwards save me; if you do not have the kindness to advise me I shall die of grief.

You must know then . . . my hand trembles, as you see, I can hardly write, my face feels on fire . . . Ah! it is indeed the red of shame. Well, I will endure it; it will be the first punishment for my fault. Yes, I will tell you everything.

You must know then that Monsieur de Valmont, who hitherto had been handing me Monsieur Danceny's letters, suddenly found that it was too difficult and wanted to have the key of my room. I can assure you I did not want to do it; but he went to the extent of writing to Danceny and Danceny wanted it too; and it hurts me so much when I refuse him anything, especially since my absence which makes him so unhappy, that I ended by consenting to it. I did not foresee the misfortune which would occur from it.

Yesterday Monsieur de Valmont made use of this key to come into my room when I was asleep; I expected it so little that he frightened me very much when he woke me up; but as he spoke to me at once, I recognised him and did not scream; and then my first idea was that he had perhaps come to bring me a letter from

Danceny. It was far from that. A moment afterwards he wanted to kiss me; and while I defended myself, as was natural, he did what I would not have him do for anything in the world . . . but, he wanted a kiss first. I had to, for what could I do? I had tried to call, but I could not and then he told me that if somebody came he could throw all the blame on me; and indeed that was very easy on account of the key. After it he did not withdraw at all. He wanted a second; and I don't know what there was about this one but it completely disturbed me; and after, it was still worse than before. Oh! This is very wrong. Afterwards . . . you will exempt me from telling you the rest; but I am as unhappy as one can be.

What I blame myself the most for, and yet must speak to you about, is that I am afraid I did not defend myself as much as I could have; I do not know how this could have happened; assuredly I do not love Monsieur de Valmont, on the contrary; and there were moments when I was as if I loved him . . . You may well suppose that did not prevent me from continuing to say no to him; but I felt I was not doing what I said; and it was as if in spite of myself; and then I was very much upset! If it is always as difficult as that to defend oneself, one must need a lot of practice! It is true that Monsieur de Valmont has a way of talking that one does not know how to answer; and will you believe it? when he went away it was as if I was sorry, and I was weak enough to consent that he should come again this evening; that troubles me even more than all the rest.

Oh! In spite of this, I promise you I shall prevent him from coming. He had scarcely gone when I felt how wrong I had been to promise it him. And I cried all the rest of the time. Above all Danceny grieved me! Every time I thought of him my tears increased until they stifled me, and still I thought of him . . . and even now you see what happens; my paper is all wet. No, I shall never be consoled were it only for his sake . . . At last I could cry no more, and yet I could not sleep a minute. This morning when

I got up and looked at myself in the mirror, it frightened me to see how changed I was.

Mamma noticed it as soon as she saw me and asked me what was the matter. I began to cry at once. I thought she was going to scold me, and perhaps that would have hurt me less; but, no, she spoke gently to me! I did not deserve it. She told me not to grieve like that! She did not know the reason for my grief. That I should make myself ill. There are moments when I wish I were dead. I could not endure it. I threw myself sobbing into her arms, saying: "Ah Mamma! your little girl is very unhappy!". Mamma could not prevent herself from crying a little and that only increased my grief; fortunately she did not ask me why I was so unhappy, for I should not have known what to say to her.

I beg you Madame, to write to me as soon as you can and tell me what I ought to do, for I have not the courage to think of anything and can do nothing but grieve. Please send me your letter by Monsieur de Valmont; but pray, if you write to him at the same time, do not mention to him what I have told you.

I have the honour to be, Madame, always with great friendship, your most humble and most obedient servant . . .

I dare not sign this letter.

From the Château de . . . , 1st of October, 17—.

LETTER XCVIII

Madame de Volanges to the Marquise de Merteuil

A few days ago, my charming friend, you asked me for consolation and advice; to-day it is my turn; and I make the same request on my own behalf which you made on yours. I am really very distressed, and I fear I have not taken the best way to avoid the grief I feel.

My daughter is the cause of my uneasiness. Since my departure. I had indeed seen that she remained sad and wretched; but I expected it and had armed my heart with the severity I considered necessary. I hoped that absence and distractions would soon destroy a love which I looked upon rather as a childish error, than as a real passion. Yet far from having improved since my sojourn here, I notice the child is more and more giving way to a dangerous melancholy; and I seriously fear her health will be affected. Particularly in the last few days she has changed visibly. Yesterday above all I was struck by her appearance and every one here was truly alarmed by it.

What proves to me still further the extent to which she is affected, is that I see her at the point of overcoming the timidity she has always showed me. Yesterday morning when I simply asked her if she was ill, she threw herself into my arms, telling me she was very unhappy; and she cried until she sobbed. I cannot tell you how much it pained me; the tears came into my eyes at once and I had only time to turn away to prevent her from seeing them. Fortunately I had the prudence not to ask her any questions and she did not dare to say anything else to me. But it is none the less clear that she is tormented by this unfortunate passion.

What course am I to take if this continues? Shall I create my daughter's unhappiness? Shall I turn against her the most precious qualities of the soul—sensibility and constancy? Is it for this I am her mother? And even if I stifled the natural sentiment which makes us desire the happiness of our children; even if I considered as a weakness what I think, on the contrary, to be the first, the most sacred of our duties; yet should I force her choice, shall I not have to answer for the disastrous results which might follow? What a use for maternal authority, to place one's daughter between crime and misery!

My friend, I shall not imitate what I have so often condemned. Doubtless I might attempt to choose a husband for my daughter: in that all I did was to help her by my experience; I was not exercising a right, I was fulfilling a duty. I should betray a duty on the contrary, if I bestowed her in defiance of an inclination I could not prevent from existing, the extent and duration of which neither she nor I can know. No, I will not allow her to marry one man and love another and I prefer to forfeit my authority rather than her virtue.

I think then that the wisest thing I can do is to withdraw from the arrangement I have made with Monsieur de Gercourt. You see the reasons for it; they seem to me more important than my promises. I go further; in the state things are now, to fulfil my engagement would be really to break it. For if I owe it to my daughter not to reveal her secret to Monsieur de Gercourt, at least I owe it to him not to abuse the ignorance in which I leave him and to do for him everything I think he would do himself, if he were informed of the situation. Shall I, on the contrary, betray him unworthily when he relies upon my faith and while he honours me by choosing me as his second mother, shall I deceive him in the choice he makes of the mother of his children? These undeniable reflections which I cannot escape, alarm me more than I can tell you.

With the misfortunes they make me dread I compare my daughter, happy with a husband her heart has chosen, only knowing her duties by the pleasure she finds in carrying them out; my son-in-law equally satisfied and congratulating himself every day upon his choice; each of them finding happiness in the other's happiness, and that of both uniting to increase mine. Should the hope of so delightful a future be sacrificed to vain considerations of prudence? And what are those which restrain me? Solely the views of interest. What advantage would it be for my daughter to have been born rich if she must none the less be the slave of wealth?

I admit that Monsieur de Gercourt is perhaps a more distinguished husband than I could have hoped for my daughter; I will even admit that I was extremely flattered by his choosing her. But after all, Danceny is as well born as he is; in personal qualities he yields to him in nothing; he has the advantage over

Monsieur de Gercourt of loving and of being loved; he is not rich, it is true, but is not my daughter rich enough for them both? Ah why should I deprive her of the delightful satisfaction of enriching the man she loves!

These marriages which are calculated instead of being matched, which are called "marriages of convenience", where everything indeed is convenient except their tastes and characters, are the most fertile source of those scandalous discords which become more frequent every day.[1] I prefer to delay; at least I shall have the time to observe my daughter whom I do not know. I have the courage to cause her a transitory grief if she will acquire through it a more solid happiness; but it is not in my heart to risk giving her up to an eternal despair.

Those are the ideas, my dear friend, which torment me, and on which I require your advice. These severe matters contrast greatly with your amiable gaiety and do not at all appear suitable to your age; but your reason is so far in advance of it! Moreover your friendship will aid your prudence; and I do not fear that either will be refused to the maternal solicitude which implores them.

Good-bye, my charming friend; never doubt the sincerity of my sentiments.

From the Château de . . . , 2nd of October, 17—.

LETTER XCIX

Vicomte de Valmont to the Marquise de Merteuil

More little events, my fair friend; but only scenes, no actions, so arm yourself with patience; take a lot of it; for while my Madame de Tourvel advances with such slow steps, your pupil retreats, and that is much worse. Well, I have the wit to amuse myself with these trifles. I am positively growing accustomed to my stay here; and I can say that I have not passed a boring moment in my old aunt's dull *Château*. Indeed, have I not enjoyment, deprivation, hope, uncertainty? What more can one have on a wider stage? Spectators? Eh! Let things go, spectators will not be lacking. Though they do not see me at work, I will show them my completed task; they will have nothing to do but admire and applaud; Yes, they will applaud; for I can at last predict with certainty the moment when my austere devotee will fall. This evening I was present at the death-agony of her virtue. Soft frailty will reign in its place. I fix the time no later than our next interview. But I hear you already crying out upon pride. To announce a victory, to boast of it beforehand! Ah well, calm yourself! To prove my modesty to you, I will begin with the story of my defeat.

Your pupil is indeed a very ridiculous little person! She is indeed a child who should be treated as such, and one would do her a favour by only giving her a child's punishment! Would you believe that after what passed between her and me the day before yesterday, after the friendly way we parted yesterday morning, when I went to her again in the evening, as she had agreed, I found the door closed from the inside? What do you say to that? We endure such puerilities sometimes on the eve, but on the day after! Is it not amusing?

Yet I did not laugh at it at first; I had never felt so much the ascendency of my character. Assuredly I went to this rendez-vous without pleasure and solely as a matter of procedure. My own bed, which I greatly needed, seemed to me at the moment preferable to anyone else's and I only left it with regret. Yet I had no sooner found an obstacle, than I burned to overcome it; I was especially humiliated to be outwitted by a child. I retired therefore in a very ill humour; and with the idea of concerning myself no further with this silly child and her affairs, I wrote her at once a note, which I meant to give to her to-day, in which I rated her at her true value. But, as people say, night brings wisdom; this morning I realised that, having no choice of distractions here, I should have to keep this one; so I suppressed the severe note.

Since I have reflected upon it, I cannot get over my having had the idea of ending an adventure before having in my hands the evidence to ruin its heroine. How far we are led astray by first thoughts! Happy is he, my fair friend, who, like you, has grown accustomed never to yield to them! In short, I put off my vengeance; I made this sacrifice to your designs on Gercourt.

Now that I am no longer angry, I only see the ridiculous side of your pupil's conduct. Indeed, I should like to know what she hopes to gain by it! For my part, I have no idea; if it were only to defend herself, you must admit that she was a little late. One day she must tell me the answer to this enigma! I very much want to know it. Perhaps it was only because she was tired? Frankly, it might be so; for doubtless she is still ignorant that the shafts of love, like the lance of Achilles, carry with them the remedy to the wounds they make. But no, from her little grimace all that day, I would wager there entered into it repentance . . . there . . . something . . . like virtue . . . Virtue! . . . Does it beseem her to have it? Ah! Let her leave virtue to the woman truly born for it, the only one who knows how to embellish it, who would make it beloved! . . . Pardon, my fair friend, but this very evening there took place between Madame de Tourvel and me the scene I am to relate to you, from which I still feel some emotion. I need to make an effort over myself to distract myself from the impression it made on me; it was as an assistance to this that I began to write to you. You must pardon something to this first moment.

For some days Madame de Tourvel and I have been in agreement upon our sentiments; we were only disputing about words. It was indeed always *her friendship* which responded to *my love;* but this conventional language did not alter the root of things; and even if we had remained there, I should perhaps have advanced less quickly, but none the less surely. Already there was no longer any question of sending me away, as she wished at first; and as to our daily interviews, if I give my attention to offering her the opportunity, she gives hers to seizing it.

As our little rendez-vous usually take place out of doors to-day's horrible weather left me no hope; I was even positively annoyed; I did not foresee how much I should profit by this mishap.

As they were unable to go out, they began to play cards after leaving the table; and as I play very little and am no longer necessary, I went up to my room with no other project but to wait there until about the end of the game.

I was returning to join the company when I met the charming woman going into her room and she, either from imprudence or frailty, said to me in her soft voice: "Where are you going? There is no one in the drawing-room." I needed nothing more, as you may believe, to make an attempt to enter her room; and I found less resistance than I expected. It is true that I took the precaution to begin the conversation at the door and to start it on indifferent topics; but we were scarcely settled when I brought in the real topic and spoke of *my love to my friend*. Although her first reply was simple, it seemed to me expressive: "Oh!" she said, "Do not let us speak of that here"; and she trembled. Poor woman! She saw herself dying.

However she was wrong to be in fear. For some time I have been certain of success one day or another, and, as I watched her use so much strength in useless combats, I resolved to spare my own and to await without effort the time when she would yield herself from very weariness. You will realise that I want a complete triumph with her and that I wish to owe nothing to opportunity. It was in accordance with this plan, and with the purpose of being pressing without engaging myself too far, that I returned to this word "love", which was so obstinately refused; since I was certain she believed me to be sufficiently ardent, I tried a more tender tone. This refusal had ceased to annoy me, it distressed me; did not my tender friend owe me some consolation?

While she was consoling me, a hand remained in mine; the pretty body leant upon my arm, and we were extremely close together. You have surely noticed in this situation how, as the defence weakens, demands and refusals take place at closer quarter; how the head turns away and the looks are lowered, while the sentences, all enunciated in a weak voice, become infrequent and broken. These valuable symptoms announce the consent of the mind in an unmistakable manner; but it has barely touched the senses at that time; I even think that it is always dangerous to attempt any too conspicuous enterprise then, because this state of abandonment never exists without a very delightful pleasure and a woman cannot be forced out of it without an ill humour on her part which infallibly turns to the profit of the defence.

But in the present case prudence was all the more necessary since I had especially to apprehend the terror which this forgetfulness of herself would not fail to cause my tender dreamer. And so I did not even exact that the admission I asked for should be put into words; a look would suffice me, a single look, and I was happy.

My fair friend, those beautiful eyes were indeed lifted to mine, that heavenly mouth even said: "Yes, I . . ." But suddenly the gaze was obscured, the voice failed, and this adorable woman fell into my arms. Scarcely had I the time to receive her when, tearing herself away with convulsive force, her eyes wandering and her hands lifted to Heaven . . . "God . . . Oh! my God, save me," she cried; and suddenly swifter than lightning, she was on her knees ten paces from me. I could hear that she was almost suffocating. I went to help her; but she, taking my hand which she bathed in tears, sometimes even embracing my knees, exclaimed: "Yes, it will be you, it will be you, who will save me! You do not wish my death, leave me; save me; leave me; in the name of God, leave me!" And these broken sentences were barely audible through her redoubled sobs. However, she held me with such strength, that I could not have gone away; then, collecting my own strength, I lifted her in my arms. At the same instant the tears ceased; she spoke no more; all her limbs stiffened and violent convulsions succeeded to this storm.

I must admit I was deeply moved and I think I should have consented to her request, even if the circumstances had not compelled me. The truth is, after I had given her some assistance, I left her as she begged me and I congratulate myself upon having done so. Already I have almost received the reward for it.

I expected that, as on the day of my first declaration, she would not show herself during the evening. But about eight o'clock she came down to the drawing-room and simply announced to the company that she had been very indisposed. Her face was dejected, her voice weak, and her behaviour composed; but her gaze was gentle and often fixed itself upon me. Her refusal to play cards having obliged me to take her place, she sat down at my side. During supper, she remained alone in the drawing-room. When we returned, I thought I noticed that she had been crying; to find out, I said to her it seemed to me she was still suffering from her indisposition; to which she obligingly replied: "This illness does not go as quickly as it comes!" Finally, when we retired, I gave her my hand; and at the door of her apartment she pressed my hand hard. It is true that this movement seemed to be almost involuntary; but so much the better, it is one more proof of my domination.

I would wager that now she is enchanted to be where she is; all the price is paid; nothing remains but to enjoy. Perhaps while I am writing to you she is already occupied with that soft idea! And even if she is occupied, on the contrary, with a new plan of defence, do we not know what becomes of all such plans? I ask you, can it be later than our next interview? I quite expect she will make some difficulties in granting it; but there! When the first step is made, do these austere prudes know how to stop? Their love is a positive explosion; their resistance gives it more strength. My shy devotee would run after me if I ceased to run after her.

In short, my fair friend, I shall very soon be with you to

demand the execution of your word. Doubtless you have not forgotten what you promised me after this success—an infidelity to your Chevalier? Are you ready? For my part, I desire it as if we had never known each other; for the rest, to have known you is perhaps a reason for desiring it the more:

> I am just, I am not being gallant.[1]

So this shall be the first infidelity I shall make my demure conquest; and I promise you to make use of the first pretext to absent myself from her for twenty-four hours. That shall be her punishment for having kept me so long away from you. Do you know that this adventure has occupied me for more than two months? Yes, two months and three days; it is true that I am including to-morrow, since it will only be really consummated then. This reminds me that Mademoiselle de B . . . resisted for three whole months. I am very glad to see that frank coquetry has more defences than austere virtue.

Good-bye, my fair friend; I must leave you, for it is very late; this letter has led me further than I reckoned; but as I am sending to Paris to-morrow, I wanted to profit by it to allow you to share your friend's joy a day sooner.

From the Château de . . . , 2nd of October, 17—.
(In the evening.)

LETTER C

The Vicomte de Valmont to the Marquise de Merteuil

My friend, I am outwitted, betrayed, lost; I am in despair: Madame de Tourvel has gone. She has gone, and I did not know it! And I was not there to oppose her departure, to reproach her with her unworthy betrayal. Ah! Do not think I should have let her go; she would have remained; yes, she would have remained, even if I had been compelled to employ violence. But in my credulous confidence, I was sleeping calmly; I was sleeping and the thunder-bolt fell on me. No, I simply cannot understand this departure; I must give up trying to understand women.

When I recollect yesterday afternoon! What am I saying? The very evening! That gentle look, that tender voice! And that pressure of the hand! And during this time she was planning to run away from me! O women, women! Complain after this, if you are deceived! Yes, every perfidy we employ is a theft from you.

What pleasure I shall have in avenging myself! I shall find this perfidious woman again, I shall reassert my power over her. If love sufficed me to find means for it, what will it not be, aided by vengeance? I shall see her again at my knees, trembling and bathed in tears, crying to me for mercy in her deceitful voice; and I shall be without pity.

What is she doing now? What is she thinking? Perhaps she congratulates herself on having deceived me; and, faithful to the tastes of her sex, finds this the most agreeable pleasure. What this boasted virtue could not achieve, the spirit of deceit obtained without effort. Fool that I was! I dreaded her modesty; I ought to have feared her bad faith.

And to be obliged to swallow down my resentment! Only to dare to show a tender pain, when my heart is filled with rage! To see myself reduced to imploring a rebellious woman, who has escaped my domination! Ought I to be humiliated to this point? And by whom? By a timid woman, who has never had any practice in defending herself. Of what use is it to me to have established myself in her heart, to have inflamed her with all the fires of love, to have brought the disturbance of her senses to the point of delirium if she is tranquil in her retirement and can take more pride in her flight to-day than I in my victories? And shall I endure this? My friend, you will not think so; you have not this humiliating idea of me.

But what fate attaches me to this woman? Are there not

a hundred others who desire my attentions? Will they not be eager to reply to them? Even if none of them was worth her, do not the attractions of variety, the charm of new conquests, the splendour of their number, offer delightful enough pleasures? Why run after her who flies from us and neglect those who offer themselves? Ah! why? . . . I do not know, but I feel it strongly.

For me there is no happiness, no rest, until I possess this woman whom I hate and whom I love with equal fury. I shall only endure my fate from the moment I am master of hers. Then I shall be calm and satisfied and I, in my turn, shall see her given up to the storms I feel at this moment; I shall create in her a thousand others as well. Hope and fear, suspicion and confidence, all the ills invented by hatred, all the good allowed by love—I want them to fill her heart, to succeed each other at my will. This time will come . . . But what work still remains! How near I was to it yesterday, and how far I am from it to-day! How am I to draw near it once more? I dare not attempt any measures; I feel that I must be calmer to adopt any course, and my blood boils in my veins.

What redoubles my torture is the calm with which everybody here replies to my questions about this event, about its cause, about all its extraordinary side . . . Nobody knows, nobody desires to know; they would scarcely have spoken of it, if I had allowed them to speak of anything else. Madame de Rosemonde, to whom I rushed this morning when I learned the news, replied with the coldness of her age that it was the natural result of the indisposition Madame de Tourvel felt yesterday; that she had feared an illness and had preferred to be in her own home. She thinks it quite simple; she would have done the same, she told me; as if there could be anything in common between them! Between her, who has nothing left but to die and the other, who is the charm and torment of my life!

Madame de Volanges, whom I suspected at first of being an accomplice, only appeared moved by the fact that she had not been consulted upon this step. I confess I am very glad that she has not had the pleasure of doing me harm. This proves to me that she does not possess the woman's confidence as much as I feared; it is one enemy the less. How glad she would be, if she knew that it was from me Madame de Tourvel fled! How puffed up with pride she would have been if it had been by her advice! How her self-importance would have redoubled! Heavens! How I hate her! Oh! I shall renew the affair with her daughter; I want to work upon her as I fancy; so I think I shall stay here some time; at least, the few reflections I have made lead me to that course.

Do you not think that after so obvious a step, my ungrateful lady must dread my presence? If the idea has occurred to her that I might follow her, she will not have failed to close her doors to me; and I no more wish to accustom her to this method than to suffer the humiliation of it myself. On the contrary, I prefer to announce to her that I am staying here; and I shall even beg her to return here; and when she is fully persuaded of my absence I shall arrive at her house; we shall see how she will sustain that adventure. But to increase its effect, it must be postponed, and I do not know yet if I shall have the patience; twenty times to-day I have had my mouth open to call for my horses. However I will control myself; I promise to wait for your reply here; I only ask you, my fair friend, not to keep me waiting.

What would annoy me most would be not to know what is happening; but my servant, who is in Paris, has the right to some access to the waiting-woman; he will be able to help me. I am sending him instructions and money. I beg you will allow me to enclose both in this letter and that you will send them to him by one of your servants, with orders to give them to him in person. I take this precaution because the rascal has a habit of never receiving the letters I write him when they order him to do something which gives him trouble, and because he does not appear to me at present to be as interested in his conquest as I should like him to be.

Good-bye, my fair friend; if you have some happy idea, some way of hastening my advance, let me know. More than once I have proved how useful your friendship can be; I feel it again now; for since I have written to you I feel calmer; at least I am speaking to someone who understands me and not to the automata near whom I have vegetated since this morning. Really, the further I go the more I am tempted to believe that you and I are the only people in the world who are worth anything.

From the Château de . . . , 3rd of October, 17—.

LETTER CI

The Vicomte de Valmont to Azolan, his servant

(Enclosed in the preceding letter)

You must be an imbecile, not to have known when you left here this morning that Madame de Tourvel was going too; or, if you knew it, not to have come to tell me. What use is it your spending my money getting drunk with the footmen and passing the time you ought to give to my service in playing the agreeable with waiting-women, if I am not better informed of what is taking place? But this is one of your negligences. I warn you that if there is one more in this affair, it will be your last in my service.

You must find out for me everything that is happening at Madame de Tourvel's; her health; if she sleeps; if she is sad or gay; if she often goes out and where she goes; if she receives company in her house and who comes to it; how she spends her time, if she is bad-tempered with her women, particularly with the woman she brought here; what she does when she is alone; if when she reads, she reads continuously or interrupts her reading to meditate; and similarly when she writes. Remember to become friendly with the servant who takes her letters to the post. Offer frequently to perform this duty in his place; and when he accepts, only post those letters which seem to you unimportant and send the others to me, especially those to Madame de Volanges, if you come across any.

Make your plans to remain for some time yet the favoured lover of your Julie. If she has another, as you thought, make her agree to share herself; and do not pique yourself upon ridiculous delicacy; you will be in the position of many others who are worth far more than you are. However, if your second becomes too importunate, if you notice, for example, that he occupies Julie too much during the day and that she is consequently less often near her mistress, get rid of him in some way; or pick a quarrel with him; do not fear for the consequences, I will support you. Above all do not leave that house. It is by assiduity that everything is seen and seen clearly. If by chance one of the servants is sent away, present yourself to take his place, as if you were no longer in my service. In this event, say that you have left me to look for a quieter and more orderly house. Try to get yourself accepted. I will continue you in my service during this time; it will be the same as it was with the Duchesse de . . .; and in the end Madame de Tourvel will reward you in the same way.

If you have enough adroitness and zeal these instructions ought to suffice you; but to make up for both I am sending you some money. The enclosed note authorises you, as you will see, to draw twenty-five *louis* from my agent; for I have no doubt you are penniless. Out of this sum you will use so much as may be necessary to persuade Julie to establish a correspondence with me. The rest will serve to buy drinks for the servants. Take care as much as possible to let this happen in the porter's lodge, so that he will be glad to see you come. But do not forget that I am not paying for your pleasures but your services.

Accustom Julie to observe everything and to report everything, even what appear to her to be details. It is better that she should write ten useless phrases than omit one important; and often what seems to be unimportant is not so. Since I must be informed at once if anything happens which appears to you to deserve attention, as soon as you receive this letter you will send Philippe on the errands horse to wait at . . .[1]; he will remain there until he receives fresh orders; he will be a relay in case of necessity. For ordinary correspondence the Post will suffice.

Be careful not to lose this letter. Read it over every day, both to make certain that you have forgotten nothing and to make certain you have not lost it. In short, do everything a man ought to do when he is honoured by my confidence. You know that if I am pleased with you, you will be so with me.

From the *Château* de . . . , 3rd of October, 17—.

LETTER CII

Madame de Tourvel to Madame de Rosemonde

You will be very surprised, Madame, when you learn I have left your house so suddenly. This step will appear very extraordinary to you; but how greatly your surprise will be increased when you know the reasons for it! Perhaps you will think that in confiding them to you I am not sufficiently respectful of the tranquillity necessary to your age; that I am even departing from those sentiments of veneration which are your due upon so many accounts? Ah! Madame, forgive me; but my heart is over-laden; it needs to pour out its grief upon the bosom of a friend who is both gentle and prudent; what other but you could it choose? Look upon me as your child. Show me the kindness of a mother; I implore it. Perhaps I have some right to it through my feelings for you.

Where is that time when, occupied solely with laudable sentiments, I was ignorant of those which bear into the soul the mortal disturbance I feel and take away the strength to combat them at the same time that they impose the duty of doing so? Ah! This fatal visit has ruined me . . .

How shall I tell you? I love, yes, I love wildly. Alas! that word which I write for the first time, that word so often asked for and never granted—I would give my life for the sweetness of being able once only to say it to him who inspires it; and yet it must be refused for ever! He will still doubt my feelings; he will think he has reason to complain of them. I am very unfortunate! Why is it not as easy for him to read in my heart as it is for him to reign over it? Yes, I should suffer less, if he knew all that I am suffering; but you, to whom I tell this, can have but a feeble idea of it.

In a few moments I am about to fly from him and to hurt him. While he still thinks me near him I shall already be far away; at the hour when I was accustomed to see him each day, I shall be in places where he has never been, where I must not permit him to come. All my preparations are made; everything is here before my eyes; I cannot rest them on anything which does not tell me of this cruel departure. Everything is ready, except me! . . . And the more my heart refuses it, the more it proves to me the necessity for submitting.

Without doubt I shall submit, it is better to die than to live guilty. Already I feel I am too much so; I have only saved my modesty, my virtue has vanished. Must I confess it? What virtue I have left I owe to his generosity. Intoxicated with the pleasure of seeing him, of hearing him, with the sweetness of feeling him near me, with the greater happiness of being able to make him happy, I was without power and without strength; I had scarcely enough to combat, not enough to resist; I shuddered at my danger without being able to escape it. Well! He saw my anguish and he pitied me. How can I not cherish him? I owe him more than life itself.

Ah! If by remaining near him I had only to tremble for my life, do not think that I should ever consent to go away. What is life to me without him; should I not be but too happy to lose it? Condemned to make him and myself unhappy for ever; to dare neither to complain nor to console him; to defend myself every day against him, against myself; to give my attention to causing his pain when I wish to consecrate it all to his happiness—to live thus, is it not to die a thousand times? Yet that will be my fate. Yet I will endure it, I shall have the courage. O you, whom I choose as my mother, receive my oath!

Receive also the oath I make you never to hide from you any of my actions; receive it, I beg you; I ask it of you as an aid I sorely need; thus, having promised to tell you everything, I shall grow accustomed to thinking myself always in your presence. Your virtue shall replace mine. Never indeed will I consent to blush at your gaze; and, restrained by this powerful curb, I shall cherish in you the indulgent friend, the confident of my frailty, and I shall still honour in you the guardian Angel who will save me from shame.

It is indeed sufficient shame to have to make this request. Fatal effect of a presumptuous self-confidence! Why did I not sooner dread this feeling which I felt growing? Why did I flatter myself that I could dominate or overcome it at will? Ah! If I had combatted it with more attention, perhaps it would have gained less power over me! Perhaps this departure would not then have been necessary; or even, in submitting to this painful course, I should have been able not to break off entirely an acquaintance which it would have sufficed to render less frequent! But to lose everything at once! And for ever! O my friend! . . . But, even as I write to you, am I not wandering into criminal wishes? Ah! Let me go, let me go, and at least let these involuntary errors be expiated by my sacrifices.

Farewell, my respected friend; love me as your daughter, adopt me as such; and be certain that, in spite of my weakness, I would rather die than render myself unworthy of your choice.

From . . . , 3rd of October, at one o'clock in the morning.

LETTER CIII

Madame de Rosemonde to Madame de Tourvel

I was more distressed, my dear Beauty, by your departure than surprised at its cause; a long experience and the interest I take in you were sufficient to enlighten me as to the state of your heart; and if I must tell you everything, your letter told me little or nothing I did not know. If I had had no information but your letter I should still not know who it is you love; for in speaking of *him* the whole time you have not written his name once. But I notice it because I recollect that was always the style of love. I see it is now as it used to be.

I did not think I should ever again be in a situation to have to return to memories so distant from me and so foreign to my years. Yet, since yesterday I have truly thought much upon them, from my desire to find something in them which might be of use to you. But what can I do except admire and pity you? I praise the wise course you have taken; but it frightens me, because I conclude you felt it necessary; and when you have reached that point it is very difficult to remain always distant from him to whom your heart always draws you near.

Yet do not be discouraged. Nothing should be impossible to your fine soul; and even if you should one day have the misfortune to succumb (which God forbid!) believe me, my dear Beauty, keep for yourself at least the consolation of having fought with all your strength. And then, what human wisdom cannot do, Divine grace operates when it pleases. Perhaps you are on the eve of receiving this aid and your virtue, tried in these terrible struggles, will emerge purer and more brilliant. Hope that you will receive to-morrow the strength you do not possess to-day. Do not expect it as a matter of course, but as an encouragement to use all your own strength.

When I leave to Providence the care of aiding you in a danger against which I can do nothing, I reserve for myself the task of supporting and consoling you as much as I am able. I shall not relieve your anguish, but I shall share it. On this basis I will gladly receive your confidences. I feel your heart must need to pour itself out. I open mine to you; age has not yet chilled it to the point of being insensible to friendship. You will find it always ready to receive you. It will be a poor relief to your pain, but at least you will not weep alone; and when this unhappy love gains too much power over you and forces you to speak of it, it is better that it should be with me than with *him*. You see I am talking as you did; and I believe that the two of us will never succeed in naming him; however, we understand each other.

I do not know if I am doing right to tell you that he appeared to be keenly affected by your departure; it would perhaps be wiser not to tell you this; but I do not like that wisdom which distresses its friends. Yet I am compelled not to speak of him much longer. My failing sight and my trembling hand do not permit me long letters when I have to write them myself.

Good-bye then, my dear Beauty, good-bye, my sweet child; yes, I gladly adopt you as my daughter, and you indeed possess all that is needed to make the pride and pleasure of a mother.

From the Château de . . . , 3rd of October, 17—.

LETTER CIV

The Marquise de Merteuil to Madame de Volanges

Indeed, my dear and good friend, I found it hard to avoid a feeling of pride when I read your letter. What! You honour me with your complete confidence! You even go so far as to ask my advice! Ah! I am very happy if I deserve this favourable opinion on your part; if I do not owe it only to the favour of friendship. But whatever the motive, it is none the less precious to my heart; and to have obtained it is in my eyes but one more reason for labouring to deserve it. I shall then (but without pretending to give you advice) tell you freely what I think. I am mistrustful of it because it differs from yours; but when I have given you my reasons you will form your judgment; and if you condemn them I accept your judgment beforehand. At least I shall have the wisdom not to think myself wiser than you.

But if on this single occasion my opinion appears preferable, the cause must be sought in the illusions of maternal love. Since this is a laudable sentiment, you must possess it. How well it can be perceived in the course you are tempted to take! It is thus, that if you ever err, it is in the choice of virtues.

It seems to me that prudence is the preferable virtue when we are disposing of the fate of others, especially when it is a matter of binding it with an indissoluble and sacred tie like that of marriage. It is then that a mother, equally wise and tender, ought, as you say so well "to help her daughter with her experience". Now, I ask you, what has she to do to achieve that? To distinguish for herself between what is pleasing and what is expedient.

Would it not be debasing maternal authority, would it not be annihilating it, to subordinate it to a frivolous inclination whose illusory power is only felt by those who dread it, which disappears as soon as it is treated with contempt? For my part, I confess I have never believed in these overpowering and irresistible passions, which by general agreement seem to have been made the excuse for our licentiousness. I cannot conceive how an inclination, which is born one moment and dies the next, can have more strength than the inalterable principles of decency, honour, and modesty; and I do not understand how a woman who betrays them can be justified by a pretended passion, any more than a thief would be by the passion for money or a murderer by that of vengeance.

Ah! Who can say she has never had to struggle? But I have always tried to convince myself that it was sufficient to want to resist in order to be able to do so; and hitherto at least my experience has confirmed my opinion. What would virtue be without the duties it imposes? Its cult is in our sacrifices, its reward in our hearts. These truths can only be denied by those whose interest it is to disregard them and who, being already depraved, hope to create a moment of illusion by trying to justify their bad conduct with bad reasons.

But can one fear this from a simple and timid child, from a child born of you, whose modest pure education can only have fortified her natural good character? Yet to this fear, which I dare to call humiliating to your daughter, you wish to sacrifice the advantageous marriage your prudence had arranged for her! I like Danceny very much; and for a long time, as you know, I have seen little of M. de Gercourt; but my friendship for the one, my indifference to the other, do not prevent me from feeling the enormous difference between them as husbands.

Their birth is equal, I admit; but one has no fortune and the other's wealth is such that even without good birth it would have sufficed to lead him anywhere. I admit that money does not make happiness; but it must be admitted also that it greatly facilitates happiness. As you say, Mademoiselle de Volanges is wealthy enough for two; and yet, the income of sixty thousand *livres* she will enjoy is already not much when she bears the name of Danceny, when she must display and keep up an establishment worthy of it. We are no longer in the times of Madame de Sévigné. Luxury absorbs everything; we deplore it, but we must imitate it; and superfluities end up by absorbing necessities.

As to the personal qualities which you consider so important, and with such good reason; assuredly M. de Gercourt is irreproachable in that respect; he has given proofs of it. I like to think, and I do indeed think that Danceny yields to him in nothing; but are we as sure of this? It is true that hitherto he has appeared free from the faults of his age and that, in spite of the current tone, he shows a taste for good company which allows one to augur well of him; but who knows if he does not owe this apparent virtue to the mediocrity of his fortune? Even if a man fears to be a rogue or debauched, he needs money to be a gambler and a libertine and he may like the faults whose excess he fears. In short, he would not be the thousandth who has kept good company solely because he could afford nothing worse.

I do not say (God forbid!) that I think all this of him; but still there is always a risk of it; and how you would blame yourself if the result were unhappy! What would you reply to your daughter if she said to you: "Mother, I was young and inexperienced; I was carried away by an error pardonable at my age; but Heaven, which had foreseen my frailty, had given me a wise mother to redress it and to preserve me. Why then, forgetting your prudence, did you consent to my misfortune? Was it for me to choose a husband when I knew nothing of the condition of marriage? Even if I wished to do so, was it not for you to oppose it? But I never had that foolish presumption. I had decided to obey you and awaited your choice with respectful obedience; I never departed from the submission which I owed you and yet to-day I endure the pain which is the lot of rebellious children. Ah! Your weakness has ruined me . . ." Perhaps her respect would stifle these complaints; but maternal love would guess them; and though your daughter's tears were concealed they would none the less flow upon your heart. Where then will you seek for consolation? Will it be in this foolish love, against which you ought to have fore-armed her and by which, on the contrary, you will have allowed yourself to be seduced?

I do not know, my dear friend, if I have too strong a prejudice against this passion; but I think it is to be dreaded, even in marriage. It is not that I have any disapproval of a soft and virtuous sentiment which should embellish the conjugal tie and soften, as it were, the duties imposed by it; but the sentiment should not form the tie; the illusion of a moment should not govern our choice for life. After all, to choose, we must compare; and how can we compare when one person occupies us entirely, when we cannot even comprehend him since we are plunged in ecstasy and blindness?

As you may suppose I have met several women infected with this dangerous disorder; and I have received the confidences of some of them. To listen to them you would think every lover was a perfect being; but these illusory perfections only exist in their imaginations. Their excited minds dream of nothing but attractions and virtues; they invest the man they prefer with them at pleasure; it is the drapery of a god, often worn by an abject model; but in any case, they have scarcely dressed him up in them when they are duped by their own handiwork and prostrate themselves to adore it.

Either your daughter does not love Danceny or she is undergoing this illusion; it is common to them both, if their love is mutual. Thus your reason for uniting them forever comes down to the certainty that they do not know each other, that they cannot know each other. But, you will say, do M. de Gercourt and your daughter know each other any better? No, doubtless they do not; but at least they do not delude themselves, they are merely ignorant of each other. What happens in a such a case between two married people, whom I suppose to be virtuous? Each of them studies the other, observes the other, seeks for and soon perceives what must be yielded to the other's tastes and wishes for their mutual tranquillity. These slight sacrifices are made without pain because they are mutual and because they have been foreseen; soon they create a mutual good-will; and habit which strengthens all inclinations it does not destroy little by little brings with it that gentle friendship, that tender confidence which, added to esteem, form in my opinion the real, the solid happiness of marriage.

The illusions of love may be sweeter; but who does not know that they are less durable? And what dangers are brought by the moment which destroys those illusions! It is then that the slightest faults appear offensive and unendurable from their contrast with the idea of perfection which had seduced us. Yet each one thinks that the other alone has changed and that he or she still possesses the worth a moment of error won for them. They are surprised at not creating the charm they no longer feel; they are humiliated; wounded vanity embitters their minds, increases their errors, produces ill-temper, gives birth to hatred; and idle pleasures are paid for by long misfortunes.

That, my dear friend, is how I think of the subject which now occupies us; I do not defend it, I merely express it; it is for you to decide. But if you persist in your opinion I ask you to let me know the reasons which overcome mine; I shall be very glad to be enlightened by you and above all to be reassured as to the fate of your charming child, whose happiness I desire most ardently, both from my friendship with her and from that which unites me to you for life.

Paris, 4th of October, 17—.

LETTER CV

The Marquise de Merteuil to Cécile Volanges

Well, little one, you are very angry and very much ashamed, and this M. de Valmont is a wicked man, is he not? Why! He dares to treat you as the woman he loves best. He teaches you what you were dying to know! Indeed, such proceedings are unpardonable. And you, for your part, wished to keep your modesty for your lover (who does not abuse it); you only cherish the pains of love, not its pleasures! Nothing could be better, and you would make an excellent figure in a novel. Passion, misfortune, above all virtue, what excellent things! In the midst of this brilliant throng one is bored sometimes, it is true, but then it is in turn boring.

Come, poor child, how she is to be pitied! She had rings under her eyes the next morning! And what will you say when they are caused by your lover? Ah! my angel, you will not always have them so; all men are not Valmonts. And then; not to dare to lift those eyes! Ah! you were quite right there; everyone would have read your adventure in them! Believe me, if that were so, our women and even our girls would have a more modest gaze.

In spite of the praises I am compelled to give you, as you see, it must still be admitted that you missed your master stroke—that was to tell your Mamma everything. You began so well! You had already thrown yourself into her arms, you were sobbing and she was crying too; what a pathetic scene! And what a pity you did not complete it! Your tender mother would have been delighted and, to aid your virtue, would have shut you up for the rest of your life; and there you could have loved as much as you liked, without rivals and without sin; you could have despaired at leisure, and Valmont assuredly would not have come to trouble your grief with annoying pleasures.

Seriously now, can you at over fifteen be as childish as you are? You are right to say you do not deserve my kindness. Yet I should like to be your friend; perhaps you will need one with the mother you have and the husband she wants to give you! But if you do not develop more, what is to be done with you? What can be hoped if that which gives girls understanding takes yours away?

If you could but persuade yourself to reason a little, you would soon find that you ought to be glad rather than to complain. But you are ashamed, and that troubles you! Hey! Calm yourself; the shame love causes is like its pain; we only feel it once. We may feign it afterwards, but we do not feel it. However, the pleasure remains, and that is indeed something. I think I could make out through your chatter that you might count it for a good deal. Come now, a little honesty. That disturbance which prevented you from "doing what you said", which made you find it so "difficult to defend" yourself, which made you "as if sorry" when Valmont went away—was it caused by shame? Or was it pleasure? "And his way of saying things which one does not know how to answer"—does that not come from "his way of pleasing"? Ah, little girl, you tell an untruth, and to your friend too. That is not right. But let us have no more of this.

What would be a pleasure for anyone, and be only a pleasure, in your situation becomes positive happiness. Indeed, placed as you are between a mother whose love is important to you and a lover by whom you desire to be loved always, how can you fail to see that the only way to obtain these opposite ends is to occupy yourself with a third? Entertained by this new adventure, you will appear to your Mamma to sacrifice to your submission to her an inclination which displeases her, while with your lover you will acquire the honour of a long resistance. While you continually assure him of your love, you will not grant him the last proofs of it. These refusals, which are not difficult in your present situation, he will place to the credit of your virtue; he will complain perhaps but he will love the more, and, to secure the double merit, in the eyes of the one of sacrificing love, in the eyes of the other of resisting it, you have to do nothing but enjoy its pleasures. O, how many women have lost their reputation who would have preserved it with care if they could have supported it by such means!

Does not the course I propose to you seem the most reasonable, as it is the pleasantest? Do you know what you have gained by your present course? Your Mamma attributes your increase of sadness to an increase of love; she is angered by it and she is only waiting to be more certain to punish you. She has just written to me; she will try everything to obtain this admission from you. She will perhaps go so far, she tells me, as to propose Danceny to you as a husband; and that to get you to speak! If you let yourself be deceived by this false tenderness, if you reply as your heart wishes, you will soon be shut up for a long time, perhaps for ever, where you can cry at pleasure over your blind credulity.

You must combat this ruse she is going to make use of against you by another. Begin by showing less sadness, by making her believe that you are thinking less of Danceny. She will be the more readily convinced of it, since it is the usual effect of absence; and she will be all the more pleased with you because it gives her an opportunity to congratulate herself on her prudence, which suggested this way. But if she retains some doubt and persists in testing you, if she speaks to you of marriage, show yourself, as a well-born girl should, completely submissive. After all, what do you risk? One husband is as good as another; and the most inconvenient is less so than a mother.

Once she is more pleased with you, your Mamma will marry you; and you will be freer in your movements and you will be able to leave Valmont for Danceny as you choose, or even to keep them both. For, notice this, your Danceny is agreeable, but he is one of those men one can have when and as long as one wishes; you can therefore be easy about him. It is not the same with Valmont; it is difficult to keep him; and it is dangerous to leave him. With him you need a lot of skill, or, if you do not possess it, great docility. But if you can succeed in attaching him to you as a friend, that would be good fortune! He would put you at once in the first rank of our fashionable women. That is the way to acquire a position in the world, not by blushing and crying as you did when the nuns made you eat your dinner kneeling.

If you are wise you will try to patch matters up with Valmont, who must be very angry with you; and since we must make up for our silly mistakes, do not be afraid to make some advances to him; you will soon learn that if men make the first advances to us, we are almost always obliged to make the second. You have a pretext for this; for you must not keep this letter and I insist that you give it to Valmont as soon as you have read it. But do not forget to re-seal it first. You must be left the merit of the measure you are going to take with him and it must not seem to have been suggested to you; and then you are the only person in the world with whom I am friendly enough to speak as I am doing.

Good-bye, my angel, follow my advice and tell me if it turns out well.

P.S. By the way, I forgot . . . One word more. Try to take more care with your style. You still write like a child. I see why that is; it is because you say everything you think and nothing you do not think. That is all very well between you and me who have nothing to hide from each other, but with everyone else! Above all with your lover! You will always seem like a little fool. You can see that when you write to someone it is for him and not for yourself; you ought then to try less to say what you are thinking than to say what will please him more.

Good-bye, my heart; I kiss you instead of scolding you in the hope that you will be more reasonable.

Paris, 4th of October, 17—.

LETTER CVI

The Marquise de Merteuil to the Vicomte de Valmont

Wonderful! Vicomte, and this time I love you madly! For the rest, after the first of your two letters the second might have been expected; so it did not surprise me; and while you were so proud of your success, while you solicited your reward and asked me if I were ready, I saw I had no such need to hurry. Yes, on my word of honour; when I read the fine account of that tender scene and that you had been so "profoundly moved"; when I saw that restraint, worthy of the best ages of our chivalry, I said twenty times: That affair will be a failure!

But it could not have been otherwise. What do you expect a poor woman to do when she yields herself and is not taken? Faith, in such cases, honour at least must be saved; and that is what your Madame de Tourvel has done. For my part I know that the course she took is really not ineffective and I propose to make use of it on my own account, at the first serious opportunity which occurs; but I promise that if the man for whom I go to the trouble does not profit by it better than you, he may certainly renounce me for ever.

There you are, absolutely reduced to nothingness, and that between two women, one of whom was already at the day after and the other was asking nothing better than to be there! Well! You will think I am boasting and that it is easy to prophesy after the event; but I swear to you that I expected it. The fact is, you have not the genius of your condition; you only know what you have learned, and you invent nothing. Thus, as soon as circumstances do not fit your usual formulas and you have to leave the beaten path, you are as taken aback as a schoolboy. So, on the one side a puerility, on the other side a return of prudery, are sufficient to disconcert you because they are not to be met with every day, and you can neither avoid nor remedy them. Ah! Vicomte, Vicomte, you teach me not to judge of men by their successes; very soon we shall have to say of you: He *was* brave on such a day. And when you have piled stupidities on stupidities, you come to me! It appears I have nothing else to do but to retrieve them. I shall soon have plenty of work, it is plain.

In any case, one of these two adventures was undertaken against my will and I shall not meddle with it; as to the other, since it was partly out of consideration for me, I make it my business. The letter I enclose, which you will read and then give to the Volanges girl, is more than sufficient to bring her back to you; but I beg you to give some attention to the child and let us together make her the despair of her mother and of Gercourt. There is no need to fear increasing the doses. I see clearly that the little person will not be frightened by it; and once we have carried out our intentions she will become what she can.

I take no further interest in her. I had some idea of making her at least a subsidiary intriguer and to take her on to play second parts under me; but I see she has not the material in her; she has a silly ingenuousness which has not even yielded to the specific you employed, which is hardly ever unsuccessful; and in my opinion, it is the most dangerous malady a woman can have. It shows especially a weakness of character, almost always incurable, which thwarts everything; so that while we are trying to fit this girl for intrigue we shall make nothing of her but a facile woman. Now, I know nothing so flat as the facility of stupidity which yields without knowing how or why, solely because it is attached and does not know how to resist. These sorts of women are absolutely nothing but pleasure machines.

You will tell me that that is the only thing to do and that it suffices for our plans. Well enough! But do not forget that very soon everybody gets to know the springs and contrivances of these machines; so, to make use of this one without danger, we must hurry, know when to stop, and then break it. After all, we shall have plenty of ways of getting rid of her, and Gercourt will always shut her up when we like to make him. Indeed, when he can no longer doubt his discomfiture, when it is public and notorious, what does it matter to us whether he revenges himself, so long as he has no consolation? What I say of the husband, you doubtless think of the mother; so it is as good as done.

The course I consider best, and have fixed on, has decided me to lead the young person rather quickly, as you will see from my letter; this makes it very important to leave nothing in her hands which might compromise us, and I beg you will give your attention to this. This precaution once taken, I will look after the moral side, the rest concerns you. If we see later that she gets over her ingenuousness, we shall always have time to change our plans. One day or other we should have had to consider what we were going to do; in any event our trouble will not be wasted.

Do you know my trouble was nearly wasted and that Gercourt's star very nearly overcame my prudence? Did not Madame de Volanges have a moment of maternal weakness? And did she not want to give her daughter to Danceny? That was what was meant by that more tender interest you noticed on "the morning after".

It is you again who would have been the cause of this wonderful master stroke! Happily, the tender mother wrote to me, and I hope my reply will disgust her with the idea. I talked so much of virtue and above all I flattered her so much that she is certain to think I am right.

I am sorry I had not time to take a copy of my letter, to edify you with the austerity of my morals. You would have seen what scorn I have for women so depraved as to have a lover! It is convenient to be a rigorist in words! It never hurts others and does not hinder us in the least . . . And then I happen to know that the good lady had weaknesses, like others, in her youth and I was not sorry to humiliate her, at least in her conscience; that consoled me a little for the praises I gave her against my own conscience. And in the same letter the idea of harming Gercourt gave me the courage to speak well of him.

Good-bye, Vicomte; I strongly approve your determination to stay where you are for some time. I have no means of hastening your advance; but I invite you to chase away boredom with our common pupil. As to what concerns me, you will see (in spite of your polite quotation) that you must wait a little longer; and no doubt you will admit that it is not my fault.

LETTER CVII

Azolan to the Vicomte de Valmont

Monsieur,

In accordance with your orders I went, as soon as I got your letter, to M. Bertrand, who handed me twenty-five *louis*, as you ordered him. I asked him for two more for Philippe, whom I had told to start at once, as Monsieur wrote me, and who had no money; but your agent would not do it, saying that he had no order for it from you. I was obliged to give them myself and Monsieur will take this into account, if he will be so good.

Philippe started yesterday evening. I warned him not to leave the inn, so that I can be sure to find him if necessary.

Immediately afterwards I went to Madame de Tourvel's house to see Mademoiselle Julie; but she had gone out, and I only spoke to La Fleur, from whom I learned nothing, because since his return he has only been in the house to meals. The second footman is doing all the waiting, and Monsieur knows that I have no acquaintance with him. But I began to-day.

I went back this morning to Mademoiselle Julie and she seemed very pleased to see me. I asked her what was the cause of her Mistress's returning; but she told me she knew nothing about it, and I think she was telling the truth. I reproached her for not having told me she was going and she assured me she only knew it the same evening when she went to help Madame to bed; so she spent the whole night in packing and the poor girl did not have two hours sleep. She only left her Mistress's room that evening after one o'clock, and when she went, her Mistress was beginning to write.

In the morning when Madame de Tourvel was leaving she gave a packet to the Concierge of the *Château*. Mademoiselle Julie does not know who it was for; she says perhaps it was for Monsieur; but Monsieur does not say anything about it.

During the whole journey Madame had a large hood over her face, so she could not be seen; but Mademoiselle Julie thinks she can be sure she often cried. She did not say a word all the way and would not stop at . . .[1], as she did when coming; which did not please Mademoiselle Julie who had had no breakfast. But, as I told her, Masters are masters.

On arriving, Madame went to bed; but she only stayed there two hours. When she got up she sent for the door-porter and told him to allow no one to come in. She made no toilet at all. She sat down at table to dine; but she only ate a little soup and left at once. Her coffee was taken up to her and Mademoiselle Julie went in at the same time. She found her Mistress arranging papers in her writing-desk and she saw they were letters. I would wager they are Monsieur's; and one of the three which came in the afternoon she kept in front of her all evening! I am sure it was another of Monsieur's again. But why did she go away like that? That surprises me. Monsieur must surely know? But it is not my business.

Madame de Tourvel went to the library in the afternoon and took two books which she brought up to her boudoir; but Mademoiselle Julie assures me she did not read them a quarter of an hour all day, and that she did nothing but read this letter, think, and lean her head on her hand. As I thought Monsieur would be glad to know what these books were, and Mademoiselle Julie did not know, I got myself taken into the library, under the pretext of seeing it. There were empty places for only two books; one is the second volume of "Christian Thoughts", and the other, the first volume of the book, has the title "Clarissa". I write it just as it was; Monsieur will perhaps know it.

Yesterday evening Madame did not take supper; she only had some tea.

She rang early this morning; she asked for her horses at once and before nine o'clock she went to the Feuillants, where she heard Mass. She wanted to go to confession, but her confessor was away and he will not be back for eight or ten days; I thought it important to tell Monsieur that.

Afterwards, she came back, had breakfast, and began to write, and remained at it until one o'clock. I soon found an opportunity to do what Monsieur most desires; and it was I took the letters to post. There was none for Madame de Volanges; but I send one to Monsieur which was for Monsieur de Tourvel; I thought that would be the most important There was one for Madame de Rosemonde as well; but I supposed Monsieur could always see it when he wanted, so I let it go. Moreover, Monsieur will soon know everything since Madame de Tourvel writes to him also. In the future I shall have all the letters Monsieur wants, for it is almost always Mademoiselle Julie who gives them to the servants, and she assures me that from friendship to me, and to Monsieur too, she will gladly do what I want.

She even refused the money I offered her; but I think Monsieur would wish to give her some little present; and if it is his wish and he will leave it to me, I shall easily find out what will please her.

I hope that Monsieur will not think I have shown any negligence in his service, and I have it upon my heart to justify myself from his reproaches. If I did not know that Madame de Tourvel was leaving, the reason is my zeal for Monsieur's service, since it is he who made me leave at three o'clock in the morning; and that was why I did not see Mademoiselle Julie the night before, as usual, having gone to sleep at the roadside inn, so as not to wake the *Château*.

As to what Monsieur says about my being often without money, first it is because I like to keep clean-looking, as Monsieur can see; and then I must keep up the honour of the cloth I wear; I know that I ought to save a little for the future; but I confide entirely in the generosity of Monsieur, who is such a good Master.

As to entering Madame de Tourvel's service, while remaining in Monsieur's, I hope Monsieur will not insist on it. It was very different with Madame la Duchesse; but assuredly I will not wear livery, and a magistrate's livery, after having had the honour to be Monsieur's servant. In anything else Monsieur may dispose of him who has the honour to be with as much respect as affection, his most humble servant.

Roux Azolan, *Chasseur.*

Paris, 5th of October, 17—, at eleven o'clock at night.

LETTER CVIII

Madame de Tourvel to Madame de Rosemonde

O my indulgent mother! How many thanks I have to give you and what need I had of your letter! I read and re-read it continually; I could not tear myself away from it. I owe to it the only less painful moments I have passed since I left. How good you are! Wisdom and virtue, then, can feel for weakness! You have pity on my misfortunes! Ah! If you knew them! . . . They are terrible. I thought I had endured the pains of love; but the inexpressible torture, the torture which must be felt for one to have any idea of, is to be separated from what one loves, and separated for ever! . . . Yes, the pain which overwhelms me to-day will return to-morrow, the day after to-morrow, all my life! Heaven! How young I am still, and how long I have yet to suffer!

To be oneself the creator of one's misery; to tear one's heart with one's own hands; and, while suffering these unendurable pains, to feel every instant that one could end them with a word and that this word is a sin! Ah! My friend! . . .

When I took this painful step of going away from him, I hoped that absence would increase my courage and my strength; how much I was deceived! On the contrary, it seems to have completed their destruction. It is true I had more to struggle against; but even in resistance everything was not deprivation; at least I saw him sometimes; often, without daring to raise my eyes to his, I felt his gaze fixed upon me; yes, my friend, I felt it, it seemed to warm my soul; and without passing through my eyes it none the less reached my heart. Now, in my painful solitude, isolated from everything that is dear to me, alone with my misfortune, every moment of my sad existence is marked by tears, and nothing softens their bitterness, no consolation mingles with my sacrifices; those I have made hitherto only serve to make more painful those I have still to make.

Yesterday I felt it very keenly. Among the letters brought to me there was one from him; I recognised it among the others when the servant was a yard away. Involuntarily I got up; I trembled, I had difficulty in hiding my emotion; and this state was not without pleasure. A moment later I was alone and this deceptive pleasure soon vanished and left me only one sacrifice the more to make. Indeed, could I open that letter, which I was burning to read? Through the fatality which pursues me, the consolations which appear to come to me, only impose new privations; and these become still more cruel, from the thought that M. de Valmont shares them.

There it is at last, that name which continually occupies me and which I have such difficulty in writing; the kind of reproach you made me really alarmed me. I beg you to believe that my confidence in you was not lessened by false shame; and why should I blush to name him? Ah! I blush for my feelings, not the person who causes them. Who else is more worthy of inspiring them! Yet I do not know why that name does not naturally come to my pen; and this time too, I had to think before putting it down. I come back to him.

You tell me that he seemed keenly affected by my leaving. What has he done? What has he said? Has he talked of returning to Paris? I beg you to dissuade him from it as much as you can. If he has judged me rightly, he ought not to be angry with me for this step; but he ought to feel too that it is a final determination. One of my greatest tortures is not knowing what he thinks. I still have his letter here . . . but you will surely agree with me that I ought not to open it.

It is only through you, my indulgent friend, that I am not entirely separated from him. I do not wish to impose upon your kindness; I understand perfectly that your letters cannot be long; but you will not refuse a word to your child; a word to support her courage, a word to console her. Good-bye, my respected friend.

Paris, 5th of October, 17—.

LETTER CIX

Cécile Volanges to the Marquise de Merteuil

It is only to-day, Madame, that I returned to M. de Valmont the letter you did me the honour to write me. I kept it four days, in spite of my terror lest someone should find it, but I hid it very carefully; and when my grief returned I locked myself in to read it.

I see that what I thought a great misfortune is scarcely one at all; and I must admit there is great pleasure in it; so I am scarcely at all unhappy. There is only the idea of Danceny which still torments me sometimes. But already there are many times when I do not think of him at all! And also M. de Valmont is very charming!

I made things up with him two days ago; it was very easy; for I had only said two words when he said that if I had anything to say to him he would come to my room at night, and I had only to say that I agreed. And then, when he came, he seemed no more angry than if I had never done anything to annoy him. He only scolded me afterwards and then very gently and in such a way . . . Just like you; which proves to me that he was friendly to me too.

I cannot tell you how many funny things he has told me which I should never have thought, especially about Mamma. You would do me a great pleasure if you would tell me whether it is all true. What is certain is that I could not keep from laughing; so that once I laughed out loud, which frightened us very much; for Mamma might have heard it; and if she had come to see what it was, what would have become of me? She would certainly have sent me back to the convent this time!

As we must be prudent and since, as M. de Valmont says, he would not risk compromising me for the world, we have agreed that henceforth he will simply come and open the door and we will go to his room. There is nothing to fear there; I went there yesterday and as I write you now I am waiting for him to come. I hope, Madame, you will not scold me now.

Yet there is one thing in your letter which surprised me; it is what you say about Danceny and M. de Valmont when I am married. I thought that one day at the Opera you said just the opposite, that when I was married I could only love my husband and that I should even have to forget Danceny; perhaps I did not hear properly and I should prefer it to be otherwise, because now I do not feel afraid of my marriage. I even want it, because I shall have more liberty; I hope that I can then arrange things so as not to think of anything but Danceny. I feel I should only be really happy with him; for the idea of him always torments me and I am not happy except when I do not think of him, which is very difficult; and as soon as I think of him, I become unhappy at once.

What consoles me a little is that you assure me Danceny will love me all the more; but are you quite sure? . . . Oh! yes, you would not deceive me. Yet it is amusing that I should love Danceny and that M.de Valmont . . . But, as you say, perhaps it is a piece of good fortune! We shall see, in any case.

I did not quite understand what you meant about my way of writing. It seems to me that Danceny likes my letters as they are. Yet I do not feel that I should tell him anything of what is going on between M. de Valmont and me; so you have no reason to fear.

Mamma has not yet spoken to me about my marriage; but let things go; when she speaks to me, I promise you I shall tell her lies, since she wants to catch me.

Good-bye, my kind friend; I thank you very much and I promise you I shall never forget all your kindness to me. I must stop now, for it is after one o'clock; M. de Valmont will be here soon.

From the Château de . . . , 10th of October, 17—.

LETTER CX

The Vicomte de Valmont to the Marquise de Merteuil

"Powers of Heaven, I had a soul for grief: give me one for joy!"[1] I think it is the tender Saint-Preux who expresses himself thus. Better gifted than he, I possess the two existences at once. Yes, my friend, I am at the same time very happy and very unhappy; and since you have my entire confidence I owe you the double account of my pains and of my pleasures.

Know then that my ungracious devotee is still severe to me. I am at the fourth returned letter. Perhaps I am wrong to say the fourth; for having guessed at the first return that it would be followed by many others, and not wishing to waste my time, I adopted the course of putting my complaints into commonplaces and adding no date; and since the second post it is always the same letter which goes to and fro; I only change the envelope. If my Beauty ends up as Beauties usually do, and softens one day, at least from lassitude, she will at last keep the missive and it will then be time to find out how things are going. You will see that with this new sort of correspondence I cannot be perfectly informed.

Yet I have discovered that the unstable person has changed her confidant; at least I have made certain that since her departure from the *Château* no letter has come from her for Madame de Volanges, while there have been two for old Rosemonde; and as she has said nothing about it to us, as she no longer opens her mouth about "her dear Beauty," whom she used to talk of ceaselessly, I concluded that it was she who now has her confidence. I presume that on the one hand, the need to speak of me, and on the other the little shame of having to make admissions to Madame de Volanges about a sentiment so long disavowed, have produced this great revolution. I am afraid I have lost by the change; for the older women get, the more harsh and severe they become; the first would have said worse things about me; but the other will say them about love; and the tender prude is much more terrified of the sentiment than of the person.

The only way for me to find out is, as you see, to intercept this clandestine commerce. I have already sent orders about it to my man-servant; and I await the execution of them daily. Until then I can do nothing except at hazard; so for the last week I have been vainly going over all the known methods, all those in novels and in my secret memoirs. I do not find one which suits either the circumstances of the adventure or the character of the heroine. The difficulty would not be to get into her house, even at night, or even to drug her and to make her a new Clarissa; but after more than two months of attentions and trouble, to have recourse to means which are not mine! To crawl servilely in the traces of others, and to triumph without glory! . . . No, she shall not have "the pleasures of vice and the honours of virtue."[2] It is not enough for me to possess her, I want her to yield to me. Now, to achieve that I must not only get into her house but come there by her consent, find her alone and with the intention of listening to me, above all I must close her eyes to the danger, for if she sees it she will either overcome it or die. But the more I know what ought to be done, the more difficulty I find in carrying it out; and even if you should laugh at me again, I must admit to you that my embarrassment increases the more I dwell on it.

I think my head would turn, but for the pleasant distractions given me by our common pupil; I owe it to her that I have done something else except write elegies.

Would you believe that the little girl was so timid that three whole days passed before your letter had produced all its effect? See how a single false idea can spoil the best natures!

It was not until Saturday that she came circling about me and stammered a few words; but spoken in so low a tone and so stifled by shame, that it was impossible for me to hear them. But the blushing they caused allowed me to guess their sense. Up till then I had remained proud; but softened by so amusing a repentance I promised to go to the pretty penitent the same evening; and this grace on my part was received with all the gratitude due to so great a benefit.

As I never lose sight of your plans or of mine, I resolve to profit by this occasion to find out exactly what the child is worth and to accelerate her education. But to carry out this labour with more liberty, I needed to change the place of our rendez-vous; for a mere closet, which separated your pupil's room from her mother's, could not inspire her with sufficient confidence to allow her to display herself freely. I had therefore promised myself to make some noise "innocently", which would cause her sufficient fear to persuade her to take a surer refuge in the future; she spared me that trouble too.

The little person laughs easily; and to encourage her gaiety I related to her, during our intervals, all the scandalous adventures which came into my head; and to make them more attractive and the better to hold her attention, I laid them all to the credit of her Mamma; whom I amused myself in this way by covering with vices and ridicule.

It was not without a reason that I made this choice; it encouraged my timid pupil better than anything else and at the same time inspired her with a most profound contempt for her mother. I have long noticed that, if this method is not always necessary for seducing a girl, it is indispensable and often the most efficacious way, when you wish to deprave her; for she who does not respect her mother will not respect herself, a moral truth which I think so useful that I was very glad to furnish an example in support of the precept.

However, your pupil, who was not thinking of morality, was suffocated with laughter every instant; and at last she once almost laughed out loud. I had no difficulty in making her believe that she had made "a terrible noise". I feigned a great terror which she easily shared. To make her remember it better I did not allow pleasure to reappear, and left her alone three hours earlier than usual; so, on separating, we agreed that from the next day onwards the meeting should be in my room.

I have received her there twice; and in this short space the scholar has become almost as learned as the master. Yes, truly, I have taught her everything, including the complaisances! I only excepted the precautions.

Occupied thus all night, I make up for it by sleeping a large part of the day; and as the present society of the *Château* has nothing to attract me, I scarcely appear in the drawing-room for an hour during the day. To-day I even decided to eat in my room and I only mean to leave it after this for short walks. These fantasies are supposed to be on account of my health. I declared that I was "overcome with vapours"; I also announced that I was rather feverish. It costs me nothing but the speaking in a slow, colourless voice. As to the change in my face, rely upon your pupil. "Love will see to that."[3] I spend my leisure in thinking of means to regain my ungracious lady, of the advantages I have lost, and also in composing a kind of catechism of sensuality for the use of my scholar. I amuse myself by naming everything by the technical word; and I laugh in advance at the interesting conversation this will furnish between her and Gercourt on the first night of their marriage. Nothing could be more amusing than the ingeniousness with which she already uses the little she knows of this speech! She does not imagine that anyone can speak otherwise. The child is really seductive! The contrast of naive candour with the language of effrontery makes a great effect; and, I do not know why, only fantastic things please me now.

Perhaps I am giving myself up too much to this, since I am wasting both my time and my health; but I hope that my feigned illness, besides saving me from the boredom of the drawing-room, may also be of some use to me with the austere devotee, whose tigress virtue is nevertheless allied with soft sensibility! I have no doubt that she already knows of this great event, and I should much like to know what she thinks of it; the more so since I should wager she will not fail to attribute the honour of it to herself. I shall regulate the state of my health according to the impression it makes on her.

You now know, my fair friend, as much about my affairs as I do. I should like soon to have more important news to tell you; and I beg you to believe that in the pleasure I promise myself, I count for a great deal the reward I expect from you.

From the Château de . . . , 11th of October, 17—.

LETTER CXI

The Comte de Gercourt to Madame de Volanges

Everything in this country, Madame, seems as if it will be tranquil; and we are daily expecting permission to return to France. I hope that you will not doubt I have still the same eagerness to return and to form the bonds which will unite me to you and to Mademoiselle de Volanges. But Monsieur le duc de . . . my cousin, to whom as you know, I have so many obligations, has just informed me of his recall from Naples. He tells me that he means to pass through Rome and to see on his way that part of Italy he does not yet know. He invites me to accompany him on this journey, which will last for six weeks or two months. I do not hide from you that it would be agreeable for me to profit by this opportunity, since I feel that once I am married it will be difficult for me to take the time for other absences than those demanded by my service. And perhaps it would be more becoming to wait until the winter for this marriage, since all my relatives will not be gathered in Paris until then, and especially Monsieur le Marquis de . . ., to whom I owe the hope of being related to you. In spite of these considerations, my plans in this matter shall be absolutely subordinated to yours. And should you prefer your first arrangement, I am ready to renounce mine. I only beg you to let me know your intentions in this matter as soon as possible. I shall await your reply here and my conduct will be guided by that alone.

I am with respect, Madame, and with all the sentiments which befit a son, your most humble, etc.

Comte de Gercourt.

Bastia, 10th of October, 17—.

LETTER CXII

Madame de Rosemonde to Madame de Tourvel

(Dictated)

I have only just received, my dear Beauty, your letter of the eleventh,[1] and the gentle reproaches it contains. Confess that you would have liked to reproach me more; and that if you had not remembered you were "my daughter", you would have really scolded me. Yet you would have been very unjust! It was the desire and hope of being able to reply to you myself which made me put it off each day, and you see that to-day I am still obliged to borrow my waiting-woman's hand. My unfortunate rheumatism has come back. It has settled this time in the right arm and I am absolutely one-armed. See what it is, young and fresh as you are, to have so old a friend! We suffer for these inconveniences.

As soon as my pain gives me a little respite I promise to talk to you at length. Meanwhile I can only tell you that I have received your two letters; that they would have increased my tender friendship for you, had that been possible; and that I shall never cease to be keenly interested in everything that touches you.

My nephew is also a little indisposed, but without any danger and without the least cause for uneasiness; it is a slight inconvenience which, in my opinion, affects his temper more than his health. We scarcely see him any more.

His retirement and your departure do not make our little circle any gayer. The Volanges child especially finds great fault with you and all day long yawns incessantly. Particularly in the last few days she has done us the honour of going fast asleep after dinner.

Good-bye, my dear Beauty; I am always your good friend, your Mamma, your sister even, if my great age permitted me that title. In short, I am attached to you by all my tenderest feelings.

(Signed Adelaide, for Madame de Rosemonde)

From the Château de . . . , 14th of October, 17—.

LETTER CXIII

The Marquise de Merteuil to the Vicomte de Valmont

I think I ought to warn you, Vicomte, that people in Paris are beginning to talk about you; that your absence has been noticed and that the cause is already guessed. Yesterday I was at a supper where many people were present; it was positively announced that you were detained in a village by a romantic and unfortunate love; immediately joy was painted on the faces of all the men who envy your successes and of all the women you have neglected. If you will take my advice, you will not allow these dangerous rumours to gain ground, and you will return at once and destroy them by your presence.

Remember that once you allow the idea that you cannot be resisted to be destroyed, you will soon find that in fact you will be resisted more easily; that your rivals will also lose their respect for you and will dare to compete with you; for which among them does not believe himself stronger than virtue? Remember especially that among the multitude of women you have claimed, all those you have not had will try to undeceive the public, while the others will attempt to abuse it. In short, you must expect to be ranked perhaps as far below your value, as you have hitherto been ranked above it.

Come back, Vicomte, and do not sacrifice your reputation to a childish caprice. You have done all we wanted with the Volanges girl; and as to your Madame de Tourvel, you will not satisfy your whim for her by remaining ten leagues distant from her. Do you think she will come to look for you? Perhaps she no longer thinks of you or only concerns herself with you to congratulate herself on having humiliated you. Here at least you might find some opportunity of reappearing with effect, and you need it; and even if you persist in your ridiculous adventure I do not see that your return can make any difference . . .; on the contrary.

Indeed, if your Madame de Tourvel "adores you"; as you have told me so often and proved to me so little; her one consolation, her only pleasure, now must be to talk about you, to know what you are doing, what you are saying, what you are thinking, down to the smallest things which concern you. These trifles gain value according to the privations one undergoes. They are the crumbs of bread falling from the rich man's table; he despises them; but the poor man eagerly gathers them and is fed by them. At present poor Madame de Tourvel receives all these crumbs; and the more of them she has, the less she will be in a hurry to indulge in an appetite for the rest.

Moreover, since you know her confidant you can have no doubt that each letter from her contains at least a little sermon and everything she thinks likely "to strengthen her modesty and fortify her virtue."[1] Then why leave the one resources for defence and the other the opportunities for harming you?

It is not that I am at all of your opinion about the loss you think you have sustained by the change of confidants. First of all, Madame de Volanges hates you and hatred is always more clear-sighted and more ingenious than friendship. All your old aunt's virtue will never bring her for a single instant to speak ill of her dear nephew; for virtue also has its weaknesses. And then your fears are founded on an absolutely false observation.

It is not true that "the older women are, the more harsh and severe they become". It is between forty and fifty that the despair of seeing their beauty fade, the rage of feeling themselves obliged to abandon the pretentions and the pleasures to which they still cling, make almost all women disdainfully prudish and crabbed. They need this long interval to make the whole of that great sacrifice; but as soon as it is consummated, they all fall into two classes.

The more numerous, that of women who have never had to their advantage anything but their face and their youth, fall into an imbecile apathy and never come out of it except to play cards and for a few devotional practices; this kind is always boring, often grumbling, sometimes a little quarrelsome but rarely spiteful. It cannot be said either that these women are or are not severe; without ideas and without existence, they repeat with indifference and without comprehension everything they hear said, and in themselves remain absolute nullities.

The other class, much more rare but really valuable, is that of women who, having had character and having not failed to nourish their minds, are able to create an existence for themselves when that of nature fails them, and adopt the course of giving to their minds all the ornaments they formerly employed for their faces. They generally have a very clear judgment, and a mind which is at once solid, gay, and gracious. They replace seductive charms by engaging kindness and still more by a cheerfulness whose charm increases in proportion to their age; in this way they succeed to some extent in becoming friends with youth and in making themselves beloved by it. But then, far from being, as you say, "harsh and severe", the habit of indulgence, their long reflections on human weakness, and above all the memories of their youth, by which alone they still hold to life, place them perhaps rather too near laxity.

I can tell you as well that as I have always sought out old women, the usefulness of whose support I early recognised, I have found among them several to whom I returned as much from inclination as from interest. I pause here, for since you now grow inflamed so quickly and so morally I should be afraid lest you might suddenly fall in love with your old aunt and bury yourself with her in the tomb where you have already lived so long. I proceed then.

In spite of the enchantment you appear to find in your little scholar, I cannot believe that she has any influence upon your plans. You found her at hand, you took her; very well! But this cannot be an inclination. To tell you the truth it is not even a complete enjoyment; you possess absolutely nothing but her person! I do not speak of her heart, about which I have no doubt you hardly concern yourself; but you do not even occupy her head. I do not know if you have noticed it, but I have the proof of it in the last letter she wrote me[2]; I send it to you for you to form your own opinion. Observe that when she speaks of you, it is always "Monsieur de Valmont"; that all her ideas, even those which start from you, always end up in Danceny, and she never calls him Monsieur, it is always "Danceny" alone. By that she distinguishes him from all others; and even as she yields herself to you, she only grows familiar with him. If such a conquest appears "seductive" to you, if the pleasures she gives "captivate you", assuredly you are modest and not hard to please! I am perfectly willing that you should keep her; that is part of my plan. But it seems to me that this is not worth putting oneself out for a quarter of an hour; you ought to have some power over her and, for example, only allow her to come back to Danceny after having made her forget him a little more.

Before I leave you and come to myself, I should still like to tell you that this method of being ill which you announce to me is very well known and quite worn out. Really, Vicomte, you are not inventive! I repeat myself sometimes also, as you will see; but I try to make up for it in the details and above all I am justified by success. I am going to try one more and to seek a new adventure. I admit it will not have the merit of difficulty; but at least it will be a diversion, and I am bored to death.

I do not know why, but since my adventure with Prévan, Belleroche has become unendurable. He has so redoubled in attentions, in tenderness, in "veneration", that I can stand it no longer. Just at first his anger appeared amusing to me; but I had to calm it, for had I let him go on I should have been compromised; and there was no way of making him listen to reason. So I adopted the plan of showing him more love in order to succeed more easily; but he took it seriously and ever since then he has wearied me with his eternal delight. I have especially noticed the insulting reliance he has upon me and the confidence with which he looks upon me as his for ever. I am positively humiliated by it. He cannot value me very highly if he thinks himself good enough to hold me! Did he not tell me recently that I shall never have loved anyone but him? Oh! I needed all my prudence not to undeceive him at once by telling him the truth. Here indeed is an amusing gentleman to have an exclusive right! I admit he is well-made and tolerably handsome; but on the whole he is really only a manual labourer in love. In short the moment has come, we must separate.

I have been trying for the last fortnight and, turn by turn, I have employed coolness, caprice, temper, quarrels; but the tenacious creature does not leave hold so easily; I must therefore adopt a more violent course and consequently I am taking him to my country house. We leave the day after to-morrow. There will only be with us a few uninterested and not very clear-sighted people, and we shall have almost as much liberty there as if we were alone. There, I shall so overburden him with love and caresses, we shall live so uniquely for one another, that I wager that, even more than I, he will desire the end of this visit which he thinks so great a happiness; and if he does not come back more weary of me than I am of him I will allow you to say that I know no more than you.

A pretence for this kind of retirement is to occupy myself seriously with my big law-suit, which will at last come up for judgment at the beginning of the winter. I am very glad of it; for it is really disagreeable to have all one's money in the air.

It is not that I am uneasy about the result; first of all I am right, all my lawyers assure me of it; and even if I were not, I should be very unskilful if I could not win a law-suit when my only adversaries are minors still in the cradle and their old guardian! But since nothing should be neglected in so important a matter, I shall in fact have two lawyers with me. Does not this journey sound gay to you? However, if it makes me win my law-suit and lose Belleroche I shall not regret my time.

And now, Vicomte, guess his successor; I give you a hundred guesses. But there! Do I not know you never guess anything? Well, it is Danceny. You are surprised, are you not? For after all I am not yet reduced to educating children! But he deserves to be an exception; he has only the graces of youth, not its frivolity. His great reserve in company is very useful in keeping away all suspicion, and he is but the more charming when he opens out in private. It is not that I have yet had any such interview with him, I am still only his confidant; but under this veil of friendship I think he has a very keen inclination for me, and I feel I am acquiring a considerable one for him. It would really be a pity that so much wit and delicacy should be sacrificed and brutalised with that little imbecile of a Volanges! I hope he deceives himself in thinking he loves her; she is so far from deserving him! It is not that I am jealous of her; but it would be a manslaughter, and I wish to save Danceny from it. I beg you then, Vicomte, to take care that he does not meet "his Cécile" (as he still has the bad habit of calling her). A first inclination has always more power than one thinks, and I should not be sure of anything if he saw her again now, especially during my absence. On my return I will take charge of everything and will answer for it.

I had thought of taking the young man with me; but I made the sacrifice to my usual prudence; and then I was afraid he might notice something between Belleroche and me, and I should be in despair if he had the least idea of what is going on. I wish at least to offer myself to his imagination pure and unspotted; such indeed as one ought to be, to be really worthy of him.

Paris, 15th of October, 17—.

LETTER CXIV

Madame de Tourvel to Madame de Rosemonde

My dear friend, I give way to my deep uneasiness; and without knowing if you are in a condition to answer me, I cannot prevent myself from questioning you. Your telling me that Monsieur de Valmont's condition is "without danger", does not leave me with as much confidence as you appear to have. It is not rare that melancholy and a disgust with the world are the preliminary symptoms of some serious illness; the sufferings of the body, like those of the mind, make us desire solitude and often we reproach a person with ill-humour when we should only pity their sickness.

I think he ought at least to consult someone. How is it, that, being ill yourself, you have not a doctor with you? I saw my doctor this morning and I will not conceal from you that I consulted him indirectly; his opinion is that with naturally active people this kind of sudden apathy should never be neglected; and, as he said afterwards, illnesses do not yield to treatment when they are not taken in time. Why do you allow someone who is so dear to you to run this risk?

My uneasiness is greatly increased since I have had no news of him for four days. Heavens! Are you not deceiving me about his condition? Why has he suddenly ceased to write to me? If it were merely the result of my persistence in sending back his letters, I think he would have adopted this course sooner. And then, without believing in presentiments, for some days I have felt a sadness which frightens me. Ah! Perhaps I am on the eve of the greatest of misfortunes!

You will not believe, and I am ashamed to tell you, how it grieves me not to receive any longer those very letters which I refused to read. At least I was sure that he was thinking of me! And I saw something which came from him. I did not open these letters, but I cried as I looked at them; my tears were sweeter and more easy; and they only partly relieve the continual oppression I have felt since my return. I beg you, my indulgent friend, write to me yourself as soon as you can, and meanwhile, let me receive news of you and of him every day.

I see that I have scarcely said a word about you; you know my sentiments, my unreserved attachments, my tender gratitude for your sympathetic friendship; you will forgive the distress I am in, my extreme pain, the dreadful torture of having to dread ills of which I am perhaps the cause. Great Heaven! That desperate idea pursues me and tears my heart; this was a misfortune I lacked, and I feel I was born to endure them all.

Good-bye, my dear friend; love me, pity me. Shall I have a letter from you to-day?

Paris, 16th of October, 17—.

LETTER CXV

The Vicomte de Valmont to the Marquise de Merteuil

It is an inconceivable thing, my fair friend, how easily as soon as people are apart, they cease to be in agreement. As long as I was near you, we had always the same perception, the same way of looking at things; and because I have not seen you for nearly three months, we no longer have the same opinion about anything. Which of us two is wrong? You certainly would not hesitate about the answer; but I, who am wiser or more polite, do not decide. I shall only reply to your letter and continue to inform you of what I am doing.

First, I have to thank you for the information you give me of the rumours about me which are circulating; but I am not yet disturbed by them; I think that very soon I am certain to have the means of ending them. Be at rest, I shall reappear in society more celebrated than ever and still more worthy of you.

I hope that I shall even gain some credit for the adventure with the Volanges girl, which you seem to think so little of; as if it were nothing to take a girl away from her beloved lover in an evening, to make use of her afterwards as much as I wanted, absolutely as my own property, and without any more difficulty to obtain from her what one does not even dare to exact from all the women who make their living by it; and that, without in the least upsetting her tender love; without making her inconstant, not even unfaithful; for, indeed, I do not even occupy her head! So that when my whim is past, I shall return her to her lover's arms, so to speak, without her having noticed anything. Is that then so ordinary a course? And then, believe me, once she leaves my hands, the principles I have given her will develop none the less; and I predict that the timid scholar will soon take her flight in a way which will do honour to her master.

But if they prefer the heroic, I shall point to Madame de Tourvel, that often quoted model of all the virtues! Respected even by our greatest libertines! To such an extent that they had even abandoned the mere idea of attacking her! I shall point to her, I say, forgetting her duties and her virtue, sacrificing her reputation and two years of modesty, to run after the happiness of pleasing me, to intoxicate herself with the happiness of loving me, thinking herself sufficiently recompensed for so many sacrifices by a word, by a look, which she even did not always obtain. I shall do more, I shall abandon her; and either I do not know this woman or I shall have no successor. She will resist the need for consolation, the habit of pleasure, even the desire for vengeance.

In short, she will have existed only for me and whether her course be more or less long, I alone shall have opened and closed the barrier. Once I have achieved this triumph, I shall say to my rivals: "Behold my work, and seek a second example in the age!"

You will ask me whence comes this excess of confidence to-day? For the past week I have been in my fair one's confidence; she does not tell me her secrets, but I surprise them. Two letters from her to Madame de Rosemonde have given me sufficient information, and I shall only read the others from curiosity. To succeed I only need to be near her, and I have found the way. I am going to make use of it immediately.

You are curious, I think . . .? But no, to punish you for not believing in my designs, you shall not know this way. Frankly, you deserve that I should withdraw my confidence from you, at least in this adventure; indeed, but for the sweet prize attached by you to this success, I would not speak to you of it again. You see that I am annoyed. However, in the hope that you will reform, I will limit myself to this slight punishment; and, returning to indulgence, I will forget my great projects for a moment to reason with you about yours.

And so you are in the country, which is as wearisome as sentiment and as dull as fidelity! And poor Belleroche! You are not content to make him drink the water of oblivion, you put him to the torture with it! How is he? How well does he endure the nausea of love? I greatly wish that it would only make him more attached to you; I am curious to see what more efficacious remedy you would come to use. Really, I pity you for being obliged to make use of this one. Only once in my life have I made love by procedure. I had certainly good grounds, since it was the Comtesse de . . .; and twenty times in her arms I have been tempted to say to her: "Madame, I renounce the place I solicited, allow me to leave that which I occupy." So, among all the women I have had, she is the only one whom it gives me pleasure to speak ill of.

As to your grounds, to tell you the truth, I think them very ridiculous; and you were right to think I should not guess the successor. What? You are giving yourself all that trouble for Danceny! Hey! My dear friend, leave him to adore "his virtuous Cecile", do not involve yourself in these children's games. Leave the school-boys to be improved by "good women"[1], or to play with school-girls at little innocent games. Why are you burdening yourself with a novice who will not know how to take you or when to leave you, and with whom you will have to do everything? I tell you seriously I disapprove of this choice, and however secret it remains it will humiliate you at least in my eyes and in your own consciousness.

You say you have a great inclination for him; come, you are surely wrong, and I think I have found the cause of your error. This wonderful disgust with Belleroche came to you at a time of famine, and as Paris offered you no choice, your fancy, always too keen, fell upon the first object you met with. But remember, when you return you can choose among a thousand; and if you dread the inaction into which you may fall by delaying, I offer myself to amuse your leisure.

Between now and your arrival, my important affairs will be settled in one way or another; and, assuredly, neither the Volanges girl nor Madame de Tourvel herself will then occupy me so much that I cannot be at your service as much as you desire. Perhaps by then I shall have already returned the little girl to the hands of her discreet lover. Without agreeing, whatever you may say, that it is not a "captivating" enjoyment, since my project is that all her life she shall keep an idea of me superior to all other men, I have adopted with her a tone which I cannot sustain for long without damaging my health; and from this moment I am only bound to her by the kind of attention we owe to family affairs . . .

You do not understand me; the truth is I am waiting for a second period to confirm my hope and to be certain I have fully succeeded in my projects. Yes, my fair friend, I have already a first indication that my scholar's husband will not run the risk of dying without posterity and that the head of the house of Gercourt will in future be only a cadet of the house of Valmont. But let me finish as I fancy an adventure I only undertook at your request. Remember, if you make Danceny unfaithful, you take away all the piquancy of this story. And then consider that when I offer myself to you as his representative I have, I think, some rights to the preference.

I rely upon it so much, that I have not shrunk from acting contrary to your views, by assisting myself in augmenting the tender passion of the discreet lover for the first and worthy object of his choice. Yesterday I found your pupil engaged in writing to him, and having first disturbed her in this sweet occupation by another still sweeter, I asked afterwards to see her letter; and since I thought it cold and constrained I pointed out to her that this was not the way to console her lover and persuaded her to write another at my dictation; in which, imitating her silly little chatter as best I could, I tried to feed the young man's love with a more certain hope. The little girl was quite delighted, she told me, to find herself speaking so well; and henceforth I am to take charge of the correspondence. What is there I shall not have done for Danceny? I shall have been at once his friend, his confident, his rival, and his mistress! and again, at this moment, I am doing him the service of saving him from your dangerous bonds; for to possess you and to lose you, is to buy a moment of happiness with an eternity of regrets.

Good-bye, my fair friend; have the courage to dispatch Belleroche as soon as you can. Leave Danceny, and prepare yourself to rediscover and to give me again the delicious pleasures of our first acquaintance.

P.S. My compliments upon the near completion of your great law-suit. I shall be very glad if this happy event takes place under my reign.

From the Château de . . . , 19th of October, 17—.

LETTER CXVI

The Chevalier Danceny to Cécile Volanges

Madame de Merteuil left this morning for the country; thus, my charming Cécile, I am deprived of the sole pleasure which was left to me in your absence, that of talking about you to your friend and mine. Recently she has allowed me to give her that title; and I adopted it the more eagerly in that it seemed to draw me nearer to you. Heaven! What an amiable woman she is! And what a flattering charm she gives to friendship! It seems as if this gentle feeling is embellished and fortified in her by all that she refuses to love. If you knew how she loves you, how it pleases her to hear me talk of you! . . . Doubtless it is this which so attaches me to her. What happiness to be able to live solely for you two, to pass continually from the delights of love to the sweetness of friendship, to consecrate all my existence to them, to be, as it were, the meeting point of your mutual attachment; and to feel always that in occupying myself with the happiness of the one I should equally be working for the happiness of the other! Love this adorable woman, love her very much, my charming friend, give an even greater value to my attachment for her by sharing it. Since I have enjoyed the charm of friendship I wish you to feel it in your turn. It seems to me I only half enjoy the pleasures I do not share with you. Yes, my Cécile, I should like to surround your heart with all the softest sentiments; I should like each of these emotions to cause you a sensation of happiness and I should still think I could never return you more than a part of the felicity I gained from you.

Why must these charming projects be nothing but a fantasy of my imagination, while reality on the contrary brings me only painful and indefinite privations? I perceive that I must abandon the hope you gave me of seeing you in the country. My only consolation is to convince myself that this is not possible for you. And you neglect to tell me so, to lament over it with me! Already my complaints on this subject have twice gone unanswered. Ah! Cécile, Cécile, I believe you love me with all the faculties of your soul, but your soul does not burn like mine! Why is it not my part to raise these obstacles? Why is it not my interest which must be safeguarded instead of yours? I could soon prove to you that nothing is impossible to love.

You do not tell me either when this cruel absence is to end; here at least I might see you, your charming glances would reanimate my drooping spirits; their touching expression would reassure my heart which sometimes needs it so much. Forgive me, Cécile; this fear is not a suspicion. I believe in your love, in your constancy. Ah! I should be too wretched if I doubted it. But so many obstacles! Continually renewed! My dear, I am sad, very sad. It seems as if Madame de Merteuil's departure has renewed in me the feelings of all my misfortunes.

Good-bye, Cécile; good-bye, my beloved. Remember that your lover is in distress, and that you alone can give him back happiness.

Paris, 17th of October, 17—.

LETTER CXVII

Cécile Volanges to the Chevalier Danceny

(Dictated by Valmont)

Do you think, my dear, that you have to scold me for me to be sad when I know that you are distressed? And do you doubt that I suffer as much as you in all your troubles? I even share those which I cause you voluntarily; and I suffer more than you when I see that you do not do me justice. Oh! That is not kind. I can see what it is that annoys you; it is because I have not replied on the last two occasions when you asked me if you could come here; but is this such an easy reply to make? Do you think I do not know that what you want is very wrong? and yet if it is so hard for me to refuse you at a distance, what would it be if you were here? And then, I should be unhappy all my life for having wished to console you for a moment.

Come, I have nothing to hide from you; here are my reasons, judge for yourself. Perhaps I should have done what you wanted, had it not been that (as I told you) this Monsieur de Gercourt who is the cause of all our grief, will not arrive here so soon; and as Mamma has showed me much more kindness for some time; and as I am as caressing to her as I can be for my part; who knows what I might be able to obtain from her? And would it not be much better if we could be happy without my having anything to reproach myself with? If I am to believe what I have often been told, men do not love their wives so much when they have loved them too much before they were their wives. This fear restrains me more than all the rest. My friend, are you not sure of my heart and will there not always be time?

Listen, I promise you that if I cannot avoid the misfortune of marrying Monsieur de Gercourt, whom I hate so much before I know him, nothing will restrain me from being yours as much as I can, and even before everything. As I only care to be loved by you, and as you will see that if I do wrong it will not be my fault, the rest is of no concern to me; provided you promise me to love me always as much as you do now. But, my dear, until then let me go on as I am; and do not ask me any more for a thing I have good reasons for not doing but which it hurts me to refuse you.

I wish also that Monsieur de Valmont were not so pressing on your behalf; it only serves to make me still sadder. Oh! You have a very good friend in him, I assure you! He does everything just as you would do it yourself. But good-bye, my dear; it was very late when I began to write to you and I have spent part of the night on it. I am going to bed to make up for lost time. I kiss you, but do not scold me again

From the Château de . . . , 18th of October, 17—

LETTER CXVIII

The Chevalier Danceny to the Marquise de Merteuil

If I am to believe my almanac, my adorable friend, you have only been absent for two days; but if I am to believe my heart it is for two centuries. Now, I have it from you yourself, that we should always believe our hearts; it is therefore quite time for you to return and all your business must be more than finished. How do you expect me to be interested in your law-suit, if, whether lost or gained, I must pay the costs, through the weariness of your absence? Oh! How I should like to quarrel! And how sad it is, with such an excellent reason for ill temper, to have no right to show it!

Yet is it not a real infidelity, a wicked betrayal, to leave your friend far from you, after you have accustomed him to depend upon your presence? However much you consult your lawyers they will not find you any justification for this wrong procedure; and then such people only give reasons, and reasons are not a sufficient response to sentiments.

For my part, you have so often told me that you were making this journey from reason, that you have altogether put me out of temper with reason. I never want to listen to it again; not even when it tells me to forget you. Yet that is a very reasonable reason; and indeed it would not be so difficult as you might believe. It would be enough merely to lose the habit of always thinking about you; and I assure you nothing here would recall you to my mind.

Our prettiest women, those who are said to be the most charming, are still so far from you that they could only give a very weak idea of you. I believe even that with practised eyes, the more one thought at first they resembled you, the more difference one would find afterwards; whatever they may do, and however much they put into it all their knowledge, they will always fail to be you and it is there that the charm lies.

Unfortunately, when the days are so long, and I am without occupation, I dream, I make castles in Spain, I create my fantasy; little by little the imagination grows warm; I wish to embellish my work; I bring together everything that can please, and at last I reach perfection; and when I am there the portrait brings me back to the model, and I am amazed to find I have only been thinking of you.

At this very moment I am again the dupe of an almost similar error. Perhaps you think I began to write to you to dwell on you? Not at all; it was to distract myself from you I had a hundred things to say to you, of which you were not the object, and which, as you know, interest me very deeply; and yet it is from these I have been distracted. Since when does the charm of friendship distract from that of love? Ah! If I looked at it very closely, perhaps I should have to blame myself a little! But hush! Let me forget this little fault for fear of falling into it again; and let my friend ignore it.

So why are you not here to reply to me, to bring me back if I wander, to talk to me of my Cécile, to increase—if that is possible—the happiness I feel in loving her by the delightful idea that it is your friend I love? Yes, I confess it, the love she inspires in me has become still more precious to me since you have condescended to receive my confidence. It gives me such pleasure to open my heart to you, to occupy yours with my feelings, to place them there without reserve! It seems to me that I cherish them the more as you deign to receive them; and then I look at you and say to myself: "It is in her that all my happiness is enclosed."

I have nothing new to tell you about my situation. The last letter I received from *her* increases and assures my hope, but still retards it. But her motives are so tender and so virtuous, that I cannot blame her or complain myself. Perhaps you do not quite understand what I am telling you; but why are you not here? Although one says everything to a friend, one dares not write everything. The secrets of love especially are so delicate that one cannot let them go thus on their own good faith. If we sometimes let them go out, we must at least not lose sight of them; we must, as it were, see them enter their new refuge. Ah! Come back, my adorable friend, you see your return is necessary. Forget the "thousand reasons" which detain you where you are, or teach me how to live away from you.

I have the honour to be, etc.

Paris, 19th of October, 17—.

LETTER CXIX

Madame de Rosemonde to Madame de Tourvel

Although I am still suffering a great deal, my dear Beauty, I am trying to write to you myself so that I can speak to you of what interests you. My nephew still retains his misanthropy. He sends very regularly each day to have news of me; but he has not come once to inquire although I sent to ask him to do so and I see no more of him than if he were at Paris. I met him this morning, however, where I hardly expected it. It was in my chapel, to which I went down for the first time since my painful indisposition. I learned to-day that for the last four days he has gone there regularly to hear Mass. God grant that this lasts!

When I entered, he came to me and congratulated me very affectionately on the better state of my health. As Mass was beginning, I cut short the conversation which I intended to renew afterwards; but he disappeared before I could join him. I will not conceal from you that I found him a little changed. But, my dear Beauty, do not make me repent my confidence in your reason, by too keen an anxiety; and above all be certain that I should rather distress you than deceive you.

If my nephew continues to avoid me, I shall take the course of going to see him in his room, as soon as I am better; and I shall try to find out the cause of this singular proceeding, in which I believe you have some part. I will let you know what I find out. I must leave you now, I cannot move my fingers any longer; and then if Adelaide knew I had been writing she would scold me all the evening. Good-bye, my dear Beauty.

From the Château de . . . , 20th of October, 17—.

LETTER CXX

The Vicomte de Valmont to Father Anselme

(Monk of St Bernard of the Convent in the rue St Honoré)

I have not the honour to be known to you, Monsieur; but I know the complete confidence which Madame de Tourvel has in you, and I know moreover how worthily that confidence is placed. I think I can address myself to you without indiscretion, to obtain an essential service, worthy indeed of your holy ministry, in which the interest of Madame de Tourvel is joined with mine.

I have in my hands certain important papers which concern her, which cannot be confided to anyone, which I must not and will not give into any hands but hers. I have no way to inform her of this, from reasons which perhaps you have learned from her but which I do not think myself justified in telling you, and which have caused her to refuse all correspondence from me; a course which I willingly admit to-day that I cannot blame, since she could not foresee the events I was myself very far from expecting; events only possible to the more than human power I am forced to recognise in them.

I beg you then, Monsieur, to be good enough to inform her of my new resolutions and to ask her on my behalf for a private interview, when I can at least partly atone for my faults by my excuses; and, as a last sacrifice, destroy before her eyes the sole existing traces of an error or of a fault which rendered me culpable towards her.

Only after this preliminary expiation shall I dare to lay at your feet the humble admission of my long misconduct; and to implore your mediation for a still more important reconciliation, happily even more difficult. Can I hope, Monsieur, that you will not refuse me such necessary and such precious care? That you will consent to support my weakness and to guide my steps in a new path which I most ardently desire to follow, but blush to admit I do not yet know?

I await your reply with the impatience of a repentance which desires to make reparation, and I beg you to believe that I am, with as much gratitude as veneration,

Your most humble, Etc.

P.S. I authorise you, Monsieur, if you think fit, to hand this letter to Madame de Tourvel, whom I shall make it a duty all my life long to respect, and in whom I shall never cease to honour her whom Heaven made use of to bring back my soul to virtue through the moving spectacle of hers.

From the Château de . . . , 22nd of October, 17—.

LETTER CXXI

The Marquise de Merteuil to the Chevalier Danceny

I have received your letter, my too young friend; but before thanking you, I must scold you, and warn you that if you do not correct yourself, you will have no more reply from me. Be ruled by me, and abandon this tone of flattery which is nothing but a mere jargon as soon as it ceases to be the expression of love. Is that the style of friendship? No, my friend, each sentiment has its befitting language, and to use another is to disguise the thought one is expressing. I know our little women understand nothing of what is said to them unless it is translated, as it were, into this customary jargon: but I confess I think I deserve to be distinguished from them by you. I am really annoyed, and perhaps more than I should be, that you have judged me so ill.

You will only find in my letter then what is lacking in yours—frankness and simplicity. I will, for example, tell you that it would give me great pleasure to see you and that I am vexed to have near me only such people as weary me, instead of those who please me; but you translate this phrase thus: "Teach me how to live away from you"; so, I suppose, that even if you were with your mistress you could not live unless I made a third. What folly! And those women, "who always fail to be me"—perhaps you find that your Cécile fails to be so! But that is where you are carried by a language which, from the abuse that is made of it to-day, is even beneath the jargon of compliments and becomes nothing more than a mere formula, in which people no more believe than in "your most humble servant"!

My friend, when you write to me, let it be to tell me your manner of thinking and feeling and not to send me phrases which I shall find, without your help, more or less well expressed in the first modern novel. I hope you will not be angry with what I have said to you, even if you should see a little ill humour in it; for I do not deny that I am a little out of temper; but to avoid even the appearance of the fault I reproach you with, I shall not tell you that this ill humour is perhaps a little increased by my absence from you. Taking it all round, it seems to me that you are preferable to a lawsuit and two lawyers, and perhaps even more so than the *attentif* Belleroche.

You see that instead of lamenting my absence, you ought to be glad of it; for I have never paid you so fine a compliment. I think that the example is infecting me and that I shall be sending you flatteries too; but no, I prefer to stick to my frankness; that alone then assures you of my tender friendship and of the interest it inspires in me. It is very delightful to have a young friend, whose heart is occupied elsewhere. It is not the habit of most women; but it is mine. It seems to me that we yield with more pleasure to a sentiment from which we can have nothing to fear; so with you I have come to the rôle of confident, perhaps rather early. But you choose your mistresses so young, that you have made me perceive for the first time that I am beginning to get old! You do well to prepare yourself thus a long career of constancy, and I hope with all my heart that it may be mutual.

You are right to give way "to the tender and virtuous motives" which, you tell me, "retard your happiness". A long defence is the sole merit left to those who do not always resist; and I should think it inpardonable in anyone but a child like the Volanges girl not to have avoided a danger, of which she was sufficiently informed by her admission of her love. You men have no idea what virtue is, and what it costs to sacrifice it! But if a woman reasons a little, she must know that, independently of the wrong she commits, a weakness is for her the greatest of misfortunes; and I cannot imagine how any woman ever allows herself to fall into it, when she has a moment to reflect.

Do not combat this idea, for it is what principally attaches me to you. You will save me from the dangers of love; and although I have hitherto been able to defend myself from them without you, I consent to be grateful for it and I shall like you better.

And now, my dear Chevalier, I pray God to have you in his holy and lofty care.

From the Château de . . . , 22nd of October, 17—

LETTER CXXII

Madame de Rosemonde to Madame de Tourvel

I hoped, my amiable daughter, that at last I should be able to calm your anxieties; and I see with grief, on the contrary, that I am about to increase them still further. But, be calm; my nephew is not in danger; it cannot even be said that he is really ill. But there is surely something extraordinary passing within him. I do not understand it; but I left his room with a feeling of sadness; perhaps even of terror, which I reproach myself for sharing with you and about which nevertheless I cannot prevent myself from talking to you. Here is an account of what happened; you may be sure that it is a faithful one; for were I to live another eighty years I should not forget the impression made upon me by this sad scene.

This morning, then, I went to my nephew's room; I found him writing and surrounded with different piles of papers, which appeared to be the object of his labour. He was so absorbed in them that I was in the middle of his room before he had even turned his head to find out who had entered. As soon as he perceived me I noticed as he rose he tried to compose his features, and perhaps it is that which made me pay more attention. He was undressed and unpowdered; but I thought him pale and wasted, and above all that his physiognomy had changed. His glance which we have seen so lively and so gay, was downcast and sad; in short, let it be said between ourselves, I should not have wished you to see him thus; for he looked very touching and very fit, I should think, to inspire that tender pity which is one of the most dangerous snares of love.

Although I was struck by my observations, I began the conversation as if I had noticed nothing. First of all I spoke to him of his health; and, without saying that it was good, he did not in so many words say that it was bad. Then I complained of his retirement, which seemed a little like an obsession, and I tried to mingle some gaiety with my little reprimand; but he only replied with an air of conviction: "It is one wrong the more I confess; but it shall be repaired with the others." His air, even more than what he said, rather disturbed my cheerfulness, and I hastened to tell him that he attached too much importance to a single reproach of friendship.

We then began to converse quietly. He told me, a little afterwards, that perhaps business, "the most important business of his life", would soon recall him to Paris; but as I was afraid of guessing it my dear Beauty, and feared this opening might lead me to a confidence I did not desire, I asked him no questions and contented myself by replying that more distraction would be useful to his health. I added that, on this occasion I would trouble him with no entreaties, since I loved my friends for their own sake; at this very simple phrase he grasped my hands, and, speaking with a vehemence I cannot convey to you, said: "Yes, aunt, love, love a nephew who respects and cherishes you, and as you say, love him for himself. Do not feel distressed for his happiness, and do not disturb by any regret the eternal calm he soon hopes to enjoy. Tell me again that you love me, that you forgive me; yes, you will forgive me; I know your goodness; but how can I hope for the same indulgence from those I have so offended?" Then he leaned over me, to hide, I think, the signs of grief which in spite of himself were revealed to me by the sound of his voice.

More moved than I can tell you, I rose hastily; and doubtless he noticed my terror; for composing himself immediately he went on: "Forgive me, Madame, forgive me, I feel that I am falling into error in spite of myself. I beg you to forget what I have said and to remember only my profound respect. I shall not fail", he added, "to come and renew my expression of it before I go." It seemed to me that this last phrase was a hint to terminate my visit; and so I went away.

But the more I reflect upon it, the less I can guess what he meant. What is this business, "the most important of his life"? Why did he ask my pardon? What was the cause of his involuntary emotion in speaking to me? I have asked myself these questions a thousand times, without being able to reply to them. I see nothing in them which touches you; yet, as the eyes of love are more clear-sighted than those of friendship, I was unwilling to leave you in ignorance of what passed between my nephew and me.

I have four times resumed the writing of this long letter, which I would make longer but for the fatigue I feel . . . Good-bye, my dear Beauty.

From the Château de . . . , 25th October, 17—.

LETTER CXXIII

Father Anselme to the Vicomte de Valmont

I have received, Monsieur le Vicomte, the letter with which you honour me, and yesterday, in accordance with your desire, I went to the person in question. I informed her of the object and the motives of the application you asked me to make to her. However attached I found her to the wise course she had first adopted, upon my pointing out that by her refusal she was perhaps placing an obstacle to your happy conversion, and thus, as it were, opposing the merciful intentions of Providence, she has consented to receive your visit on condition that it is the last, and has charged me to tell you that it will be at her house on Thursday next, the twenty-eighth. If this day does not suit you, will you let her know and appoint another. Your letter will be received.

But, Monsieur le Vicomte, let me ask you not to delay without very good reason, so that you can the sooner and the more entirely give yourself up to the praiseworthy designs of which you inform me. Remember that he who delays to profit by the moment of grace, runs the risk of its being withdrawn from him; that if the goodness of God is infinite his use of it is nevertheless regulated by justice; and that there may come a moment when the God of mercy is changed into a God of vengeance.

If you continue to honour me with your confidence, I beg you to believe that all my exertions are at your command, as soon as you desire; however considerable my occupations may be, my most important business will always be to carry out the duties of the holy ministry, to which I am particularly devoted;

and the fairest moment of my life is that when I shall see my efforts prosper through the blessing of the Almighty. Weak sinners that we are, we can do nothing of ourselves! But the God who calls you back to Him can do everything; and we both owe it equally to His goodness, you the constant desire of uniting yourself to Him, and I the means of guiding you there. It is with His aid, that I hope soon to convince you that only our holy religion can give, even in this world, the solid and durable happiness we vainly search for in the blindness of human passions.

I have the honour to be, with respectful esteem, etc.

Paris, 25th of October, 17—.

LETTER CXXIV

Madame de Tourvel to Madame de Rosemonde

In the midst of the surprise which was caused me, Madame, by the news I learned yesterday, I do not forget the satisfaction it must cause you, and I hasten to give you information of it. Monsieur de Valmont is now concerned neither with me nor with his love; and only desires to repair by a more edifying life the faults, or rather the errors, of his youth. I was informed of this great event by Father Anselme, to whom he applied for direction in the future, and also to procure him an interview with me, the principal object of which I think is to return me my letters which he has hitherto retained, in spite of the contrary request I had made him.

Doubtless I cannot but applaud this happy change, and congratulate myself upon it, if, as he says, I have furthered it to some extent. But why should I have to be the instrument, and why should it cost me a life-time's tranquillity? Could Monsieur de Valmont's happiness only come about through my misfortune? Oh! My indulgent friend, forgive me for this complaint. I know it is not for me to fathom the decrees of God; but while

I ceaselessly ask Him, and ask Him always in vain, for strength to overcome my unhappy love, He squanders it upon one who did not ask for it, and leaves me helpless, entirely surrendered to my weakness.

But I must stifle this guilty murmur. Do I not know that the prodigal son on his return obtained more grace from his Father than the son who had never been away? What account have we to demand from Him who owes us nothing? And even if it were possible that we should have some rights with Him, what could mine be? Should I boast a virtue, which already I only owe to Valmont? He saved me, and I dare to complain of suffering for him! No; my sufferings will be dear to me, if his happiness is their reward. Doubtless he had to come back to the common Father in his turn. The God who created him must cherish His work. He did not create that charming being, to make him only a castaway. It is for me to bear the pain of my bold imprudence; should I not have felt that, since it was forbidden me to love him I ought not to have allowed myself to see him?

My sin or my misfortune is that I refused too long to see this truth. You are my witness, my dear and worthy friend, that I submitted to this sacrifice as soon as I recognised its necessity; but the one thing lacking to make it complete was that Monsieur de Valmont should not share it. Shall I confess to you that this idea is now my greatest torment? Unendurable pride, which softens the ills we feel by those we make others suffer! Ah! I will overcome this rebellious heart, I will accustom myself to humiliations.

It is with the especial purpose of achieving this that I have consented to receive the painful visit of Monsieur de Valmont on Thursday next. There, I shall hear him tell me himself that I am nothing to him, that the faint and passing impression I had made upon him is entirely effaced! I shall see his gaze rest upon me without emotion, while the fear of revealing my own will make me keep my eyes lowered. I shall receive from his indifference those very letters which he refused so long to my reiterated request; he will return them to me as useless objects which no longer interest him; and as my trembling hands receive this shameful packet, they will feel that it is given to them by a calm, firm hand! And then, I shall see him go away. Go away for ever, and my gaze will follow him, will not see his turned back upon me!

And I was reserved for such humiliation! Ah! At least let me make it useful to myself, by letting it fill me with the sentiment of my weakness . . . Yes, I will carefully keep those letters he no longer cares to retain. I will impose upon myself the shame of re-reading them every day, until my tears have washed out the last traces of them; and I shall burn his as being infected with the dangerous poison which has corrupted my soul. Oh! What then is love, if it makes us regret even the danger to which it exposes us; above all if we may dread to feel it still even when we no longer inspire it! Let me fly from this disastrous passion, which leaves no choice between shame and misery, and often unites them both; and at least let prudence replace virtue.

How far off Thursday still is! Why can I not consummate this painful sacrifice at this instant, and forget at once both its cause and its object! This visit troubles me; I regret having promised it. Ah! Why does he need to see me again! What are we to each other now? If he has offended me, I pardon him. I even congratulate him on the desire to repair his wrongs; I praise him for it. I will do more, I will imitate him; and, seduced by the same errors, I will be saved by his example. But when his project is to fly from me, why does he begin by seeking me! Is not the most urgent matter for each of us to forget the other? Ah! Doubtless, and henceforth it shall be my one anxiety.

If you permit it, my amiable friend, I shall come and occupy myself with this difficult labour near you. If I need aid, perhaps even consolation, I wish to receive them only from you. You alone can understand me and can speak to my heart. Your precious friendship will fill my whole existence. Nothing will appear difficult to me to second the solicitude you might give me. I shall owe you my tranquillity, my happiness, my virtue; and the fruit of your kindnesses to me will be to have made me worthy of them at last.

I think I have wandered a good deal in this letter; I presume so, at least, from the agitation I have been in all the time I was writing to you. If it contains some sentiments at which I might have to blush, hide them with your indulgent friendship. I commit myself entirely to it. I do not wish to conceal any of the emotions of my heart from you.

Farewell, my respectable friend. In a few days I hope to inform you of the day of my arrival.

Paris, 25th October, 17—.

END OF PART THREE

PART IV

LETTER CXXV

The Vicomte de Valmont to the Marquise de Merteuil

She is conquered, that proud woman who dared to think she could resist me! Yes, my friend, she is mine, entirely mine; after yesterday she has nothing left to grant me.

I am still too full of my happiness to be able to appreciate it, but I am astonished by the unsuspected charm I felt. Is it true then that virtue increases a woman's value even in her moment of weakness? But let me put that puerile idea away with old woman's tales. Does one not meet almost everywhere a more or less well-feigned resistance to the first triumph? And have I not found elsewhere the charm of which I speak? But yet it is not the charm of love; for after all, if I have sometimes had surprising moments of weakness with this woman resembling that pusillanimous passion, I have always been able to overcome them and return to my principles. Even if yesterday's scene carried me, as I now think, a little further than I intended; even if I shared for a moment the agitation and ecstasy I created; that passing illusion would now have disappeared, and yet the same charm remains. I confess I should even feel a considerable pleasure in yielding to it, if it did not cause me some anxiety. Shall I, at my age, be mastered like a schoolboy, by an involuntary and unsuspected sentiment? No; before everything else, I must combat and thoroughly examine it.

And perhaps I have already a glimpse of the cause of it! At least the idea pleases me and I should like it to be true.

Among the crowd of women for whom I have hitherto filled the part and functions of a lover, I had never met one who had not at least as much desire to yield to me as I had to bring her to doing so; I had even accustomed myself to call "prudes" those who only came half way to meet me, in opposition to so many others whose provocative defence never covers but imperfectly the first advances they have made.

Here on the contrary I found a first unfavourable prejudice, supported afterwards by the advice and reports of a woman who hated me but who was clear-sighted; I found a natural and extreme timidity strengthened by an enlightened modesty; I found an attachment to virtue, controlled by religion, able to count two years of triumph already, and then remarkable behaviour inspired by different motives but all with the object of avoiding my pursuit.

It was not then, as in my other adventures, a merely more or less advantageous capitulation which it is more easy to profit by than to feel proud of; it was a complete victory, achieved by a hard campaign and decided by expert manoeuvres. It is therefore not surprising that this success, which I owe to myself alone, should become more valuable to me; and the excess of pleasure I feel even yet is only the soft impression of the feeling of glory. I cling to this view which saves me from the humiliation of thinking I might depend in any way upon the very slave I have enslaved myself; that I do not contain the plenitude of my happiness in myself; and that the faculty of enjoying it in energy should be reserved to such or such a woman, exclusi[illegible] all others.

These sensible reflections shall guide my conduct in this important occasion; and you may be sure I shall not allow myself to be so enchained that I cannot break these new ties at any time, by playing with them, and at my will. But I am already talking to you of breaking off and you still do not know the methods by which I acquired the right to do so; read then, and see what wisdom is exposed to in trying to help folly. I observed my words and the replies I obtained so carefully that I hope to be able to report both with an exactness which will content you.

You will see by the copies of the two enclosed letters[1] who was the mediator I chose to bring me back to my fair one and what zeal the holy person used to unite us. What I must still inform you of—a thing I learned from a letter intercepted as usual—is that the fear and little humiliation of being abandoned had rather upset the austere devotee's prudence and had filled her heart and head with sentiments and ideas which, though they lacked common sense, were none the less important. It was after these preliminaries, necessary for you to know, that yesterday, Thursday, the 28th, a day fixed beforehand by the ungrateful person herself, that I presented myself before her as a timid and repentant slave, and left a crowned conqueror.

It was six o'clock in the evening when I arrived at the fair recluse's house, for, since her return, her doors have been shut to everyone. She tried to rise when I was announced; but her trembling knees did not allow her to remain in that position: she sat down at once. As the servant who had brought me in had some duty to perform in the room, she seemed to be made impatient by it. We filled up this interval with the customary compliments. But not to lose any time, every moment of which was precious, I carefully examined the locality; and there and then I noted with my eyes the theatre of my victory. I might have chosen a more convenient one, for there was an ottoman in the same room. But I noticed that opposite it there was a portrait of the husband; and I confess I was afraid that with such a singular woman one glance accidently directed that way might destroy in a moment the work of so many exertions. At last we were left alone and I began.

After having pointed out in a few words that Father Anselme must have informed her of the reasons for my visit, I complained of the rigorous treatment I had received from her; and I particularly dwelt upon the "contempt" which had been shown me. She defended herself, as I expected; and as you would have expected too, I founded the proofs of it on the suspicion and fear I had inspired, on the scandalous flight which had followed upon them, the refusal to answer my letters, and even the refusal to receive them, etc., etc. As she was beginning a justification (which would have been very easy) I thought I had better interrupt; and to obtain forgiveness for this brusque manner I covered it immediately by a flattery: "If so many charms have made an impression on my heart", I went on, "so many virtues have made no less a mark upon my soul. Seduced no doubt by the idea of approaching them I dared to think myself worthy of doing so. I do not reproach you for having thought otherwise; but I am punished for my error." As she remained in an embarrassed silence, I continued: "I desired, Madame, either to justify myself in your eyes or to obtain from you forgiveness for the wrongs you think I have committed, so that at least I can end in some peace the days to which I no longer attach any value since you have refused to embellish them."

Here she tried to reply, however: "My duty did not permit me." And the difficulty of finishing the lie which duty exacted did not permit her to finish the phrase. I therefore went on in the most tender tones: "It is true then that it was from me you fled?" "My departure was necessary." "And what took you away from me?" "It was necessary." "And for ever?" "It must be so."

I do not need to tell you that during this short dialogue the tender prude was in a state of oppression and her eyes were not raised to me.

I felt I ought to animate this languishing scene a little; so, getting up with an air of pique, I said: "Your firmness restores me all of my own. Yes, Madame, we shall be separated, separated even more than you think; and you shall congratulate yourself upon your handiwork at leisure." She was a little surprised by this tone of reproach and tried to answer. "The resolution you have taken [illegible]" said she. "Is only the result of my despair", I replied with vehemence. "It is your will that I should be unhappy; I will prove to you that you have succeeded even beyond your wishes." "I desire your happiness," she answered. And the tone of her voice began to show a rather strong emotion. So, throwing myself at her feet, I exclaimed in that dramatic tone of mine you know: "Ah! cruel woman, can there exist any happiness for me which you do not share? Ah! Never! Never!" I confess that at this point I had greatly been relying on the aid of tears; but either from a wrong disposition or perhaps only from the painful and continual attention I was giving to everything, it was impossible for me to weep.

Fortunately I remembered that any method is equally good in subjugating a woman, and that it sufficed to astonish her with a great emotion for the impression to remain both deep and favourable. I made up therefore by terror for the sensibility I found lacking; and with that purpose, only changing the inflection of my voice and remaining in the same position, I continued: "Yes, I make an oath at your feet, to possess you or die." As I spoke the last words our eyes met. I do not know what the timid person saw or thought she saw in mine, but she rose with a terrified air, and escaped from my arms, which I had thrown round her. It is true I did nothing to detain her; for I have several times noticed that scenes of despair carried out too vividly become ridiculous as soon as they become long, or leave nothing but really tragic resources which I was very far from desiring to adopt. However, as she escaped from me, I added in a low and sinister tone, but loud enough for her to hear: "Well, then! Death!"

I then got up; and after a moment of silence I cast wild glances upon her which, however much they seemed to wander, were none the less clear-sighted and observant. Her ill-assured bearing, her quick breathing, the contraction of all her muscles, her trembling half-raised arms, all proved to me that the result was such as I had desired to produce; but, since in love nothing is concluded except at very close quarters, and we were rather far apart, it was above all things necessary to get closer together. To achieve this, I passed as quickly as possible to an apparent tranquillity, likely to calm the effects of this violent state, without weakening its impression.

My transition was: "I am very unfortunate. I wished to live for your happiness, and I have disturbed it. I sacrifice myself for your tranquillity, and I disturb it again." Then in a composed but constrained way: "Forgive me, Madame; I am little accustomed to the storms of passions and can repress their emotions but ill. If I am wrong to yield to them, remember at least that it is for the last time. Ah! Calm yourself, calm yourself, I beseech you." And during this long speech, I came gradually nearer. "If you wish me to be calm", said the startled beauty, "you must yourself be more tranquil." "Well, then, I promise it," said I. And I added in a weaker voice: "If the effort is great, at least it will not be for long. But", I went on immediately in a distraught way, "I came, did I not, to return you your letters? I beg you, deign to receive them back. This painful sacrifice remained for me to accomplish; leave me nothing that can weaken my courage." And taking the precious collection from my pocket, I said: "There it is, that deceitful collection of assurances of your friendship! It attached me to life," I went on. "So give the signal yourself which must separate me from you for ever."

Here the frightened Mistress yielded entirely to her tender anxiety. "But, M. de Valmont, what is the matter, and what do you mean? Is not the step you are taking to-day a voluntary one? Is it not the fruit of your own reflections? And are they not those which have made you yourself approve the necessary course I adopted from a sense of duty?" "Well", I replied, "that course decided mine." "And what is that?" "The only one which, in separating me from you, can put an end to my own." "But tell me, what is it?" Then I clasped her in my arms, without her defending herself in the least; and, judging from this forgetfulness of conventions, how strong and powerful her emotion was, I said, risking a tone of enthusiasm: "Adorable woman, you have no idea of the love you inspire; you will never know to what extent you were adored, and how much this sentiment is dearer to me than my existence! May all your days be fortunate and tranquil; may they be embellished by all the happiness of which you have deprived me! At least reward this sincere wish with a regret, with a tear; and believe that the last of my sacrifices will not be the most difficult for my heart. Farewell."

While I was speaking, I felt her heart beating violently; I observed the change in her face; I saw above all that she was suffocated by tears but that only a few painful ones flowed. It was at that moment only that I feigned to go away; but, detaining me by force, she said quickly: "No, listen to me." "Let me go," I answered. "You will listen to me, I wish it". "I must fly from you, I must." "No", she cried. At this last word she rushed or rather fell into my arms in a swoon. As I still doubted of so lucky a success, I feigned a great terror; but with all my terror I guided, or rather, carried her towards the place designed beforehand as the field of my glory; and indeed she only came to her senses submissive and already yielded to her happy conqueror.

Hitherto, my fair friend, I think you will find I adopted a purity of method which will please you; and you will see that I departed in no respect from the true principles of this war, which we have often remarked is so like the other. Judge me then as you would Frederic or Turenne. I forced the enemy to fight when she wished only to refuse, battle; by clever manoeuvres I obtained the choice of battle-field and of dispositions; I inspired the enemy with confidence, to overtake her more easily in her retreat; I was able to make terror succeed confidence before joining battle; I left nothing to chance except from consideration of a great advantage in case of success and from the certainty of other resources in case of defeat; finally, I only joined action when I had an assured retreat by which I could cover and retain all I had conquered before. I think that is everything that can be done; but now I am afraid of growing softened in the delights of Capua, like Hannibal. This is what happened afterwards.

I expected so great an event would not take place without the usual tears and despair; and if I noticed at first a little more confusion, and a kind of interior meditation, I attributed both to her prudishness; so, without troubling about these slight differences which I thought were purely local, I simply followed the high-road of consolations; being well persuaded that, as usually happens, sensations would aid sentiment, and that a single action would do more than all the words in the world, which, however, I did not neglect. But I found a really frightening resistance, less from its excess than from the manner in which it showed itself.

Imagine a woman seated, immovably still and with an unchanging face, appearing neither to think, hear, nor listen; a woman whose fixed eyes flowed with quite continual tears which came without effort. Such was Madame de Tourvel while I was speaking; but if I tried to recall her attention to me by a caress, even by the most innocent gesture, immediately there succeeded to this apparent apathy, terror, suffocation, convulsions, sobs, and at intervals a cry, but all without one word articulated.

These crises returned several times and always with more

strength; the last was so violent that I was entirely discouraged and for a moment feared I had gained a useless victory. I fell back on the usual commonplaces, and among them was this: "Are you in despair because you have made me happy?" At these words the adorable woman turned towards me; and her face, although still a little distraught, had yet regained its heavenly expression. "Your happiness!" said she. You can guess my reply. "You are happy then?" I redoubled my protestations. "And happy through me?" I added praises and tender words. While she was speaking, all her limbs relaxed; she fell back limply, resting on her armchair; and abandoning to me a hand which I had dared to take, she said: "I feel that idea console and relieve me."

You may suppose that having found my path thus, I did not leave it again; it was really the right, and perhaps the only, one. So when I wished to attempt a second victory I found some resistance at first, and what had passed before made me circumspect; but having called to my aid that same idea of happiness I soon found its results favourable. "You are right", said the tender creature, "I cannot endure my existence except as it may serve to make you happy. I give myself wholly up to it; from this moment I give myself to you and you will experience neither refusals nor regrets from me."

It was with this naive or sublime candour that she surrendered to me her person and her charms, and that she increased my happiness by sharing it. The ecstasy was complete and mutual; and, for the first time, my own out-lasted the pleasure. I only left her arms to fall at her knees, to swear an eternal love to her; and, I must admit it, I believed what I said. Even when we separated, the idea of her did not leave me and I had to make an effort to distract myself.

Ah! Why are you not here to balance the charm of this action by that of its reward? But I shall lose nothing by waiting, shall I? And I hope I can consider as a thing agreed on between us the pleasant arrangement I proposed to you in my last letter. You see I carry things out and that, as I promised you, my affairs have gone well enough for me to be able to give you part of my time. Make haste then to get rid of your heavy Belleroche, abandon the whining Danceny, and concern yourself with me. What can you be doing in the country that you do not even reply to me? Do you know I should like to scold you for it? But happiness inclines us to indulgence. And then I do not forget that by returning to my place among your numerous suitors I must again submit to your little caprices. But remember, the new lover wishes to lose none of his old rights as a friend.

Good-bye, as of old . . . Yes, good-bye, my angel! I send you all love's kisses.

P.S. Do you know that Prévan, after a month's imprisonment, has been forced to leave his Corps? It is the news of all Paris to-day. Really, he is cruelly punished for a fault he did not commit, and your success is complete!

Paris, the 29th of October, 17—.

LETTER CXXVI

Madame de Rosemonde to Madame de Tourvel

I should have answered you sooner, my dear child, if the fatigue of my last letter had not brought back my pains, which have again deprived me of the use of my arm during the past few days. I was very anxious to thank you for the good news you gave me of my nephew, and equally anxious to send you my sincere congratulations on your own account. We cannot help but see in this a stroke of Providence which, by moving the one, has saved the other. Yes, my dear Beauty, God wished only to try you and helped you at the moment when your strength was exhausted; and, in spite of your little murmuring, I think you have some thanks to return Him. It is not that I am unaware that it would have been pleasanter for you if this resolution had come to you first and Valmont's had only been the result of it; it even seems, humanly speaking, that the rights of our sex would have been better preserved, and we are unwilling to lose any of them! But what are these slight matters compared with the important objects which are attained? Do we see a man who escapes from a shipwreck complain that he did not have a choice of ways?

You will soon feel, my dear daughter, that the pain you dread will disappear of itself; and even if it remains always and entire you will none the less feel that it is easier to endure than remorse and self-contempt. It would have been useless for me to speak to you with this apparent severity earlier; love is an independent sentiment, which prudence may enable us to avoid but which she cannot overcome, and, once love is born, it only dies of a natural death or from a complete lack of hope. You are in the latter position and this gives me the courage and the right to tell you my opinion freely. It is cruel to frighten a despairing invalid, who can endure nothing but consolations and palliatives, but it is wise to enlighten a convalescent on the dangers he has run, to give him the prudence he needs and submission to advice which may still be necessary to him.

Since you have chosen me for your doctor, I speak to you as such and tell you that the little discomforts you now feel (which perhaps require a few medicines) are yet nothing in comparison with the alarming illness whose cure is now assured. Then as your friend, as the friend of a reasonable and virtuous woman, I take it upon myself to add that this passion, in itself so unfortunate, became even more so through its object. If I am to believe what I am told, my nephew, whom I confess I love to a point of weakness and who, indeed, possesses many praise-worthy qualities and great charm, is not without danger for women and has wronged many, indeed prizes almost equally the seducing and the ruining of them. I believe you would have converted him. Doubtless no one was ever more worthy of doing so; but so many others have flattered themselves in the same way and their hope has been deceived, that I prefer you should not have been reduced to this resource.

Think now, my dear Beauty, that instead of all the dangers you would have had to run, you will have the quiet of your own conscience, your own tranquillity, as well as the satisfaction of having been the principal cause of Valmont's conversion. For my part, I have no doubt that it was largely due to your courageous resistance, and that a moment of weakness on your part might perhaps have left my nephew in a perpetual state of infidelity. I like to think so and wish to see you think the same; you will find in it your first consolation and I new reasons for loving you still more.

I expect you here in a few days, my charming daughter, as you have promised me. Come and seek calm and happiness in the very place you lost them; above all come and rejoice with your tender mother that you have kept so well the promise you made her, that you would never do anything which was unworthy of her and of you!

From the Château de . . . , 30th October, 17—.

LETTER CXXVII

The Marquise de Merteuil to the Vicomte de Valmont

If I did not answer your letter of the 19th, Vicomte, the reason was not that I lacked time; it is merely that it annoyed me and that I thought it devoid of common sense. I thought I could not do better than leave it in oblivion; but since you return to it and seem to value the ideas it contained while you take my silence for consent, I must plainly tell you my opinion.

I may sometimes have claimed to take the place of a harem by myself, but it has never suited me to be part of one. I thought you knew that. Now at least, when you know it, you will easily see how ridiculous your proposal seemed to me. What I! I sacrifice an inclination, a new inclination, to concern myself with you? And concern myself how? By waiting my turn, like a submissive slave, for the sublime favours of your "Highness!" When, for example, you wished to distract yourself for a moment from "this unsuspected charm" which you have only experienced with the "adorable, the heavenly" Madame de Tourvel, or when you are afraid of compromising in the eyes of the "delightful Cécile" the superior idea you are very glad to have her retain of you: then, condescending to me, you will come to seek pleasures, less keen indeed, but without consequence; and your precious favours, although rather rare, will quite suffice for my happiness!

Assuredly, you are rich in your good opinion of yourself; but apparently I am not so in modesty; for, however much I examine myself, I cannot think I have fallen so low as that. Very likely I am wrong, but I warn you that I am so in many other matters.

I am especially wrong in thinking that "the schoolboy, the whining" Danceny, entirely occupied with me, sacrificing to me, without even making a merit of it, a first passion, even before it has been satisfied, and loving me as men love at his age, could, in spite of his twenty years, work more effectively than you for my happiness and my pleasures. I take it upon me to add that if I had the caprice to give him a rival, it would not be you, at least at this moment.

And for what reasons? you will ask. But, in the first place, there might not be any; for the caprice which might make you preferred, might just as well exclude you. But for politeness' sake I am willing to tell the motives of my decision. It seems to me that you would have too many sacrifices to make me; and I, instead of feeling grateful as you would not fail to expect, should be capable of thinking that you owed me even more! You see that our way of thinking keeps us so far apart that we cannot draw near each other in any way; and I fear it will take a long time, a very long time, before my feelings change. I promise to let you know when I change. Until then, take my advice, make other arrangements, and keep your kisses, you have so many better ways of bestowing them!

"Good-bye, as of old", you say? But of old, it seems to me, you thought more of me; you did not destine me to play third parts; and above all you waited until I said yes before being certain of my consent. Excuse me then if, instead of saying "Good-bye, as of old", I say to you "Good-bye, as at present".

Servant, Vicomte.

From the Château de . . . , 31st of October, 17—.

LETTER CXXVIII

Madame de Tourvel to Madame de Rosemonde

It was only yesterday, Madame, that I received your delayed reply. It would have killed me there and then, if my existence were still my own; but it belongs to another; and that other is M. de Valmont. You see I hide nothing from you. Even if you should now think me unworthy of your friendship, I would rather lose it than hold it by deceit. All I can tell you is that I was placed by M. de Valmont between his death or his happiness, and I chose the latter. I neither boast of it nor blame myself for it; I simply say what is.

After that, you will easily imagine what an effect your letter and the severe truths it contained had upon me. Yet do not think it created any regret in me or can ever make me alter my feelings or my conduct. It is not that I do not pass through cruel moments; but when my heart is most torn, when I feel I can no longer endure my torture, I say to myself: Valmont is happy; and everything vanishes before that idea, or rather it changes everything into pleasure.

I have then consecrated myself to your nephew; for his sake

I am ruined. He has become the sole centre of my thoughts, of my feelings, of my actions. As long as my life is necessary to his happiness, it will be precious to me, and I shall think it a fortunate one. If some day he should think otherwise . . . he will hear no complaint and no reproach from me. I have dared already to fix my eyes upon that fatal moment and my mind is made up.

You can now see how I must be affected by the fear you seem to have that one day M. de Valmont might ruin me; for before he wishes to do so, he must have ceased to love me; and what effect on me can be made by vain reproaches I shall not even hear? He alone shall be my judge. As I shall only have lived for him, my memory will rest with him; and if he is forced to recognise that I loved him, I shall be sufficiently justified.

You have now read my heart, Madame. I prefer the misfortune of losing your esteem by my frankness to that of rendering myself unworthy of it by the degradation of a lie. I felt I owed this complete confidence to your former kindnesses to me. To add a word more might make you suspect that I have the pride to claim it still when, on the contrary, I do myself justice by abandoning any claim to it. I am with respect, Madame, your most humble and most obedient servant.

Paris, 1st of November, 17—.

LETTER CXXIX

The Vicomte de Valmont to the Marquise de Merteuil

Tell me, my fair friend, what is the reason for the tone of bitterness and banter which reigns in your last letter? What is this crime I have committed, apparently without knowing it, which makes you so out of humour? You blame me because I appeared to count upon your consent before I had obtained it; but I thought that what might appear presumption with other people, could never be taken for anything but confidence between you and me; and since when has that feeling harmed friendship and love? By adding hope to desire I only yielded to a natural impulse, which makes us always place the happiness we seek as near as possible; and you took the result of my eagerness for that of pride. I know very well that custom has introduced a respectful doubt into this matter; but you know it is only a form, a protocol; and I think I had every reason to believe that these minute precautions were not necessary between us.

It even seems to me that this frank and free proceeding, when it is founded upon old acquaintance, is very preferable to the insipid flattery which so often makes love tasteless. And perhaps the value I find in this manner only comes from that I attach to the happiness of which it reminds me; but, for that very reason, it would be still more disagreeable for me to have you think otherwise.

But that is the only fault I know of myself; for I do not imagine that you can have seriously believed there is a woman in the world whom I should prefer to you; and still less that I should have rated you as low as you feign to believe. You examined yourself in this matter, you tell me, and you do not think you have fallen so low as that. So I think and that simply proves that your mirror is faithful. But might you not have concluded more easily and with more justice that I certainly never had such an opinion of you?

I vainly look for a reason for this strange idea. But it seems to me that it comes more or less from the praises I have allowed myself to give other women. At least, so I infer from your affectation in dwelling upon the epithets of "adorable", "heavenly", "delightful", which I used in speaking to you of Madame de Tourvel or the Volanges girl. But do you not know that these words, taken at hazard rather than on reflection, do not so much express one's opinion of the person as the situation one is in when speaking? And if, at the very moment when I was so keenly affected by one or the other, I still desired you no less; if

I gave you a marked preference over both of them, since I could not renew our first *liaison* except at the expense of both of them; I do not think there is any great reason for blaming me here.

It would not be more difficult for me to justify myself on the "unsuspected charm", by which you also seem to be a little annoyed; for, first of all, it does not follow that it is stronger because it was unsuspected. Ah! Who could surpass you in the delicious pleasures which you alone can make always new and always more intense? I merely wished to say it was of a kind I had not before experienced, but without intending to classify it; and I added, as I repeat today, that, whatever it may be, I shall combat and overcome it. I shall do this with even more zeal, if I can think that this slight labour is a homage I can offer you.

As to little Cécile—I see no use in talking about her to you. You have not forgotten that I took charge of the child at your request, and only await your permission to get rid of her. I may have mentioned her ingenuousness and her freshness; I may even for a moment have thought her "delightful", because one is always more or less pleased with one's own work; but she certainly has not enough consistency in any way to hold one's attention at all.

And now, my fair friend, I appeal to your justice, to your first favours to me; to the long and perfect friendship, to the complete confidence, which have so long bound us to each other: have I deserved the harsh tone you take towards me? But how easy it will be for you to compensate me for it whenever you choose! Say but a word, and you will see if all the charms and all the delights keep me here—not a day, but a minute. I shall fly to your feet and your arms, and I will prove to you a thousand times and in a thousand ways that you are, that you always will be, the true ruler of my heart.

Good-bye, my fair friend; I expect your reply with great eagerness.

Paris, 3rd of November, 17—.

LETTER CXXX

Madame de Rosemonde to Madame de Tourvel

And why, my dear Beauty, do you no longer wish to be my daughter? Why do you seem to declare that all correspondence between us is to be broken off? Is it to punish me for not having guessed something which was contrary to all probability? Or did you suspect me of having voluntarily distressed you? No, I understand your heart too well to believe that it thinks of mine in this way. So the pain your letter caused me is much less relative to myself than to you!

O my friend! It is with pain I say it, but you are far too worthy of being loved, for love ever to make you happy. Ah! What really delicate and sensitive woman has not found misfortune in the very sentiment which promised her so much happiness! Do men know how to appreciate the women they possess?

Many of them are honourable in their behaviour and constant in their affection but, even among those, how few are able to place themselves in unison with our hearts! Do not think, my dear child, that their love is like ours. They feel indeed the same ecstasy; they often put more vehement emotion into it; but they do not know that anxious eagerness, that delicate solicitude, which in us produce those tender and continuous attentions whose sole end is always the beloved person. Man enjoys the happiness he feels, woman the happiness she gives. This difference, so essential and so rarely noticed, has a very perceptible influence on the whole of their respective behaviour. The pleasure of one is the satisfaction of desires, the pleasure of the other is above all to create them. To please is for him only a method of success; but for her it is success itself. And coquetry, for which women are so often blamed, is nothing but the excess of this way of feeling, and by that very fact proves its reality. Finally, that exclusive desire which particularly characterises love is, in man, only a preference which at most serves to increase a pleasure which another object would perhaps weaken but not destroy; while with women it is a profound sentiment which not only annihilates every other desire but, being stronger than nature itself and out of its control, makes them feel nothing but repugnance and disgust even where it seems pleasure should be created.

And do not think that the more or less numerous objections which may be quoted can be successfully opposed to these general truths! They are supported by public opinion which in men alone distinguishes infidelity from inconstancy—a distinction which they are proud of when they ought to feel humiliated; for our sex this distinction has only been adopted by those depraved women who are its shame, women to whom any way seems good if they can hope by it to be saved from the painful feeling of their baseness.

I thought, my dear Beauty, it might be useful for you to have these reflections to oppose to the deceitful ideas of perfect happiness with which love never fails to abuse our imagination; a fallacious hope to which we still cling even when forced to abandon it, the loss of which irritates and multiplies the griefs, already but too real, inseparable from a strong passion! This occupation of softening your troubles or diminishing their number is the only one I can and will fulfil at this moment. In ills without remedy, advice can only be directed to the regimen. All I ask of you is to remember that to pity an invalid is not to blame him. Ah! What are we, to blame one another? Let us leave the right of judging to Him who reads in our hearts; and I even dare to think that in his paternal eyes a host of virtues can atone for one weakness.

But, I beg you, dear friend, avoid especially these violent resolutions which are less a sign of strength than of complete discouragement; do not forget that, by rendering another the possessor of your existence (to use your own expression) you cannot have frustrated your friends of what they possessed before, which they will continue to claim.

Good-bye, my dear daughter; think sometimes of your tender mother and believe that you will always and above everything be the object of her dearest thoughts.

From the Château de . . . , 6th of November, 17—

LETTER CXXXI

The Marquise de Merteuil to the Vicomte de Valmont

Well and good, Vicomte, I am more pleased with you this time than before; but now let us discuss matters in a friendly way; I hope to convince you that the arrangement you appear to desire would be a real folly for you as well as for me.

Have you not yet noticed that pleasure, which is indeed the sole motive for the union of the two sexes, is not sufficient to form a bond between them? And that, though it is preceded by the desire which brings together, it is none the less followed by the disgust which repels? It is a law of nature that love alone can change; and is one able to feel love at will? Yet it is always necessary, and this would be really a difficulty had people not perceived that it was enough if love existed on one side. In this way the difficulty was halved, and without much loss; indeed, one enjoys the happiness of loving, the other that of pleasing, a little less animated it is true, but I add to it the pleasure of deceiving, which makes a balance; and thus everything goes well.

But tell me, Vicomte, which of us two will undertake to deceive the other! You know the story of the two rogues who recognised each other when gambling: "We shall achieve nothing" they said, "let us each pay half for the cards"[1]; and they gave up the game. Take my advice, let us follow this prudent example, and not lose together a time we could employ so much better elsewhere.

To prove to you that your interest influences this decision as much as my own and that I am not acting from ill humour or caprice, I will not refuse you the reward agreed upon between us; and I am quite certain that for one evening only we shall satisfy each other, and I have no doubt that we shall be able to embellish it in such a way that we shall see it end with regret. But do not forget that this regret is necessary to happiness; and however pleasant our illusion may be, do not let us think it can be durable.

You see that I fulfil my promise in my turn and that before you have carried out your part; for I was to have the first letter from the heavenly prude and yet, either you still value it or you have forgotten the conditions of a bargain, which perhaps interests you less than you would like to persuade me; but I have received nothing, absolutely nothing. Yet either I am wrong or the tender devotee must write often; for what does she do when she is alone? She surely has not the wit to amuse herself. If I wished, then, I might make you a few little reproaches; but I leave them in silence to atone for a little ill-humour I may have had in my last letter.

And now, Vicomte, I have nothing else to say but to make you a request which is as much for your sake as for mine; this is to put off a moment, which perhaps I desire as much as you do but which I think ought to be delayed until my return to town. On the one hand, we shall not have the necessary liberty here, for it would only need a little jealousy to attach to me more firmly than ever this dull Belleroche, who is now only hanging by a thread. He is already in the state of having to lash himself up to the point of loving me; he is in such a state that I put as much malice as prudence into the caresses with which I overwhelm him. But at the same time you will see that this would not be a sacrifice to make you! A mutual infidelity will make the charm much more powerful.

Do you know I sometimes regret that we are reduced to these resources. In the time when we loved each other, for I think it was love, I was happy; and you too, Vicomte! . . . But why trouble about a happiness which can never return? No, whatever you may say, it is impossible. First of all, I should exact sacrifices which you would not or could not make me and which I may not deserve; and then, how is one to retain you? Oh! No, no, I will not even consider the idea; and in spite of the pleasure I feel in writing to you, I prefer to leave you abruptly. Good-bye, Vicomte.

From the Château de . . . , 6th of November, 17—.

LETTER CXXXII

Madame de Tourvel to Madame de Rosemonde

I am touched by your kindness to me, Madame; I should yield entirely to it were I not restrained to some degree by the fear of profaning it by the acceptance. Why must it be that when I find it so precious I feel at the same time I am not worthy of it? Ah! I shall at least dare to express my gratitude to you; above all I shall admire that indulgence of virtue which only knows our faults to pity them, whose powerful charm keeps so sweet and so strong a domination over our hearts even beside the charm of love.

But can I still deserve a friendship which no longer suffices for my happiness? I say the same of your advice; I feel its value but I cannot follow it. And how could I not believe in perfect happiness when I feel it at this very moment? Yes, if men are as you say, they must be avoided, they are hateful; but how far is Valmont from resembling them! If, like them, he has that violence of passion, which you call vehement emotion, is it not exceeded in him by his delicacy! O, my friend! You speak of sharing my troubles, rejoice at my happiness; I owe it to love and how much more to the object of that love who increases its value! You love your nephew, you say, perhaps to the point of weakness? Ah! If you knew him as I do! I love him with idolatry, and even then much less than he deserves. No doubt he has been carried away into a few errors, he admits it himself; but who has ever known real love as he does? What more can I say to you? He feels it as he inspires it.

You will think that this is "one of the deceitful ideas with which love never fails to deceive our imagination"; but in this case why should he become more tender, more ardent, now he has nothing more to obtain? I will confess I had noticed in him formerly an appearance of reflection and reserve which rarely left him and often brought me back, in spite of myself, to the false and cruel impressions I had been given of him. But since he has been able to yield unconstrainedly to the emotions of his heart he seems to guess all the desires of mine. Who knows whether we were not born for each other! If this happiness were not reserved for me—to be necessary to his! Ah! If it is an illusion, may I die before it ends. But no; I want to live to cherish him, to adore him. Why should he cease to love me? What other woman could make him happier than I do? And I feel it myself; this happiness one creates is the strongest bond, the only one which really binds. Yes, it is this delicious feeling which ennobles love, which, in a way, purifies it, and makes it really worthy of a tender and generous soul, like Valmont's.

Good-bye, my dear, my respectable, my indulgent friend. It would be vain for me to try to write longer to you; this is the time he promised to come and every other idea abandons me. Forgive me! But you desire my happiness and it is so great at this moment that I am scarcely capable of feeling it all.

Paris, 7th of November, 17—.

LETTER CXXXIII

The Vicomte de Valmont to the Marquise de Merteuil

And what, my fair friend, are these sacrifices you think I would not make, the reward of which would be to please you? Only let ıow them and if I hesitate to offer them to you, I give you ıission to refuse the homage of them. Hey! What have you beeıı thinking of me for some time if, even in a mood of indulgence, you doubt my feelings and my energy? Sacrifices I would not or could not make! So you think me in love, subjugated? And the value I put upon the victory you suspect me of attaching to the person? Ah! Thanks be to Heaven, I am not yet reduced to that and I offer to prove it to you. Yes, I will prove it to you, even if it must be at the expense of Mme de Tourvel. After that you cannot surely have any doubts.

I think I may give some time, without compromising myself, to a woman who at least has the merit of belonging to a class one rarely meets. And perhaps the dead season in which this adventure happened made me give myself up to it more completely; and even now, when the full rush of society has hardly begun, it is not surprising that it absorbs me almost entirely. But remember it is hardly a week ago that I enjoyed the fruit of three months of trouble. I have so often stayed longer with someone who was worth less and had not cost me so much! . . . And you never drew any conclusions from them about me.

And then, would you know the real cause of the ardour I put into it? Here it is. The woman is naturally timid; at first, she continually doubted her happiness, and this doubt was enough to disturb her, so that I am scarcely beginning to be able to notice how far my power extends in this class. Yet it is a thing I was curious to know, and the opportunity is not so easily found as people think.

First, for a great many women pleasure is always pleasure, and is never anything more; whatever title may be given us, to them we are never anything but factors, mere agents whose sole merit is activity, and among whom he who does most, does best.

In another class, perhaps the most numerous to-day, the celebrity of the lover, the pleasure of taking him away from another woman, the fear of his being taken away in turn, preoccupy women almost entirely; we have a share, to a gre... or less extent, in the kind of happiness they enjoy, but it is rather a matter of the circumstances than the person. It comes to them by way of us and not from us.

For my observations I needed to find a delicate and sensitive woman who would make love her sole interest and who in love itself would see only her lover; a woman whose emotion, far from following the ordinary track, would start from the heart and reach the senses; whom I have seen, for example (and I am not speaking of the first day) emerge from pleasure in tears and a moment later regain delight through a word which replied to her soul. And then, it was necessary for her to add to all this the natural candour which has become insurmountable from the habit of yielding to it, which does not allow her to dissimulate any of the feelings of her heart. Now, you will admit that such women are rare; and I am ready to believe that but for her I should never have met with one perhaps.

It would not be surprising then if she should hold me longer than another woman would; and since the experiment I wish to make with her demands that I render her happy, perfectly happy, why should I refuse to do so, especially when it aids and does not hinder me? But because the mind is engaged, does it follow that the heart is enslaved? Certainly not. So the value I do not deny that I set on this adventure will not prevent my seeking others or even my sacrificing it to more agreeable ones.

I am so free that I have not even neglected the Volanges girl, for whom I care so little. Her mother is bringing her back to town in three days' time; yesterday I arranged my communications; some money to the porter, a few compliments to his wife, arranged the whole matter. Can you imagine that Danceny did not find out so simple a way? And yet people say love makes them ingenious! On the contrary those it dominates it makes stupid. And I am unable to protect myself from it! Ah! You need not be uneasy. In a few days I shall weaken, by sharing it elsewhere, the possibly too lively impression I have received; and if sharing it with one only is not enough, I shall share it with several.

I shall be quite ready, however, to hand the schoolgirl over to her discreet lover as soon as you think fitting. It seems to me you have no further reason to prevent it, and I agree to render poor Danceny this signal service. Indeed it is the least I can do for him in exchange for all those he has done me. At the present moment he is in a great state of anxiety to know whether Mme. de Volanges will receive him; I calm him as best I can by assuring him I will make him happy on the first opportunity; and meanwhile I continue to take charge of his correspondence which he wishes to continue on the arrival of "his Cécile". I have six letters from him already and I shall have one or two more before the happy day. He must be a very unoccupied young man!

But let us leave this childish couple and come back to ourselves; let me concentrate solely on the delightful hope your letter gave me. Yes, of course, you will hold me and I should not forgive you for doubting it. Our ties were loosened, not broken; and the so-called breaking-off was merely an error of our imagination; our feelings, our interests, remained just as united as ever. Like a traveller who returns home undeceived, I shall realise, like him, that I had left happiness to pursue hope and I shall say with Harcourt:

"The more foreigners I saw, the more I loved my country."[1] Do not oppose any longer the idea, or rather, the sentiment which brings you back to me; and after having tried all pleasures in our different paths, let us enjoy the happiness of feeling that none of them is comparable with that we have experienced, which we shall once more find still more delicious!

Good-bye, my charming friend. I consent to wait until you return; but hasten it, and do not forget how much I desire it.

Paris, 8th of November, 17—.

LETTER CXXXIV

The Marquise de Merteuil to the Vicomte de Valmont

Really, Vicomte, you are like children, in whose presence one must never say anything and to whom one can never show anything without their wanting to get hold of it once! A mere idea comes to me and I warn you I do not intend to dwell on it; and, because I speak of it to you, you take advantage of this to bring back my attention to it, to hold me to it when I wish to be diverted from it and, as it were, to make me share your absurd desires in spite of myself! Is it generous of you to leave me to bear the whole burden of prudence? I tell you again, and I repeat it to myself still more often, the arrangement you propose is really impossible. Even if you put into it all the generosity you show at the moment, do you think I have not any delicacy myself, that I would accept sacrifices which would injure your happiness?

Now, Vicomte, have you really any illusions about the sentiment which attaches you to Madame de Tourvel? It is love, or love never existed; you may deny it in a hundred ways but you prove it in a thousand. For example, what subterfuge is this you make use of against yourself (for I think you are sincere with me) which makes you attribute to a desire for observation the wish to keep this woman, a wish you can neither hide nor combat? Would one not say you had never made another woman happy, perfectly happy? Ah! If you doubt that, you have very little memory! But no, it is not that. Your heart simply abuses your mind and makes it accept bad reasons; but I, who have a great interest in not allowing myself to be deceived, am not so easily contented.

Thus, in noticing your politeness which made you carefully suppress all the words which you thought had displeased me, I yet perceived that you had none the less preserved the same ideas, perhaps without realising it. Indeed, it is no longer the adorable, the heavenly Madame de Tourvel, but "an amazing woman, a delicate and sensitive woman", and that to the exclusion of all others; "a rare woman", such that "a second one like her could never be met with". It is the same with that unsuspected charm which is not "the more powerful". Well! So be it; but since you have never found it until now, it is to be supposed that you would never find it again in the future, and the loss you would make would be just as irreparable. Vicomte, either these are the certain symptoms of love or one must give up ever looking for them.

Rest assured, this time I am speaking to you without ill humour. I have promised myself not to be out of humour again; I have realised but too well that it may become a dangerous snare. Take my advice, let us be friends and remain as we are. Be thankful to me only for my courage in defending myself; yes, my courage, for it is needed sometimes even to avoid taking a course one knows to be the wrong one.

It is therefore not with the purpose of bringing you round to my opinion by persuasion that I reply to the request you make about the sacrifices I should exact, and you could not make. I use the word "exact" on purpose, because I am sure that in a moment you will indeed think me too exacting; but so much the better! Far from being annoyed by your refusal, I shall thank you for it. Come, I will not dissimulate with you, though perhaps I need it.

I should exact then—see the cruelty of it!—that this rare, this astonishing Madame de Tourvel be no more to you than an ordinary woman, a woman such as she really is; for we must not deceive ourselves; this charm we think we find in others exists in us, and love alone embellishes so much the beloved person. Perhaps you would make the effort to promise me, to swear to me that you would do what I ask, however impossible it may be; but, I confess, I should not believe mere words, I could only be persuaded by your whole conduct.

That is not all; I should be capricious. I should not pay any attention to the sacrifice of little Cécile which you offer me so graciously. On the contrary I should ask you to continue this painful service until further orders from me; either because I should like to abuse my power or because from indulgence or justice it would suffice me to control your sentiments without thwarting your pleasures. In any case, I should want to be obeyed; and my orders would be very harsh!

It is true that I should then think myself obliged to thank you and—who knows?—perhaps even to reward you. For example, I should certainly cut short an absence which would become unendurable to me. I should see you at last, Vicomte, and I should see you . . . how? . . . But remember this is nothing more than a conversation, a mere account of an impossible project, and I do not want to forget it alone . . .

Do you know I am a little anxious about my law-suit. I wanted to find out exactly what my rights are; my lawyers quote me a few laws and a great many "precedents" as they call them; but I do not see as much reason and justice in it. I am almost at the point of regretting that I did not agree to a compromise. However, I reassure myself by remembering that the solicitor is sharp, the barrister eloquent, and the client pretty. If these three methods do not succeed, the whole course of affairs would have to be changed, and what would become of respect for ancient customs!

This law-suit is now the only thing which keeps me here. Belleroche's is over; nonsuited, each side to pay its own costs. He is now regretting this evening's ball—it is indeed the regret of an idler! I shall return him complete liberty when I get back to town. I make him this painful sacrifice and console myself for it by the generosity he perceives in it!

Good-bye, Vicomte, write to me often; the recital of your pleasure will compensate me to some extent for the boredom I endure.

From the Château de . . . , 11th of November. 17—.

LETTER CXXXV

Madame de Tourvel to Madame de Rosemonde

I am trying to write to you without knowing if I shall be able to do so. Ah! Heaven, when I think that in my last letter it was the excess of my happiness which prevented my continuing it! And now it is the excess of my despair which overwhelms me, which leaves me no strength except to feel my pain, and takes away from me the strength to express it.

Valmont . . . Valmont loves me no more, he never loved me. Love does not disappear in this way. He deceives me, he betrays me, he outrages me. I endure all the misfortunes and humiliations which can be gathered together, and they come upon me from him.

And do not think this is a mere suspicion; I was so far from having any! I have not the happiness to be able to doubt. I saw him; what can he say to justify himself? . . . But what does it matter to him! He will not even attempt it . . . Wretch that I am! What will my tears and reproaches matter to him? What does he care for me! . . .

It is true that he has sacrificed me, surrendered me . . . and to whom? . . . a vile creature . . . But what am I saying? Ah! I have lost even the right to scorn her. She has betrayed fewer duties, she is less guilty than I. Oh! How painful is grief when it is founded on remorse! I feel my tortures increase. Farewell, my dear friend; however unworthy of your pity I have rendered myself, you will yet have some pity for me if you can form any idea of what I am suffering.

I have just re-read my letter, and I see that it tells you nothing; I will try to have the courage to relate this cruel event. It was yesterday; for the first time since my return I was to dine out. Valmont came to see me at five; never had he seemed so tender. He let me know that my plan of going out vexed him, and, as you may suppose, I soon decided to stay at home. However, suddenly, two hours later his manner and tone changed perceptibly. I do not know if anything escaped me which could offend him; but in any case, a little later he pretended to recollect an engagement which forced him to leave me and he went away; yet it was not without expressing very keen regrets which seemed to me tender and which I then thought sincere.

Left to myself, I thought it more polite not to avoid my first engagement since I was free to carry it out. I completed my toilet and entered my carriage. Unhappily my coachman took me by the Opera and I found myself in the confusion of the exit; a few steps in front of me in the next line I saw Valmont's carriage. My heart beat at once, but it was not from fear; and the one idea which filled me was the desire that my carriage would go forward. Instead of that, his was forced to retire, and stopped opposite mine. I leaned forward at once; what was my astonishment to see at his side a woman of ill-repute, well known as such! I leaned back, as you may suppose; this alone was enough to rend my heart; but, what you will scarcely believe, this same creature, apparently informed by an odious confidence, did not leave her carriage window, kept looking at me, and laughed so loudly it might have made a scene.

In the state of prostration I was in I allowed myself to be driven to the house where I was to sup, but it was impossible to remain there; every moment I felt ready to faint and I could not restrain my tears.

When I returned I wrote to M. de Valmont and sent him my letter immediately; he was not at home. Desirous, at any price, to emerge from this state of death or to have it confirmed for ever, I sent the man back with orders to wait, but my servant returned before midnight and told me that the coachman, who had come back, had said that his master would not be in that night. This morning I thought there was nothing to do but to ask for my letters once more and to request him never to come to my house again. I gave orders to that effect; but doubtless they were useless.

It is nearly midday; he has not yet arrived and I have not even received a word from him.

And now, my dear friend, I have nothing more to add; you know all about it and you know my heart. My one hope is that I shall not be here long to distress your tender friendship.

Paris, 15th of November, 17—.

LETTER CXXXVI

Madame de Tourvel to the Vicomte de Valmont

Doubtless, Monsieur, after what happened yesterday you will not expect to be received in my house again and, doubtless, you have no great wish to do so! The object of this note, then, is not so much to ask you not to come again as to request from you once more those letters which ought never to have existed; letters which, if they may have amused you for a moment as proofs of the delusion you created, cannot but be indifferent to you now that it is dissipated and now that they express nothing but a sentiment you have destroyed.

I recognise and confess I was wrong to give you a confidence by which so many other women before me have been victimised; in this I blame myself alone; but I thought at least I did not deserve from you that I should be abandoned to contempt and insult. I thought that in sacrificing myself to you, in losing for you alone my rights to the esteem of others and of myself, that I might expect not to be judged by you with more severity than by the public, whose opinion still recognises an immense difference between a weak woman and a depraved woman. These wrongs, which would be wrongs to anyone, are the only ones of which I speak. I am silent upon those of love; your heart would not understand mine. Farewell, Monsieur.

Paris, 15th of November, 17—.

LETTER CXXXVII

The Vicomte de Valmont to Madame de Tourvel

Your letter, Madame, has only just been handed to me; I shuddered on reading it, and it has left me scarcely strength to reply. What a dreadful idea of me you hold! Ah! No doubt I have done wrong, and such wrong that I shall never forgive myself for it in my life, even if you should hide it with your indulgence. But how far from my soul those wrongs you reproach me with have ever been! What I! I humiliate you! I degrade you! When I respect you as much as I cherish you, when my pride dates from the moment when you thought me worthy of you! Appearances have deceived you; and I confess they may have been against me; but why have you not in your heart that which should combat them? Why was it not revolted at the mere idea that it could have any reason to complain of mine? Yet you believed this! So you not only thought me capable of this atrocious madness, but you even feared you had made yourself liable to it by your favours to me. Ah! Since you feel yourself so degraded by your love I am then very vile in your eyes?

Oppressed by the painful feeling this idea causes me, I am wasting in resenting it the time I should employ in destroying it. I will confess all; but another consideration still restrains me. Must I then relate facts I wish to annihilate and fix your attention and mine upon a moment of error which I should like to atone for by the remainder of my life, whose cause I have yet to comprehend, whose memory must for ever bring me humiliation and despair? Ah! If by accusing myself I must excite your anger, you will at least not have to seek far for your vengeance; it will suffice you to leave me to my remorse.

Yet—who would believe it?—the first cause of this event is the all-powerful charm I feel in your presence. It was that which made me forget too long an important engagement which could not be postponed. When I left you it was too late and I did not find the person I was looking for. I hoped to find him at the Opera and this step was equally fruitless. Emilie, whom I found there, whom I knew at a time when I was far from knowing either you or love; Emilie had no carriage and asked me to take her to her home close at hand. I saw no objection and consented. But it was then that I met you and I felt at once that you would be tempted to think me guilty.

The fear of displeasing or distressing you is so powerful in me that it must have been, and was indeed, speedily noticed. I confess even that it made me request the woman not to show herself; this precaution of delicacy turned against my love. Accustomed, like all those of her condition, never to be certain of a power which is always usurped, except through the abuse they allow themselves to make of it, Emilie took care not to allow so striking an opportunity to escape. The more she saw my embarrassment increase the more pains she took to show herself; and her silly mirth—I blush to think that you could have believed for a moment that you were its object—was only caused by the cruel anxiety I felt, which was itself the result of my respect and my love.

So far I am certainly more unfortunate than guilty; and these wrongs "which would be wrongs to anyone and are the only ones of which you speak" cannot be blamed upon me since they do not exist. But it is in vain that you are silent upon the wrongs of love; I shall not keep the same silence about them; too great an interest compels me to break it.

In my shame at this inconceivable aberration I cannot recall its memory without extreme pain. I am deeply convinced of my errors and would consent to bear their punishment or to await their forgiveness from time, from my unending affection, and from my repentance. But how can I be silent when what I have to say is important to your sensibility?

Do not think I am looking for a roundabout way to excuse or palliate my error; I confess myself guilty. But I do not confess, I will never confess that this humiliating fault can be regarded as a wrong to love. Ah! What can there be in common between a surprisal of the senses, between a moment of self-forgetfulness soon followed by shame and regret, and a pure sentiment which can only be born in a delicate soul, can only be sustained there by esteem, and whose fruit is happiness! Ah! Do not profane love thus! Above all, fear to profane yourself by collecting under the same point of view what can never be confounded. Let vile and degraded women dread a rivalry they feel can be established in spite of them, let them endure the torments of a jealousy which is both cruel and humiliating; but you, turn your eyes away from objects which sully your gaze and, pure as the Divinity itself, punish, like it, the offence without resenting it.

But what punishment can you inflict upon me which could be more painful than that I feel, which can be compared with the regret of having displeased you, to the despair of having distressed you, to the crushing idea of having rendered myself less worthy of you? You are thinking how to punish me! And I ask you for consolation; not that I deserve it, but because it is necessary to me and can only come to me from you.

If, suddenly forgetting my love and yours and setting no value upon my happiness, you wish, on the contrary, to give me over to an eternal pain, you have the right to do so; strike; but if, being more indulgent or more tender, you remember still the sweet feelings which united our hearts—that pleasure of the soul, always reborn and always more deeply felt; those days, so sweet, so happy, which each of us owes to the other; all those treasures of love which love alone procures—perhaps you may prefer the power of recreating them to that of destroying them. What more shall I say? I have lost everything, and lost it through my own fault; but I can regain all by your benefaction. It is for you to decide now. I add only one word. Yesterday you swore to me that my happiness was certain as long as it depended on you! Ah! Madame, will you abandon me to-day to an eternal despair?

Paris, 15th of November, 17—.

LETTER CXXXVIII

The Vicomte de Valmont to the Maquise de Merteuil

I persist, my fair friend; no, I am not in love; and it is not my fault if circumstances force me to play the part. Only consent and come back; you will soon see for yourself that I am sincere. I gave proofs of it yesterday and they cannot be destroyed by what is happening to-day.

I went to see the tender prude and went without having any other engagement; for the Volanges girl, in spite of her age, was to spend the whole night at Madame V . . .'s precocious ball. At first, lack of occupation made me desire to prolong this evening with her and I had even, with this purpose, exacted a small sacrifice; but scarcely was it granted when the pleasure I promised myself was disturbed by the idea of this love which you persist in believing of me, or at least in accusing me of; to such an extent that I felt no other desire than that of being able at the same time to assure myself and to convince you that it was a pure calumny on your part.

I therefore adopted a violent course; and on some slight pretext I left my Beauty in great surprise and doubtless even more distress. For myself, I went off calmly to meet Emilie at the Opera; and she can inform you that no regret troubled our pleasures until we separated this morning.

Yet I had a fair enough cause for anxiety if my perfect indifference had not preserved me from it; for you must know that I was barely four houses from the Opera, with Emilie in my carriage, when the austere devotee's carriage came up exactly opposite mine and a block kept us for nearly ten minutes beside one another. We could see each other as plainly as at midday and there was no way to escape.

But that is not all; it occurred to me to confide to Emilie that this was the woman of the letter. (Perhaps you will remember that jest and that Emilie was the writing-table[1]). She had not forgotten it; she is a merry creature; and she had no rest until she had observed at her ease "that virtue", as she called it, and that with peals of laughter outrageous enough to provoke a temper.

That is still not all; did not the jealous creature send to my house that very evening? I was not there; but, in her obstinacy, she sent there a second time with orders to wait for me. As soon as I had decided to remain with Emilie I sent back my carriage with no orders to the coachman except to come for me there next morning; when he got back he found the messenger of love, and thought it quite simple to say that I was not returning that night. You can guess the effect of this news and that when I returned I found my dismissal expressed with all the dignity demanded by the situation.

So this adventure which you think interminable might have been ended this morning, as you see; if it is not ended the reason is not, as you will think, that I set any value on its continuance; the reason is that, on the one hand, I did not think it decent to allow myself to be deserted; and, on the other hand, that I wished to reserve the honour of this sacrifice for you.

I have therefore replied to the severe note by a long sentimental letter; I gave lengthy reasons and I relied on love to get them accepted as good. I have just received a second note, still very rigorous and still confirming the eternal breach, as was to be expected; but its tone was not the same. She will above all things not see me; this determination is announced four times in the most irrevocable manner. I concluded I ought not to lose a moment before presenting myself. I have already sent my servant to get hold of the door-porter, and in a moment I shall go myself to have my pardon signed; for in wrongs of this kind there is only one formula which gives a general absolution and that can only be obtained in person.

Good-bye, my charming friend; I am now going to attempt this great event.

Paris, 15th of November, 17—.

LETTER CXXXIX

Madame de Tourvel to Madame de Rosemonde

How I blame myself, my tender friend, for having spoken to you of my passing troubles too much and too soon! It is because of me that you are now in distress; the grief which came to you from me still lasts while I am happy. Yes, all is forgotten, forgiven; let me express it better, all is retrieved. Calm and bliss have succeeded grief and anguish. O joy of my heart, how shall I express you! Valmont is innocent—a man who loves so much cannot be guilty. He had not done me the heavy, offensive injuries for which I blamed him with such bitterness; and if I needed to be indulgent in one point, had I not my own injustices to repair?

I will not tell you in detail the facts or reasons which justify him; perhaps the mind would not thoroughly appreciate them; the heart alone is capable of feeling them. Yet if you suspect me of weakness, I shall appeal to your own judgment in support of mine. For men, you said yourself, infidelity is not inconstancy.

It is not that I do not feel that this distinction, which opinion authorises in vain, wounds my susceptibility; but how can mine complain when Valmont's susceptibility suffers even more? Do not think that he pardons or can console himself for this very fault which I forgive; and yet, how he has retrieved this little error by the excess of his love and of my happiness!

Either my felicity is greater or I am more conscious of its value since I feared I had lost it; in any case I can say that if I felt I had the strength to endure again distress as cruel as that I have just passed through, I should not think I was buying too dearly this increase of happiness I have since enjoyed. O! my tender mother, scold your inconsiderate daughter for having troubled you overmuch by her hastiness; scold her for having judged rashly and calumniated him whom she ought never to have ceased to adore; but as you recognise that she is imprudent, see that she is happy, and increase her joy by sharing it.

Paris, 16th of November, 17—, in the evening.

LETTER CXL

The Vicomte de Valmont to the Marquise de Merteuil

How does it happen, my fair friend, that I have had no answer from you? Yet it seems to me that my last letter deserved a reply and, though I ought to have received it three days ago, I am still waiting for it! I am vexed, to put it at the lowest; so I shall not talk to you about my important affairs at all.

That the reconciliation had its full result; that instead of reproaches and suspicion, it produced only new affection; that I it was who received excuses and compensation due to my mistrusted candour; of all this I will not say a word; and, but for the unforeseen occurrence of last night, I should not write to you at all. But as this concerns your pupil and as she will probably not be in a position herself to inform you of it at least for some time, I have undertaken this task.

For reasons which you will guess, or which you will not guess, Madame de Tourvel has not occupied me so much for the last few days, and as these reasons could not exist with the Volanges girl, I became more assiduous with her. Thanks to the obliging porter, I had no obstacle to overcome; and your pupil and I were leading a regular and convenient life. But habit leads to carelessness; the first days we could not take too many precautions for our safety; we trembled even behind bolts. Yesterday, an incredible carelessness caused the accident I am about to tell you; and if, for myself, I escaped with nothing but the fright, it cost the little girl more dearly.

We were not asleep, but we were in the repose and abandonment which follow pleasure, when we heard the door of the room suddenly open. I leaped for my sword at once, both for my own defence and for that of our common pupil; I advanced and saw nobody; but the door was actually open. Since we had a light I investigated and found no living soul. Then I remembered that we had forgotten our usual precautions; and doubtless the door had been only pushed to or partly closed and had opened of itself.

I returned to calm my timid companion, but did not find her in the bed; she had fallen or had hidden herself between it and the wall; in any case, she was lying there unconscious and with no movement but rather violent convulsions. Imagine my embarrassment! However, I succeeded in getting her back to bed and in bringing her back to consciousness; but she had hurt herself in her fall and very soon felt the results of it.

Pains in the loins, violent colics, still less uncertain symptoms soon enlightened me as to her condition; but to tell her what it was I had first to inform her of the state she was in before, for she did not suspect it. Never perhaps until now has a girl kept so much innocence while doing so effectually everything necessary to get rid of it! Oh! She is not one to waste her time in reflection!

But she wasted a lot of time in lamenting and I felt something had to be done. I finally agreed with her that I would go at once to the Physician and the Surgeon of the family, that I would tell them to come to her and at the same time confide everything to them under the seal of secrecy; that she would ring for her waiting-woman; that she would or would not take the woman into her confidence, as she chose; but that she would send for assistance and especially forbid that Madame de Volanges should be awakened—a delicate and natural consideration on the part of a daughter who did not wish to cause her mother anxiety.

I carried out my two errands and two confessions as quickly as I could and then went home and have not since gone out; but the Surgeon, whom I knew before, came to me at midday to give me news of the invalid. I was not wrong; but the hopes that if no accident happens, nothing will be noticed in the house. The waiting-woman is in the secret; the Physician has given some name to the illness; and this affair will be arranged like a hundred others, unless it is afterwards useful to us to have it talked about.

But is there still any common interest between you and me? Your silence might make me think so; I should even have ceased to believe there was, if my desire for it did not make me seek all means of retaining hope of it.

Good-bye, my fair friend; I kiss you, with a grudge.

Paris, 21st of November, 17—.

LETTER CXLI

The Marquise de Merteuil to the Vicomte de Valmont

Heavens, Vicomte, how you worry me with your obstinacy! What does my silence matter to you? Do you think that I keep silence because I lack reasons to defend myself? Ah! How much better it would be! But no, it is only that it is painful to me to tell them to you.

Tell me the truth; are you deluding yourself or are you trying to deceive me? The difference between your words and your actions leaves me no choice except between these two sentiments; which is the true one? What do you expect me to say to you when I do not know what to think myself?

You appear to make a great merit of your last scene with

Madame de Tourvel; but what does it prove for your system or against mine? Assuredly I never told you that you loved the woman enough not to be unfaithful to her, not to seize all opportunities which might appear pleasant or easy; I even felt that it would be almost the same to you to satisfy with another woman, the first who came to hand, the very desires which she alone could create; and I am not surprised that, from a libertinage of mind which it would be wrong to deny you, you should have once done from design what you have done on a thousand other occasions from opportunity. Who does not know that it is the mere run of the world and the custom of all of you from the scoundrel to the "creatures"![1] He who abstains from it to-day is considered romantic; and I do not think that is the fault I blame you for.

But what I said, what I thought, what I still think, is that you are none the less in love with your Madame de Tourvel; not indeed with a very pure or very tender love, but with the kind of love you can feel; the kind of love, for example, which makes you think a woman possesses the charms or qualities she does not possess, which puts her in a class apart and ranks all others in a second order, which holds you still to her even when you insult her; such a love in fact as I suppose a Sultan might feel for a favourite Sultana, which does not prevent his often preferring a mere odalisque to her. My comparison seems to me all the more accurate because, like a Sultan, you are never a woman's lover or friend, but always her tyrant or her slave. So I am quite sure you humiliated yourself, degraded yourself, to return to this fair creature's good graces! And you were but too happy to have succeeded, so that as soon as you thought the moment had come to obtain your forgiveness, you left me "for this great event".

Even in your last letter, if you do not speak to me wholly of this woman it is because you do not want to tell me anything of "your important affairs"; they seem to you so important that to be silent about them appears to you a kind of punishment to me. And after a thousand proofs of your decided preference for another, you ask me calmly if there is still "a common interest between you and me!" Be careful, Vicomte! If I once reply, my answer will be irrevocable; and to fear to do so now, is perhaps already to say too much. So I absolutely will not speak of it.

All I can do is to tell you a story. Perhaps you will not have time to read it or time to give it enough attention to understand it properly? It is for you to choose. At worse, it will only be a story wasted.

A man I know had entangled himself, like you, with a woman who did him very little honour. At intervals he had indeed the wit to see that sooner or later the adventure would do him harm; but although he blushed for it, he had not the courage to break away. His embarrassment was the greater because he had boasted to his friends that he was entirely free and he realised that one's ridiculousness increases in proportion as one denies it. Thus he passed his life, continually doing foolish things and continually saying afterwards: "It is not my fault." This man had a woman friend who was for a moment tempted to exhibit him to the public in this state of intoxication and thus to make his ridiculousness perpetual; but yet, more generous than malignant, or perhaps from some other motive, she wished to try one last means to be able in any event to say like her friend: "It is not my fault." She therefore sent him without any other remark the following letter, as a remedy whose application might be useful to his disease.

> One grows weary of everything, my angel, it is a law of Nature; it is not my fault.
>
> If therefore I am weary to-day of an adventure which has wholly preoccupied me for four mortal months, it is not my fault.
>
> If, for example, I had just as much love as you had virtue

> (and that is surely saying a lot) it is not astonishing that one should end at the same time as the other. It is not my fault.
> From this it follows that for some time I have been deceiving you; but then your pitiless affection forced me, as it were, to do so! It is not my fault.
> To-day, a woman I love madly insists that I sacrifice you to her. It is not my fault.
> I realise that this is a fine opportunity of crying out upon perjury; but if Nature has only given men assurance, while she gave women obstinacy, it is not my fault.
> Take my advice, choose another lover, as I have chosen another mistress. This is good advice, very good; if you think it bad, it is not my fault.
> Farewell, my angel, I took you with pleasure, I abandon you without regret; perhaps I shall come back to you. So goes the world. It is not my fault.

This is not the moment, Vicomte, to tell you the result of this last effort and what followed upon it; but I promise to tell you in my next letter. You will also find in it my "ultimatum" on the renewal of the treaty which you propose to me. Until then, good-bye and nothing more . . .

By the way, thank you for your details about the Volanges girl; it is a matter to reserve until the day after the marriage for the Scandalmonger's Gazette. Meanwhile, I send you my complimentary condolences on the loss of your posterity. Good night, Vicomte.

From the Château de . . . , 24th of November, 17—.

LETTER CXLII

The Vicomte de Valmont to the Marquise de Merteuil

Faith, my fair friend, I am not sure whether I have misread or misunderstood your letter, the story you relate and the little epistolary model which accompanied it. What I can tell you is that the letter seemed to me original and likely to make an effect; so I simply copied it out, and still more simply I sent it to the heavenly Madame de Tourvel. I did not waste a moment, for the tender missive was sent off yesterday evening. I preferred it thus, because first of all I had promised to write to her yesterday; and then too because I thought she would not have too much time if she took all night to meditate and consider "this great event", even if you should a second time reproach me for that expression.

I hoped to be able to send you this morning my beloved's answer; but it is nearly midday and I have not yet received anything. I shall wait until five o'clock; and if I have no news then I shall go for them in person; for especially in such proceedings it is only the first step which is troublesome.

As you may suppose, I am now very eager to learn the end of the story of the man you know who is so violently suspected of being unable to sacrifice a woman when necessary. Has he not reformed? And will his generous woman friend not show him some favour?

I am not less anxious to receive your "ultimatum", as you call it so politely! Above all I am curious to know whether you will still see love in this latest step of mine. Ah! No doubt, there is a great deal! But for whom? However, I do not mean to lay stress on anything, and I await everything from your goodness.

Farewell, my charming friend; I shall not close this letter until two o'clock in the hope that I may be able to enclose the desired reply.

Two o'clock in the afternoon.

Still nothing, and I have no time; I have not time to add a word; but this time will you still refuse the tenderest kisses of love?

Paris, 27th of November, 17—.

LETTER CXLIII

Madame de Tourvel to Madame de Rosemonde

The veil is torn, Madame, the veil upon which was painted the illusion of my happiness. The disastrous truth enlightens me and allows me to see nothing but a certain and near death, the path to which is laid between shame and remorse. I shall follow it . . . I shall cherish my tortures if they shorten my existence. I send you the letter I received yesterday; I will add no reflections upon it, they are carried in it. It is no longer the time to complain; there is nothing to do but suffer. It is not pity I need, but strength.

Receive, Madame, the only farewell I shall make and grant my last prayer; it is to leave me to my fate, to forget me entirely, to feel as if I were no longer on the earth. There is a limit in misery after which friendship itself increases our sufferings and cannot heal them. When wounds are mortal all aid becomes inhuman. Every feeling but that of despair is foreign to me. Nothing now can befit me save the profound night where I go to bury my shame. There I shall weep my errors, if I can still weep! Since yesterday I have not shed a tear. My broken heart grants me none.

Farewell, Madame. Do not reply to me. I have sworn upon that cruel letter never to receive another.

Paris, 27th of November, 17—.

LETTER CXLIV

The Vicomte de Valmont to the Marquise de Merteuil

Yesterday, at three o'clock in the afternoon, growing impatient at receiving no news, I presented myself at the house of the abandoned fair one; I was told she had gone out. In this phrase I only saw a refusal to receive me which neither vexed nor surprised me; and I left in the hope that this step would at least force so polite a woman to honour me with a word of reply. My desire to receive it made me return home expressly about nine o'clock, and I found nothing. Surprised by this silence, which I did not expect, I ordered my servant to collect information and to find out if the affectionate creature were dead or dying. At length, when I returned, he informed me that Madame de Tourvel had indeed gone out at eleven o'clock in the morning, that she had driven to the Convent of . . ., and that at seven o'clock in the evening she had sent back her carriage and her servants, with the message that she was not to be expected home. Certainly, she is doing the right thing. The convent is the proper refuge for a widow; and if she persists in so praiseworthy a resolution I shall add to all the obligations I already owe her that of the celebrity which will follow this adventure.

Some time ago I told you that, in spite of your anxieties, I should only reappear on the world's stage shining with a new lustre. Now let those malignant critics who accused me of a romantic and unlucky love show themselves, let them break off with more rapidity and brilliance; but no, let them do better, let them present themselves as consolers, the way is marked out for them. Well! Let them only attempt the course I have run through completely, and if one of them obtains the least success, I yield him first place. But they will all find that when I give my attention to it, the impression I leave is ineffaceable. Ah! This one will certainly be so; and I should count all my other triumphs as nothing, if ever I should have a rival preferred by this woman.

The course I have adopted flatters my self-love, I admit; but I am sorry she found sufficient strength to separate herself from me so completely. There will then be obstacles between us other than those I have placed there myself! What! If I wished to return to her, she would be in a position not to allow it; what am I saying? not to desire it, no longer to make it her supreme happiness! Is that the way to love? And do you think, my fair friend, that I ought to endure it? For example, am I not able, and would it not be better, to bring the woman back to the point of foreseeing the possibility of a reconciliation, which people always desire as long as they hope? I might attempt this without attaching any importance to it, and consequently without offending you. On the contrary! It would be a mere attempt which we would make together; and even if I should succeed, it would only be one way more of repeating at your will a sacrifice which seems to be pleasing to you. And now, my fair friend, I still have to receive the reward and all my wishes are for your return. Come back quickly and find once more your lover, your pleasures, your friends, and the regular course of adventures.

The Volanges girl's adventure has turned out excellently. Yesterday, when my anxiety did not allow me to stay anywhere, I went among other expeditions to call on Madame de Volanges. I found your pupil already in the drawing-room, still in her invalid's dress, but in full convalescence, and only the fresher and more attractive for it. In a similar case you women would remain a month on your sofas; faith, long live girls! Really she made me want to find out if her cure is complete.

I have still to tell you that the little girl's accident almost drove your "sentimental" Danceny mad. First, it was with grief; then, it was with joy. "His Cécile" was ill! You can guess how he lost his head in such a misfortune. Three times a day he sent for news and did not let a day pass without calling himself; finally he sent a handsome letter to the Mamma asking permission to come and congratulate her on the convalescence of so dear an object, and Madame de Volanges consented; so that I found the young man installed there as of old, except for a little familiarity which he dared not yet take.

It is from him that I learned these details; for I left at the same time as he did and I made him chatter. You have no idea of the effect this visit made upon him. It is impossible to describe his joy, his desires, his transports. I like these great emotions and made him lose his head entirely by assuring him that in a very few days I would enable him to see his fair one still more intimately.

In fact, I have decided to hand her back to him as soon as I have made my experiment. I want to devote myself entirely to you; and then would it be worth while for your pupil to be mine also if she were to be unfaithful only to her husband? The master-stroke is to be unfaithful to her lover and particularly to her first lover! For I cannot reproach myself with having uttered the word love to her.

Good-bye, my fair friend; return as quickly as possible to enjoy your power over me, to accept my submission, and to give me its reward.

Paris, 28th of November, 17—.

LETTER CXLV

The Marquise de Merteuil to the Vicomte de Valmont

Seriously, Vicomte, you have deserted Madame de Tourvel? You sent her the letter I composed for her! Really, you are charming, and you have surpassed my expectation! I admit freely that this triumph flatters me more than all those I have obtained up till now. You will perhaps think I value this woman very highly after having formerly rated her so low; not at all; I have not obtained this advantage over her, but over you; that is the amusing thing and it is really delicious.

Yes, Vicomte, you loved Madame de Tourvel very much and you still love her; you love her like a madman; but because I amused myself by making you ashamed of it, you have bravely sacrificed her. You would have sacrificed a thousand rather than endure one jest. Where vanity will take us! The wise man is indeed right when he says that it is the enemy of happiness.

Where would you be now if I had wanted to do more than play you a trick? But I am incapable of deceiving, as you well know; and even if you should reduce me in turn to despair and a convent, I will run the risk and yield to my conqueror.

Yet if I capitulate, it is from the merest weakness; for, if I wanted, how many cavils I should still have to raise! And perhaps you would deserve them? For example, I marvel at the skill or clumsiness with which you calmly propose that I should let you patch things up with Madame de Tourvel. It would suit you very well, would it not, to take the credit for breaking off without losing the pleasures of possession? And as this apparent sacrifice would no longer be one for you, you make me an offer to repeat it at my will! By this arrangement the heavenly devotee would still think herself the one choice of your heart, while I should pride myself upon being the preferred rival; we should both be deceived, but you would be pleased, and what matters the rest?

It is a pity that with so much talent for planning you have so little for execution; and that by one unconsidered step you have placed an invincible obstacle between you and what you most desire.

What! You think of making things up and you could have written that letter I sent! You must have thought me very clumsy in my turn! Ah! Believe me, Vicomte, when a woman strikes at another woman's heart, she rarely fails to find the sensitive place, and the wound is incurable. When I struck her, or rather when I guided your blows, I did not forget that this woman was my rival, that for a moment you had thought her preferable to me and, in short, that you had placed me beneath her. If I am deceived in my vengeance, I consent to put up with the mistake. So, I agree to your trying every means; I even invite you to do so and promise not to be angry at your success, if you manage to obtain any. I am so much at ease upon this matter, that I shall not concern myself with it further. Let us speak of other things.

For example, the Volanges girl's health. You will give me precise news of it on my return, will you not? I shall be very glad to have it. After that, it is for you to judge whether it suits you better to hand the little girl back to her lover, or to try to become a second time the founder of a new branch of the Valmonts, under the name of Gercourt. This idea seemed rather amusing to me and, while I leave the choice to you, I still request you not to make up your mind definitely until we have talked it over together. This is not putting you off for an indefinite time, because I shall be back in Paris immediately. I cannot tell you which day positively, but be certain that you shall be the first to be informed of my arrival.

Good-bye, Vicomte; in spite of my quarrels, my tricks, and my reproaches, I still love you very much and I am preparing to prove it to you. Until we meet, my friend.

From the Château de . . . , 29th of November, 17—.

LETTER CXLVI

The Marquise de Merteuil to the Chevalier Danceny

At last I am leaving here, my young friend, and tomorrow evening I shall be back in Paris. In the midst of all the confusion which a journey brings with it I shall receive no one. However, if you have any very urgent confidence to make me, I will except you from the general rule; but I shall except no one else; so I beg you to keep my arrival a secret. Even Valmont will not know it.

If anyone had told me a little time ago that you would soon have my complete confidence, I should not have believed it. But yours has impelled mine. I should be tempted to think that you had put adroitness into it, perhaps even seduction. That would be very wrong, at the least! But it would not be dangerous now; you have indeed other things to do! When the heroine is on the stage nobody cares about the confidante.

So you have had no time to inform me of your new successes? When your Cécile was away, the days were not long enough to listen to your tender complaints. You would have made them to the echoes, if I had not been there to hear them. When later she was ill, you still even honoured me with an account of your anxieties; you needed someone to tell them to. But now that she whom you love is in Paris, that she is well, and especially since you see her sometimes, she suffices for everything and your friends are now nothing to you.

I do not blame you; it is the fault of your youth. Do we not know that from Alcibiades down to you, young people have never wanted friendship except in their troubles? Happiness sometimes makes them indiscreet but never confiding. I will say with Socrates: "I like my friends to come to me when they are unhappy"; but as he was a philosopher, he could get on without them when they did not come. In that respect, I am not quite so wise as he, and I felt your silence with all a woman's weakness.

But do not think me exacting; I am far from being that! The same feeling which makes me notice these deprivations makes me endure them with courage, when they are the proof or the cause of my friend's happiness. I do not count upon seeing you tomorrow evening then, unless love leaves you free and unoccupied, and I forbid you to make me the least sacrifice.

Good-bye, Chevalier; it will be a great delight to me to see you again; will you come?

From the Château de . . . , 29th of December, 17—.

LETTER CXLVII

Madame de Volanges to Madame de Rosemonde

You will certainly be as distressed as I am, my excellent friend, when you hear of Madame de Tourvel's condition; she has been ill since yesterday; her illness came upon her suddenly and shows such grave symptoms that I am really alarmed.

A burning fever, a violent and almost continual delirium, a thirst which nothing can quench, are all that can be observed. The doctors say they cannot yet prognosticate anything and the treatment will be the more difficult because the invalid obstinately refuses any assistance whatever, to such an extent that she had to be held down by main force to be bled; and twice afterwards force has had to be used to replace her bandage which in her delirium she kept wanting to tear off.

You who saw her, as I did, so delicate, so timid, and so gentle, can you imagine that four persons could hardly restrain her and that she becomes inexpressibly furious if one attempts to point anything out to her? For my part, I am afraid it is something more than delirium and that it is a real mental alienation.

What happened yesterday increases my fears in this respect.

On that day she arrived at the Convent of . . . at eleven o'clock in the morning with her waiting-woman. Since she was brought up in that house and has the habit of going to it sometimes, she was received in the usual way, and appeared calm and well to everybody. About two hours later she enquired if the room she occupied as a schoolgirl was vacant, and when she was told it was, she asked to see it again; the Prioress accompanied her with several other nuns. She then declared that she had come back to live in this room which (she said) she ought never to have left; and she added that she would only leave it "for death"; that was her expression.

At first they did not know what to say; but after their first astonishment, they pointed out that as she was a married woman she could not be received there without a special permission. This reason and a thousand others had no effect on her; from that moment she persisted not only that she would not leave the convent but even her room. Finally, growing weary of the discussion, at seven o'clock in the evening they consented that she should spend the night there. They sent away her carriage and her servants and put off taking any steps until the next day.

I am assured that during the whole evening her air and behaviour had nothing disordered about them, both were composed and thoughtful; only, four or five times she fell into a reverie so profound that each time she could not be drawn out of it by being spoken to; and each time, before coming out of it, she lifted both hands to her head which she appeared to clasp tightly; whereupon one of the nuns who were present asked her if she were suffering in her head and she gazed at her a long time before answering, saying at last: "The pain is not there." A moment later she asked to be left alone and requested that she should be asked no more questions.

Everyone retired, except her waiting-woman, who luckily was to sleep in the same room, from lack of other accommodation.

According to the girl's account, her mistress was quite calm until eleven o'clock at night. She then said she wished to go to bed; but, before she was entirely undressed, she began to walk up and down her room with a great deal of motion and frequent gestures. Julie, who had been a witness of what had happened during the day, did not dare say anything to her, and waited in silence for nearly an hour. At last, Madame de Tourvel called twice in quick succession; she had barely time to rush forward when her mistress fell into her arms, saying: "I am exhausted." She let herself be guided to her bed, would not take anything nor allow any assistance to be sent for. She simply had water placed beside her and ordered Julie to go to bed.

Julie declares that she did not go to sleep until two o'clock and that during that time she heard neither movement nor complaints. But she says she was awakened at five o'clock by her mistress' speaking in a loud voice, that she then asked if she needed anything and, obtaining no reply, found a light and took it to Madame de Tourvel's bed but Madame de Tourvel did not recognise her and, suddenly breaking off the incoherent words she was saying, exclaimed: "Let me be left alone, let me be left in darkness; it is darkness which befits me." I have myself noticed to-day that she often repeats this phrase.

Julie made this apparent order an excuse to go for other people and help, but Madame de Tourvel refused both with the wildness and delirium which have returned to her so often since.

The difficulty in which the whole convent was placed by this determined the Prioress to send for me at seven o'clock in the morning . . . It was not yet daylight. I hastened thither at once. When I was announced to Madame de Tourvel, she seemed to regain consciousness and replied: "Ah! Yes, let her come in." But when I went up close to her bed, she gazed steadfastly at me, took my hand quickly and pressed it, and said to me in a clear but melancholy voice: "I am dying because I did not trust in you." Immediately afterwards she hid her eyes and returned to her most frequent expression: "Let me be left alone, &c"; and she lost all consciousness.

This remark she made me and some others which escaped her in delirium make me fear this cruel illness has a still more cruel cause. But we must respect our friend's secret and content ourselves with pitying her misfortune.

The whole of yesterday was equally stormy, and was divided between outbursts of terrifying delirium and moments of swooning exhaustion, the only moments when she takes any rest. I did not leave her bed-side until nine o'clock at night and I am returning there this morning to spend the day. I shall assuredly not abandon my unhappy friend; but what is distressing is her persistence in refusing all attention and help.

I send you last night's bulletin which I have just received; as you will see, it is the reverse of consoling. I shall be careful to send them all to you.

Good-bye, my excellent friend, I must now go to the invalid.

My daughter, who is happily nearly well again, presents her respects to you.

Paris, 29th of November, 17—.

LETTER CXLVIII

The Chevalier Danceny to the Marquise de Merteuil

O you whom I love! O you whom I adore! O you who began my happiness! O you who completed it! Sensitive friend, tender lover, why does the memory of your pain come to trouble the charm I feel? Ah! Madame, compose yourself, it is friendship that asks this of you. O my friend! Be happy, that is the prayer of love.

Ah! What reproaches have you to make yourself? Believe me, your susceptibility misleads you. The regrets it causes you, the faults of which it accuses me, are equally illusory; and I feel in my heart that there has been no other seducer between us but love. Fear no longer to give yourself up to the feelings which you inspire, to let yourself be imbued with all the passions you create. What! Can our hearts be less pure because they were late enlightened? No, indeed. On the contrary, it is seduction, which, never acting except from design, can organise its advance and its methods and foresee events from afar. But true love does not allow us to meditate and reflect thus; it diverts us from our thoughts by our feelings; its power is never stronger than when it is unknown; and it is in silence that love binds us by ties which it is equally impossible to perceive and to break.

Thus, even yesterday, in spite of the keen emotion caused me by the idea of your return, in spite of the extreme pleasure I felt in seeing you again, I still thought I was called and guided by friendship alone; or rather, I was so entirely given up to the soft feelings of my heart that I gave very little attention to distinguishing their origin or cause. Like me, my tender friend, you felt without realising it that despotic charm which delivered up our souls to the sweet influences of affection; and we both only recognised love when we emerged from the ecstasy into which God had plunged us.

But that alone justifies us instead of condemning us. No, you have not betrayed friendship, neither have I abused your trust. It is true we were both in ignorance of our feelings; but we experienced the illusion only, we did not try to create it. Ah! We should not complain of it, we should only think of the happiness it gives us; let us not disturb it by unjust reproaches, let us concern ourselves with increasing it even more by the charm of trust and confidence. O! My friend! How precious that hope is to my heart! Yes, henceforth freed from all fear and entirely given up to love, you will share my desires, my transports, the delirium of my senses, the ecstasy of my soul; and every moment of our happy days will be marked by a new pleasure.

Farewell, you whom I adore! I shall see you this evening, but shall I find you alone? I dare not hope it. Ah! You do not desire it as much as I.

Paris, 1st of December, 17—.

LETTER CXLIX

Madame de Volanges to Madame de Rosemonde

I hoped almost all day yesterday, my excellent friend, that I should be able to give you this morning more favourable news of our dear invalid's health; but after yesterday evening that hope has vanished and I have nothing left but the regret of losing it. An occurrence, quite unimportant in appearance, but very cruel from the results it has had, rendered the invalid's condition at least as unfavourable as it was before, if it has not made it worse.

I should have understood nothing of this sudden change, had I not yesterday received our unhappy friend's complete confidence. Since she did not conceal from me that you are also informed of all her misfortunes, I can speak to you without reserve about her sad situation.

Yesterday morning, when I reached the Convent, I was told that the invalid had been asleep for more than three hours, and her sleep was so profound and so calm that for a time they were afraid it was a swoon. Some time later she awoke and herself drew the bed-curtains. She looked at us with an air of surprise, and, as I got up to go to her, she recognised me, called me by name, and asked me to come to her. She left me no time to put any question to her, but asked me where she was, what she was doing there, if she were ill, and why she was not at home. At first I thought this was a new attack of delirium, of a calmer kind than the others; but I noticed that she quite understood my replies. She had indeed recovered her reason, but not her memory.

She questioned me in great detail about everything which had happened to her since she had been at the convent, to which she did not remember coming. I answered her truthfully, only suppressing what I thought might frighten her too much; and when in turn I asked her how she felt, she replied that she was not in any pain at the moment, but that she had been very harassed in her sleep and that she felt tired. I urged her to compose herself and not to speak much; after which, I partly closed the curtains, leaving them a little apart, and sat down at her bed-side. At the same time she was offered some soup, which she accepted and enjoyed.

She remained in this state for about half an hour, during which she only spoke to thank me for the care I had taken of her; and she put into her thanks the charm and grace which you will remember she had. After this, she remained absolutely silent for some time and then broke her silence by saying: "Ah! yes, I remember coming here", and a moment afterwards she exclaimed sorrowfully: "My friend, my friend, pity me; I have recovered all my miseries." As I then went towards her, she seized my hand, leaned her head against it, and went on: "Great Heaven! Why can I not die?" Her look, even more than her words, moved me to tears; she noticed them from my voice and said: "You pity me! Ah! If you knew!" . . . And then she interrupted herself: "Let us be left alone, I will tell you everything."

As I think I hinted to you, I had already some suspicions of what the subject of this confidence might be; and as I feared that this conversation (which I foresaw would be long and sad) might perhaps be harmful to our friend's condition, I refused at first, on the pretext that she needed rest; but she insisted and I yielded to her request. As soon as we were alone, she told me everything which you have heard from her, for which reason I shall not repeat it to you.

At last, when she was speaking to me of the cruel way in which she had been sacrificed, she added: "I thought I was quite certain to die of it, and I had the courage to do so; but what is impossible to me is to survive my misery and my shame." I tried to combat this discouragement or rather this despair with the weapons of religion, hitherto so powerful over her; but I soon felt I had not strength enough for these august functions, and I limited myself to proposing that Father Anselme should be sent for, since I know he has her entire confidence. She consented and even seemed to wish it greatly. He was sent for and came at once. He remained for a very long time with the invalid and, on coming out, said that if the doctors were of the same opinion he thought that the ceremony of the sacraments might be postponed, and that he would return the next day.

It was then about three o'clock in the afternoon, and until five our friend was fairly composed, to such an extent that we had all regained hope. Unfortunately, a letter was then brought to her. When they tried to give it to her, she said at first that she would receive none, and nobody insisted. But from that moment she appeared more agitated. Very soon afterwards she asked where this letter came from. It was not stamped. Who had brought it? Nobody knew. On whose behalf it had been delivered? The door-keepers had not been told. She then remained silent for some time; after which she began to speak, but her disconnected words only showed us that the delirium had returned.

However, there was again an interval of calm, until at length she asked to be given the letter which had been brought for her. As soon as she looked at it, she exclaimed: "From him! Great Heaven!" And then in a weaker voice: "Take it away, take it away." She had the bed-curtains closed at once and forbade anyone to approach; but almost immediately afterwards we were compelled to return to her. The delirium had returned with more violence than ever and it was accompanied by truly frightful convulsions. These symptoms did not cease all evening, and this morning's bulletin informs me that the night has been no less stormy. In short, she in such a condition that I am surprised she has not already succumbed to it, and I will not conceal from you that I have very little hope remaining.

I suppose this unfortunate letter is from M. de Valmont; but what can he dare say to her now? Forgive me, my dear friend, I restrain myself from any remarks; but it is very cruel to see perish so cruelly a woman until now so happy and so worthy of being so.

Paris, 2nd of December, 17—.

LETTER CL

The Chevalier Danceny to the Marquise de Merteuil

While awaiting the pleasure of seeing you, my tender friend, I abandon myself to the pleasure of writing to you; and I charm away the regret of being absent from you by occupying myself with you. It is a true delight to my heart to trace my feelings, to remember yours; and in this way the very time of deprivation still offers a thousand precious treasures to my love. Yet, if I am to believe you, I shall obtain no reply from you; this very letter will be the last and we shall deprive ourselves of a familiar intercourse which, in your opinion, is dangerous and "unnecessary to us". Assuredly I shall believe you, if you persist; for what is there you can wish that I should not also wish for that very reason? But before you decide finally, will you not allow us to discuss it together?

You alone must judge the matter of danger; I can calculate nothing and limit myself to begging you to take care of your safety, for I cannot be calm if you are anxious. In this respect, it is not we two who are one, it is you who are both of us.

It is not the same "about the necessity"; here we can only have the same thought; and if we differ in opinion, it can only be from lack of explaining ourselves or of understanding each other. This is what I think I feel.

Doubtless, a letter appears not very necessary when we can see each other freely. What can it say which a word, a look, or even silence do not express a hundred times better? This seemed so true to me that when you spoke to me of not writing any longer, the idea glided easily over my soul; perhaps the idea incommoded it, but did not make an effect on it. It was much the same as when I wish to set a kiss upon your heart and meet a ribbon or a gauze—I merely put it aside and yet do not feel there was an obstacle.

But since then, we have separated; and as soon as you were no longer present, the idea of letters returned to torment me. Why, I asked myself, this additional privation? What! Because we are apart, have we nothing more to say to each other? I will suppose that we are favoured by circumstances and that we spend a whole day together; must we take time to talk from that of enjoyment? Yes, of enjoyment, my tender friend; for beside you even moments of rest still provide delicious enjoyment. But at last, whatever the time may be, we end by separating; and then one is so lonely! It is then that a letter is so precious; if one does not read it, at least one looks at it . . . Ah! Certainly, one can look at a letter without reading it, as it seems to me that at night I should still have some pleasure in touching your portrait . . .

Your portrait, did I say? But a letter is the soul's portrait. It is not like a cold image, with its stagnation, so remote from love; it lends itself to all our emotions; turn by turn it grows animated, it enjoys, it rests . . . Your feelings are all so precious to me—will you deprive me of the means of collecting them?

Are you then certain that the necessity for writing will never torment you? If in solitude your heart dilates or is oppressed, if an emotion of joy passes to your soul, if an involuntary sadness should trouble it for a moment, would you not pour out your happiness or your grief in the bosom of your friend? Will you have a feeling which he does not share? Will you then leave him to wander dreamily and in solitude far from you? My friend . . . my tender friend! But it is for you to decide. I only wished to argue, not to seduce you; I have given you only reasons, I dare to think that I should have been more potent by using entreaties. If you persist, I shall try not to be afflicted; I shall make efforts to say to myself what you would have written me; but there! You would say it better than I and I should above all take more pleasure in hearing it.

Good-bye, my charming friend; the hour at last approaches when I can see you; I leave you hastily to come to meet you the sooner.

Paris, 3rd of December, 17—.

LETTER CLI

The Vicomte de Valmont to the Marquise de Merteuil

No doubt, Marquise, you do not think me so little experienced as to believe that I could be deceived about the private conversation

I found you in, and about "the extraordinary chance" which brought Danceny to your house! Your practised face was indeed able to take on marvellously the expression of calm and serenity and you betrayed yourself by none of those phrases which sometimes escape owing to disorder or regret. I even agree that your docile looks served you perfectly and that if they could have made themselves believed as easily as they made themselves understood, I should have been far from feeling or retaining the least suspicion and should not have doubted one moment the extreme annoyance caused you by "this importunate third party". But, if you do not want to display such great talents in vain, if you wish to obtain the success you count on, if you wish to produce the illusion you tried to create, you must first of all train your novice of a lover with more care.

Since you are beginning to take charge of educations, teach your pupils not to blush and grow disconcerted at the least pleasantry, not to deny so vehemently about one woman the same things they deny so languidly about all others. Teach them also to be able to hear their mistress praised without thinking themselves obliged to do the honours of it; and, if you allow them to gaze at you in company, let them at least know beforehand how to disguise that look of possession which is so easily recognised and which they so clumsily confuse with the gaze of love. Then you can let them appear in your public exercises without their conduct doing any harm to their modest schoolmistress; and I myself, but too happy to contribute towards your celebrity, promise you to make and publish the programmes of this new college.

But I must admit I am surprised that you should have tried to treat me like a schoolboy. Oh! How soon I should be avenged with any other woman! With what pleasure I should do so! And how easily it would surpass the pleasure she would think she had made me miss! Yes, it is indeed with you alone that I can prefer reparation to vengeance; and do not think I am restrained by the least doubt, by the least uncertainty; I know everything.

You have been four days in Paris; and each day you have seen Danceny and you have seen him alone. To-day even, your door was still shut; and to prevent my coming in upon you, your porter only lacked an assurance equal to your own. But you wrote me that I should be the first to be informed of your arrival, of this arrival whose date you could not tell me, though you wrote to me on the eve of your departure. Will you deny these facts or will you try to excuse yourself? Either is equally impossible; and yet I still restrain myself! See then your power; but, take my advice, be content with having tested it and do not abuse it any longer. We both know each other, Marquise; that ought to be enough for you.

You are going out all day tomorrow you said? Very good, if you do really go; and you will realise that I shall know it. But then, you will come back in the evening; and to effect our difficult reconciliation we shall not have too much time between then and the next morning. Let me know if we are to make our numerous and mutual atonements at your house or "at the other". Above all, no more Danceny. Your silly head was filled with the idea of him and it is possible for me not to be jealous of this delirium of your imagination; but remember that from this moment on, what was merely a fancy would become a marked preference. I do not think I was made for that humiliation and I do not expect to receive it from you.

I even hope that this sacrifice will not appear to be one to you. But even if it cost you something, I think I have given you quite a good example! I think a beautiful and sensitive woman, who only lived for me, who at this very moment is perhaps dying of love and regret, is well worth a schoolboy who (if you like) lacks neither face nor wit, but who so far has neither experience nor stability.

Good-bye, Marquise; I say nothing of my feelings for you.

All I can do at this moment is not to scrutinize my heart. I await your reply. When you make it, remember, remember carefully that the more easy it is for you to make me forget the offence you have given me, the more a refusal on your part, a mere delay, would engrave it upon my heart in permanent characters.
Paris, 3rd of December, 17—.

LETTER CLII

The Marquise de Mertcuil to the Vicomte de Valmont

Pray be careful, Vicomte, and treat my extreme timidity with more caution! How do you expect me to endure the crushing idea of incurring your indignation and, especially, not to succumb to the fear of your vengeance? The more so, since, as you know, if you do anything cruel to me it would be impossible for me to revenge it! Whatever I said, your existence would not be less brilliant or less peaceful. After all, what would you have to fear? To be obliged to leave the country—if you were given time to do so! But do not people live abroad as they do here? After all, provided the Court of France left you in peace at the foreign court you resided at, it would only mean for you that you had changed the scene of your triumphs. After this attempt to make you cool again with these moral considerations, let me come back to our affairs.

Do you know why I never remarried, Vicomte? It was certainly not because I could not find advantageous matches; it was solely because I would not allow anyone the right to criticize my actions. I am not even afraid of being unable to carry out my wishes, for I should always have achieved that at length; but it would have annoyed me if there had been anyone who had the right even to complain of them; in short, I wanted to deceive only for my pleasure, not from necessity. And then you write the most marital letter one could ever behold! You speak to me of nothing but wrongs on my side and forgiveness on yours! But how can one fail a person to whom one owes nothing? I cannot conceive how!

Come, what is it all about? You found Danceny at my house and it displeased you? Very well; but what could you deduce from this? Either that it was the result of chance, as I told you; or the result of my wish, as I did not tell you. In the first case, your letter is unjust; in the second, it is ridiculous; it was not worth the trouble of writing! But you are jealous and jealousy does not reason. Well! I will reason for you.

Either you have a rival or you have not. If you have one, you must please if you wish to be preferred to him; if you have not a rival, you must still please in order to avoid having one. In either case, you have the same line of conduct to follow; so, why torment yourself? Why, especially, torment me? Are you now unable to be the more charming of the two? Are you no longer sure of your successes? Come, Vicomte, you wrong yourself. But, it is not that; for your own sake I do not want you to give yourself so much trouble. You want my favours less than you want to abuse your power. Ah, you are ungrateful! There! I believe that is sentiment! And if I went on a little further this letter might become very tender; but you do not deserve it.

You do not deserve either that I should justify myself. To punish you for your suspicions, you shall keep them; so I shall say nothing about the date of my return or about Danceny's visits. You gave yourself a lot of trouble to obtain information about it, did you not? Well, are you any better off? I hope you found a great deal of pleasure in it; for my part, it did no harm to my pleasure.

All I can say in answer to your threatening letter, then, is that it has neither the gift of pleasing me nor the power of intimidating me; and that at the moment I could not be less disposed to grant your demands.

To accept you as you show yourself to-day would positively

be showing you a real infidelity. It would not be returning to my former lover; it would be taking a new one who is far from being worth the other. I have not so far forgotten the first as to deceive myself thus. The Valmont I loved was charming. I will even admit that I have not met a more delightful man. Ah! I beg you, Vicomte, to bring him to me if you meet him; he will always be well received.

But warn him that in any case it cannot be to-day or tomorrow. His twin-brother has done him a little harm; and if he pressed me too much I should be afraid of making a mistake; or perhaps I have made arrangements with Danceny for those two days? And your letter informs me that you do not joke when someone breaks his word. So you see what you have to expect.

But what does it matter to you? You can always avenge yourself on your rival. He will do nothing worse to your mistress than you do to his, and, after all, is not one woman worth as much as another? They are your own principles. Even she who is "tender and sensitive, who only lived for you, and who would die at last of love and regret" would none the less be sacrificed to the first fancy, to the fear of being laughed at for a moment; and you expect people to put themselves out for you! Ah! That is not just.

Good-bye, Vicomte; make yourself pleasant once more. Why, I ask nothing better than to think you charming; and as soon as I am sure of it, I promise to prove it to you. Positively I am too good.

Paris, 4th of December, 17—.

LETTER CLIII

The Vicomte de Valmont to the Marquise de Merteuil

I answer your letter at once; I shall try to make myself clear, which is not easy with you when once you have made up your mind to misunderstand.

Long phrases were not necessary to show that, since each of us possesses all that is needed to ruin the other; we have an equal interest in treating each other with mutual consideration; so, that is not the question. But between the violent course of ruining each other and the obviously better course of remaining united as we have been, or becoming still more so by renewing our former affair—between these two courses, I say, there are a thousand others which might be adopted. It was therefore not ridiculous to tell you and is not ridiculous to repeat to you that from this day on I shall be either your lover or your enemy.

I am perfectly aware that this choice incommodes you; that you would much prefer to evade it; and I am not ignorant that you never liked to be placed between yes and no; but you must be aware also that I cannot let you out of this narrow circle, without running the risk of being tricked; and you ought to have foreseen that I would not endure that. It is for you to decide now; I can leave you the choice, but not remain in uncertainty.

I only warn you that you will not impose upon me with your reasons, good or bad; nor will you seduce me any the more by the few flatteries with which you attempt to dress up your refusals; in fact the moment has come to be frank. I ask nothing better than to give you an example of frankness and I tell you gladly that I prefer peace and union, but if one or the other must be broken, I think I have the right and the means.

I add to this that the least obstacle presented on your part will be taken on my part as a real declaration of war; you see that the reply I ask from you needs neither long nor fine phrases. A word will do.

Paris, 4th of December, 17—.

The Marquise de Merteuil's reply, written at the end of the above letter:

Very well! War.

LETTER CLIV

Madame de Volanges to Madame de Rosemonde

The bulletins will inform you better than I can do, my dear friend, of our poor invalid's unhappy condition. I am entirely occupied with the attention I give her and only take time to write to you when there are other events besides those of the illness. Here is one, which I certainly did not expect. It is a letter I received from M. de Valmont, who has been pleased to choose me as a confident and even as a mediator between himself and Madame de Tourvel, for whom he enclosed a letter in the one to me. I sent back the one when I replied to the other. I send you the latter and I think you will be of my opinion, that I could not do and ought not to do what he asks me. Even if I had wanted to do it, our unfortunate friend is not in a state to understand me. Her delirium is continuous. But what do you say to this despair of M. de Valmont's? Is one to believe in it, or does he only want to deceive everyone and that until the end?[1] If he is sincere this time, he may well tell himself that he has made his own happiness. I think he will not be very pleased with my reply; but I confess that everything which fixes my attention on this unhappy adventure makes me more and more indignant with the author of it.

Good-bye, my dear friend; I am now going back to my sad duties, which become far more so from the little hope I have of their proving successful. You know my feelings for you.

Paris, 5th of December, 17—.

LETTER CLV

The Vicomte de Valmont to the Chevalier Danceny

I have twice been to your house, my dear Chevalier; but since you have abandoned your role of lover for that of a ladies' man it has become impossible to find you, as one might expect. Your man-servant tells me, however, that you come home at night; that he had orders to wait for you; but I, who know your projects, know very well that you only return for a moment, to adopt the custom of the thing, and that you immediately start again on your victorious career. Well and good; I can do nothing but applaud; but perhaps this evening you may be tempted to change its direction. You only know half your affairs yet; I must tell you about the other half and then you will decide. But take the time to read my letter. It will not distract you from your pleasures, since on the contrary its only object is to give you a choice between them.

If I had had your complete confidence, if I had learned from you that part of your secrets which you have left me to guess, I should have been informed in time; my zeal would have been less clumsy and would not impede your progress to-day. But let us start from the point where we are. Whatever course you adopt, the worst you do will make someone else happy.

You have a rendez-vous for to-night, have you not? With a charming woman whom you adore? For at your age, what woman does one not adore, at least for the first week! The setting of the scene will add even more to your pleasures. A delicious "little house", "which has only been taken for you" will embellish the pleasure with the charms of liberty and of mystery. All is agreed; you are expected; and you are burning to go! That is what we both know, although you have told me nothing. Now, here is what you do not know, which I must tell you.

Since my return to Paris I have busied myself with the means of bringing you and Mademoiselle de Volanges together; I had promised it to you; and the very last time I spoke to you of it I had reason to think from your replies, I might say from your ecstasies, that I was busying myself with your happiness. I could not succeed unaided in this rather difficult enterprise; but after having prepared the means, I left the rest to the zeal of your young mistress. In her love she has found resources which were lacking to my experience; in short, your misfortune has willed it that she should succeed. She told me this evening that for the last two days all obstacles have been overcome and that your happiness only depends upon yourself.

For the last two days also she has been flattering herself that she would give you this news herself, and, in spite of her Mamma's absence, you would have been received; but you did not even call! And to tell you everything, the little person (either from caprice or reason) seemed to me a little annoyed by this lack of eagerness on your part. At all events, she succeeded in bringing me to her and made me promise to give you as soon as possible the letter I enclose with this. From her eagerness I would wager that it is concerned with a rendez-vous for this evening. However that may be, I promised upon honour and friendship that you should have the tender missive to-day and I cannot and will not break my word.

And now, young man, what are you going to do? You are placed between coquetry and love, between pleasure and happiness; what will your choice be? If I were speaking to the Danceny of three months ago, even to the Danceny of a week ago, I should be certain of his heart and of what he would do; but the Danceny of to-day is snatched at by women, runs after adventures, and has become, as always happens, something of a scoundrel; will he prefer a very timid girl who has nothing on her side but her beauty, her innocence and her love to the charms of a perfectly "accustomed woman?"

For my part, my dear fellow, it seems to me that even with your new principles (which I confess are also mine to some extent) circumstances would make me decide for the young mistress. First, it is one more, and then there is the novelty and again the fear of losing the fruit of your exertions by neglecting to gather it; for, in this respect, it would really be a lost opportunity, especially for a first weakness; in such a case, it often needs only a moment of ill-humour, a jealous suspicion, even less, to frustrate the fairest triumph. Drowning virtue sometimes clutches at a straw and, once it has escaped, is on its guard and is not easily surprised.

What do you risk on the other side? Not even a breaking off; at most a quarrel where the pleasure of a reconciliation is purchased by a few attentions. What course can be taken by a woman who has already yielded, except that of indulgence? What would she gain by severity? The loss of her pleasures, with no profit to her pride.

If, as I suppose, you adopt the course of love (which also seems to me that or reason) I think it would be prudent not to send excuses about the missed rendez-vous; let her simply wait for you; if you try to give a reason, she may be tempted to verify it. Women are curious and obstinate; everything might be discovered; as you know, I have just been an example of this myself. But if you leave hope, which will be supported by vanity, it will not be lost until long after the time for enquiries has past; then tomorrow you can choose the insurmountable obstacle which detained you; you can have been ill, dead if necessary, or anything else which might have reduced you to despair, and everything will be made up.

For the rest, whichever way you decide, I only ask you to let me know; and as I have no interest in the matter, I shall approve whatever you do. Good-bye, my dear friend.

I must just add that I regret Madame de Tourvel; I am in despair at being separated from her; I would pay with one half my life the happiness of devoting the other half to her. Ah! Believe me, we are only happy through love.

Paris, 5th of December, 17—.

LETTER CLVI

Cécile Volanges to the Chevalier Danceny

(Enclosed with the preceding letter)

How does it happen, my dear, that I have ceased to see you at a time when I have not ceased to desire it? Do you not want it any more as much as I? Ah! Now I am very sad! Sadder then when we were quite separated. The grief which I endured through others now comes to me from you, and that is much worse.

For some days Mamma has been out, as you know; and I hoped you would try to profit by this time of freedom; but you do not even think of me; I am very miserable! You told me so often that I did not love as much as you! I knew to the contrary, and this is the proof of it. If you had come to see me, you would have seen me; for I am not like you, I only think of what will unite us. You do not deserve that I should tell you anything of what I have arranged for it, which I took so much trouble about; but I love you too much and I want to see you too much to be able to prevent myself from telling you. And then, I shall see afterwards if you really love me!

I have so arranged it that the porter is on our side and he has promised me that every time you come, he will let you come in as if he did not see you; and we can rely on him, for he is a very honest man. The only other difficulty is that you must not let yourself be seen in the house; and that will be quite easy if you only come in the evening when there will be nothing at all to fear. For example, since Mamma has been away all day she has been going to bed at eleven o'clock each evening; so we should have plenty of time.

The porter says that when you want to come in this way, instead of knocking at the door, you must just tap on his window and he will let you in at once; you will easily find the little stairway and, as you cannot have any light, I will leave my bedroom door ajar which will give you a little light. You must be very careful not to make a noise, especially when passing the side door into Mamma's room. It does not matter about my waiting-woman's door, because she has promised me she will not wake up; and she is a very good girl! And it will be the same when you leave. Now, we shall see whether you will come.

Heaven, why does my heart beat so fast as I write to you! Will some misfortune fall on me, or is it the hope of seeing you which so upsets me! What I do feel is that I never loved you so much and never desired so much to tell you so. Come, my friend, my dear friend; let me repeat to you a hundred times that I love you, that I adore you, that I shall never love anyone but you.

I found a way to let M. de Valmont know I had something to say to him; and, as he is a very good friend, he will surely come tomorrow and I shall beg him to give you this letter at once. So I shall expect you tomorrow evening and you will not fail to come unless you want your Cécile to be very miserable . . .

Good-bye, my dear; I kiss you with all my heart.

Paris, 4th of December, 17—. in the evening.

LETTER CLVII

The Chevalier Danceny to the Vicomte de Valmont

Do not doubt either my heart or my actions, my dear Vicomte; how could I resist a desire of my Cécile's? Ah! It is she, she only, whom I love, whom I shall love for ever! Her ingenuousness, her affection, have a charm for me, a charm from which I may have been weak enough to allow myself to be distracted, but which nothing will ever efface. I was engaged in another adventure, as it were without realising it, and often the memory of Cécile has troubled me in the most delightful pleasures; and perhaps my heart never paid her truer tribute than at the very moment when I was unfaithful to her. But let us spare her susceptibility, my friend, and hide my errors from her; not to deceive her, but to save her from distress. Cécile's happiness is my most ardent desire; I should never forgive myself for a fault which cost her a tear.

I know I deserved your jest about what you call my new principles; but you may believe me, I am not guided by them at this moment; and tomorrow I am determined to prove it. I shall go and confess to her who caused my error and who has shared it; I shall say to her: "Read in my heart; it has the most tender friendship for you; friendship united with desire is so like love! . . . We have both been deceived; but though I am liable to error I am incapable of bad faith." I know this woman friend of mine; she is as honourable as she is indulgent; she will do more than forgive me, she will approve of what I do. She has often reproached herself for betraying friendship; her delicacy often frightened her love; she is wiser than I and she will strengthen in my soul these useful fears which I rashly attempted to suppress in hers. I shall owe it to her that I am better, as I shall owe it to you that I am happier. O! my friend, share my gratitude! The idea that I owe my happiness to you increases its value.

Good-bye, my dear Vicomte. The excess of my joy does not prevent me from thinking of your troubles and sharing them. Why can I not be of use to you! Does Madame de Tourvel remain inexorable? I hear too she is very ill. Heavens, how I pity you! May she regain both health and indulgence and make you happy for ever! These are the wishes of friendship; I dare to hope they will be granted by love.

I should like to talk longer with you; but time presses and perhaps Cécile is waiting for me already.

Paris, 5th of December, 17—.

LETTER CLVIII

The Vicomte de Valmont to the Marquise de Merteuil

(First thing in the morning)

Well, Marquise, how are you after last night's pleasures? Are you not a little tired? Confess now, Danceny is charming! The young man performs prodigies! You did not expect that of him, did you? Come now, I estimate myself justly, such a rival deserved that I should be sacrificed to him. Seriously, he has many good qualities! But, especially, how much love, constancy, delicacy! Ah! If you are ever loved by him as his Cécile is, you will have no rivals to fear; he proved it to you last night. Perhaps another woman by dint of coquetry might take him away from you for a moment—a young man never knows how to resist provocative advances—but, as you see, one word from the beloved person is sufficient to dissipate this illusion; so you have only to become that person to be perfectly happy.

Certainly, you will not make any mistake in the matter; you have too sure a tact for there to be any fear of this. But the friendship which unites us, as sincere on my part as it is well recompensed on yours, made me desire last night's proof for your sake; it is the work of my zeal; it succeeded; but, pray, no thanks, it is not worth the trouble; nothing could have been easier.

After all, what did it cost me? A slight sacrifice and a little skill. I consented to share with the young man his mistress's favours; but then he had quite as much right to them as I, and I cared so little for them! The letter sent him by the young person was dictated by me; but it was only to gain time, since we had a better use for it. As to the letter I sent with it—oh! it was nothing, practically nothing; a few friendly ideas to guide the new lover's choice; but upon honour, they were unnecessary; to tell you the truth, he did not hesitate a moment.

And then, in his candour, he is to come to you to-day to tell you everything; and that will certainly be a great pleasure for you! He will say to you: "Read in my heart"; he writes me this; and, as you see, that patches everything up. I hope that when you read what he wants in his heart you will perhaps also read that such young lovers have their dangers; and in addition that it is better to have me for a friend than an enemy.

Good-bye, Marquise, until the next occasion.

Paris, 6th of December, 17—.

LETTER CLIX

The Marquise de Merteuil to the Vicomte de Valmont

(A note)

I do not like people who add paltry jokes to paltry proceedings; it is neither my way nor to my taste. When I have reason to complain of someone, I do not jest; I do something better; I avenge myself. However pleased with yourself you may be now, do not forget that this is not the first time you have congratulated yourself too soon and alone in the hope of a triumph which escapes you at the very moment you rejoice over it. Good-bye.

Paris, 6th of December, 17—.

LETTER CLX

Madame de Volanges to Madame de Rosemonde

I am writing to you from the bedroom of your unfortunate friend, whose condition is still about the same. This afternoon there is to be a consultation of four doctors. Unhappily, as you know, this is more often a proof of danger than a means of aid.

It appears, however, that she recovered partial consciousness last night. The waiting-woman told me this morning that about midnight her mistress called her, desired to be left alone with her and dictated quite a long letter. Julie added that, while she was making the envelope, Madame de Tourvel's delirium returned, so that the girl did not learn to whom the letter was to be addressed. I was surprised at first that the letter itself was not enough to tell her; but on her replying that she was afraid of making a mistake and that her mistress had ordered her to send it at once, I took it upon myself to open the letter.

I found the enclosed writing, which indeed is not addresed to anyone but to too many. I think, however, that our unhappy friend at first wished to write to M. de Valmont; and that, without realising it, she yielded to the disorder of her ideas. However this may be, I felt that the letter ought not to be sent to anyone. I am forwarding it to you because you will see from it, better than I could tell you, what are the thoughts which occupy our invalid's head. As long as she remains so deeply affected I shall have hardly any hope. It is difficult for the body to regain health when the mind is so disordered,

Good-bye, my dear and excellent friend. I am glad that you are far away from the sad spectacle I have continually before my eyes.

Paris, 6th of December, 17—.

LETTER CLXI

Madame de Tourvel to . . .

(Dictated by her and written by her waiting-woman)

Cruel and malevolent being, will you never grow weary of persecuting me? Is it not enough for you that you have tormented me, degraded me, debased me, that you wish to ravish from me even the peace of the grave? What! In this dwelling place of darkness in which I have been forced by ignominy to bury myself, is pain without cessation, is hope unknown? I do not implore a mercy I do not deserve; I will suffer without complaint if my sufferings do not exceed my strength. But do not make my tortures unendurable. Leave me my grief, but take from me the cruel memory of the treasures I have lost. When it is you who ravished them from me, do not again draw their agonising image before my eyes. I was innocent and at peace; it is because I saw you that I have lost my peace of mind; it is by listening to you that I became criminal. You are the author of my sins; what right have you to punish them?

Where are the friends who cherished me, where are they? My misfortune terrifies them. None dares to approach me. I am crushed and they leave me without aid! I am dying and none weeps for me. All consolation is refused me. Pity stays on the brink of the gulf into which the criminal plunges. He is torn by remorse and his cries are not heard!

And you, whom I have outraged; you, whose esteem adds to my torture; you, who alone have the right to avenge yourself, what are you doing so far from me? Come, punish a faithless wife. Let me suffer deserved torments at last. Already I should have submitted to your vengeance, but courage failed me to confess my shame to you. It was not dissimulation, it was shame. At least may this letter tell you my repentance. Heaven took up your cause and avenges you for an injury you did not know. It is Heaven which bound my tongue and restrained my words; it feared you might have pardoned a fault it wished to punish. It has removed me from your indulgence which would have wounded its justice.

Pitiless in its vengeance, it has delivered me up to him who ruined me. It is at once through him and by him that I suffer. I try to fly him, in vain, he follows me; he is there; he besets me continually. But how different he is from himself! His eyes only express hatred and scorn. His mouth only utters insult and blame. His arms embrace me only to rend me. Who will save me from his barbarous fury?

But what! It is he . . . I am not deceived; I see him again. O! My charming love! Receive me into your arms; hide me in your bosom; yes, it is you, it is indeed you! What disastrous illusion made me mistake you? How I have suffered in your absence! Let us not separate again, let us never be separated. Let me breathe. Feel my heart, feel how it beats! Ah! It is no longer fear, it is the sweet emotion of love. Why do you refuse my tender caresses? Turn that soft gaze upon me! What are those bonds you try to break? Why do you prepare that equipment of death? What can have so altered those features? What are you doing? Leave me; I shudder! God! It is that monster again! My friends, do not abandon me. You who call upon me to fly, help me to combat him; and you, more indulgent, you who promised to lessen my pain, come nearer to me. Where are you both? If I am not allowed to see you again, at least answer this letter; let me know that you still love me.

Leave me, cruel one! What new frenzy animates you? Are you afraid some gentle sentiment might pierce to my soul? You redouble my tortures; you force me to hate you. Oh! How painful hate is! How it corrodes the heart which distills it! Why do you persecute me? What more can you have to say to me? Have you not made it impossible for me to listen to you, impossible to reply to you? Expect nothing more of me. Farewell, Monsieur.

Paris, 5th of December, 17—.

LETTER CLXII

The Chevalier Danceny to the Vicomte de Valmont

I am informed, Monsieur, of your conduct towards me. I know as well that, not content with basely duping me, you have not feared to boast of it, to congratulate yourself on it. I have seen the proof of your betrayal written by your own hand. I confess it broke my heart and that I felt some shame at having contributed so much myself to your odious abuse of my blind confidence; yet I do not envy you this shameful advantage; I am only curious to know if you will keep all other advantages over me. I shall learn this, if, as I hope, you will be good enough to meet me tomorrow morning, between eight and nine o'clock, at the Bois de Vincennes gate; Village of Saint-Mandé. I shall take care to bring there everything necessary for the explanations I yet have to make with you.

The Chevalier Danceny.

Paris, 6th of December, 17—, in the evening.

LETTER CLXIII

M. Bertrand to Madame de Rosemonde

Madame,

It is with great regret that I carry out the sad duty of announcing to you an event which will cause you such cruel grief. Allow me first of all to exhort you to that pious resignation which everyone has so often admired in you and which alone can enable us to endure the ills which strew our miserable life.

Your nephew . . . Heaven! Must I so afflict so respectable a lady! Your nephew has had the misfortune to die in a duel he had this morning with the Chevalier Danceny. I am entirely ignorant of the cause of the quarrel; but it appears, from the note I found in the Vicomte's pocket, which I have the honour to send you; it appears, I say, that he was not the aggressor. And it was he whom Heaven allowed to die!

I was waiting for M. le Vicomte at his house at the very time when he was brought home. Imagine my terror at seeing your nephew carried in by two of his servants, soaked in his own blood. He had two sword wounds in his body and was already very weak. M. Danceny was there too and was even weeping. Ah! No doubt he must weep; but it is too late to shed tears when one has caused an irreparable misfortune!

I lost control of myself; and in spite of my humble rank I told him what I thought of him. But it was there that M. le Vicomte showed himself truly great. He ordered me to be silent; and he took the hand of his murderer, called him his friend, embraced him in our presence and said to us: "I order you to treat this gentleman with all the respect due to a brave and gallant man." Moreover, in my presence, he handed to him a voluminous mass of papers with which I am not acquainted but to which I know he attached great importance. Then he desired that they should be left alone for a time. Meanwhile I sent for aid, spiritual and temporal; but alas! the ill was beyond remedy. Less than half an hour afterwards M. le Vicomte lost consciousness. He could receive nothing but the extreme unction; and the ceremony was barely over when he breathed his last sigh.

Ah Heavens! When I received into my arms at his birth this precious support of so illustrious a family, could I foresee that he would die in my arms, and that I should have to weep his death? A death so early and so unhappy! My tears flow in spite of me; I beg your pardon, Madame, for daring thus to mingle my grief with yours; but in all ranks we have a heart and sensibility; and I should be very ungrateful if I did not mourn all my life a Lord who was so kind to me, who honoured me with so much confidence.

Tomorrow, after the body is taken away, I shall have seals put on everything and you can rely entirely upon my services. You know, Madame, that this unfortunate event puts an end to the entail and leaves your disposal entirely free. If I can be of any use to you, I beg you will be good enough to send me your orders; I will give all my attention to carrying them out punctually.

I am with the deepest respect, Madame, your most humble, etc., etc.

Bertrand

Paris, 7th of December, 17—.

LETTER CLXIV

Madame de Rosemonde to M. Bertrand

I have just received your letter, my dear Bertrand, and learn from it the terrible event of which my nephew has been the unhappy victim. Yes, certainly I have orders to give you and it is only with them that I concern myself with anything but my mortal grief.

M. Danceny's note, which you sent me, is a very convincing proof that he provoked the duel, and it is my intention that you should lodge a complaint against him in my name. By forgiving his enemy, his murderer, my nephew satisfied his natural generosity; but I must avenge at once his death, humanity, and religion. The severity of the laws cannot be too much excited against this remnant of barbarity which still infects our customs and I do not think that this is a case when we are commanded to forgive injuries. I expect you to carry out this business with all the zeal and activity of which I know you capable and which you owe to the memory of my nephew.

Before everything else, you will be careful to see M. le Président de . . .[1] on my behalf and to confer with him. I am not writing to him, I am in haste to give myself up entirely to my grief. You will make my excuses to him and show him this letter.

Farewell, my dear Bertrand; I praise you and thank you for your right feelings and, for life, I am yours, etc.

From the Château de . . . , 8th of December, 17—.

LETTER CLXV

Madame de Volanges to Madame de Rosemonde

I know you are informed, my dear and excellent friend, of your recent loss; I knew your affection for M. de Valmont and I share very sincerely the afflication you must feel. I am truly pained to have to add new regrets to those you already feel; but alas!

there is nothing left but tears for you to give our unfortunate friend. We lost her yesterday at eleven o'clock in the evening. By the fatality attached to her lot which seemed to mock at all human prudence, the short time she survived M. de Valmont was long enough for her to learn of his death; and, as she said herself, to succumb under the weight of her misfortunes only when the measure was filled.

You have already heard that for the last two days she was absolutely unconscious; yesterday morning, when her doctor came, and we both went over to her bed, she recognised neither of us and we could not obtain a word or the least sign from her. Well, we had scarcely returned to the fire-place where the doctor was telling me of the sad happening of M. de Valmont's death, when the unfortunate woman regained her senses, either because nature alone produced this change or because the repetition of the words "M. de Valmont" and "death" may have recalled to the invalid the only ideas which have so long occupied her.

However this may be, she suddenly opened her bed-curtains, exclaiming: "What! What is it you say? M. de Valmont is dead?" I hoped to make her think she was mistaken and I assured her at first that she had misunderstood us; but far from being convinced thus, she compelled the doctor to repeat the cruel announcement and when I tried once more to dissuade her, she called me to her and said in a low voice: "Why try to deceive me? Was he not already dead for me!" I was forced to yield then.

Our unhappy friend listened to the account at first calmly enough, but very soon she interrupted, saying: "Enough, I know enough." She immediately requested that her curtains might be closed; and when the doctor then tried to give her his professional services, she would never allow him to come near her.

As soon as he had gone, she sent away her nurse and her waiting-woman as well; and when we were alone together she asked me to help her to get upon her knees in bed and to support her. There she remained for some time in silence with no other expression than that of her tears which flowed abundantly. At last, clasping her hands and lifting them to Heaven, she said in a weak but fervent voice: "Almighty God, I submit myself to your justice; but pardon Valmont. May my misfortunes, which I know I deserved, not be blamed upon him and I will bless your mercy!" I have allowed myself, my dear and excellent friend, to dwell upon these details concerning a matter which I know must renew and aggravate your grief, because I have no doubt that Madame de Tourvel's prayer will bring great consolation to your spirit.

After our friend had spoken these few words, she fell back in my arms; she was scarcely back in bed when she fell into a long swoon, which yielded at length to the usual remedies. As soon as she regained consciousness, she asked me to send for Father Anselme, and added: "He is now the only doctor I need; I feel that my woes will soon be over." She complained greatly of a feeling of oppression and spoke with difficulty.

A little time after she sent me by her waiting-woman a casket (which I send to you) which she said contained papers of hers, charging me to send them to you immediately after her death.[1] Afterwards she spoke to me of you and of your friendship for her, as much as her condition allowed her and with great tenderness.

Father Anselme arrived about four o'clock and remained nearly an hour alone with her. When we returned, the invalid's face was calm and serene; but it was easy to see that Father Anselme had wept a great deal. He remained to be present at the last ceremonies of the Church. This spectacle, always so imposing and painful, became more so from the contrast between the sick person's calm resignation and the profound grief of her venerable confessor, who burst into tears beside her. The emotion became general; and she whom we all wept for was the one person who did not weep.

The remainder of the day was spent in the customary prayers which were only interrupted by the invalid's frequent swoons. Finally, about eleven o'clock at night, she seemed to me more oppressed and in greater pain. I put out my hand to find her arm; she still had strength enough to take it and put it on her heart. I could not feel it beating; and in fact our unhappy friend expired at that very moment.

You remember, my dear friend, that on your last journey here less than a year ago, we were talking of a few persons whose happiness seemed more or less assured and we dwelt with complaisance on the fate of this same woman whose misfortunes and death we weep to-day! So many virtues, praiseworthy qualities and charms; so gentle and so easy a character; a husband whom she loved and by whom she was adored; a society which pleased her and which she delighted; good looks, youth, wealth; so many advantages united have been lost through a single imprudence! O Providence! Doubtless we must adore your decrees, but how incomprehensible they are! I restrain myself, I am afraid of increasing your sadness by yielding to my own.

I must leave you and go to my daughter who is a little indisposed. When she heard this morning of the sudden death of these two persons of her acquaintance, she became unwell and I sent her to bed. But I hope this slight indisposition will have no evil results. At her age, they are un-accustomed to grief and its impression is sharper and deeper. No doubt this active sensibility is a praiseworthy quality, but how much one learns to dread it after what one sees every day! Good-bye, my dear and excellent friend.

Paris, 9th of December, 17—.

LETTER CLXVI

M. Bertrand to Madame de Rosemonde

Madame,

In consequence of the orders you did me the honour of sending me, I had the honour to see M. le Président de . . ., and

I handed him your letter, informing him at the same time that, in accordance with your orders, I should do nothing except upon his advice. That respectable Magistrate commands me to inform you that the complaint you intend to lodge against M. le Chevalier Danceny would equally compromise the memory of your nephew and that his honour would inevitably be stained by the sentence of the Court, which undoubtedly would be a great misfortune. His opinion is that no step whatever should be taken; or, if one must be taken, it should be to try to prevent the Public Prosecutor from taking notice of this unfortunate adventure, which has already become too public.

These remarks seemed to me full of wisdom and I have determined to await fresh orders from you.

Permit me, Madame, to ask you to be so kind, when you send me your orders, as to add a word on the state of your health, on whose account I dread extremely the result of so many griefs. I hope you will pardon this liberty in consideration of my attachment and my zeal.

I am with respect, Madame, your, &c.

Paris, 10th of December, 17—.

LETTER CLXVII

Anonymous to the Chevalier Danceny

Monsieur,

I have the honour to inform you that this morning, at the Public Prosecutor's office, His Majesty's legal advisers discussed the meeting which took place a few days ago between you and M. le Vicomte de Valmont, and there is reason to fear that the Public Prosecutor may lodge a complaint. I thought this warning might be useful to you, either for you to put in motion your protectors to prevent these unpleasant consequences; or, in case you are unable to achieve this, for you to be in a position to look after your personal safety.

If you will allow me to advise you, I think you would do well for a short time to show yourself less than you have been accustomed to do recently. Although this sort of affair is usually treated with indulgence, yet this form of respect is nevertheless due to the law.

This precaution becomes the more necessary since I have heard that a Madame de Rosemonde, said to be an aunt of M. de Valmont, wishes to lodge a complaint against you, in which case the Public Prosecutor could not refuse her demand. It might be useful if you could find someone to speak to this lady.

Private reasons prevent me from signing this letter. But I hope that, although you do not know from whom it comes, you will not render less justice to the sentiment which dictated it.

I have the honour to be, etc.

Paris, 10th of December, 17—.

LETTER CLXVIII

Madame de Volanges to Madame de Rosemonde

Very surprising and annoying rumours, my dear and excellent friend, are being spread here concerning Madame de Merteuil. I am certainly very far from believing them and I would wager that it is only a dreadful calumny; but I know too well how the least probable spitefulness gains credit and how hard it is to erase the impression it leaves, not to be very alarmed by these, however easily I think they may be destroyed. But it was very late yesterday when I heard of these horrors which are only beginning to be repeated; and this morning, when I sent to Madame de Merteuil's house, she had gone to the country where she is to remain for two days. I could not hear where she had gone. Her second waiting-woman, whom I sent for to speak to me told me that her mistress had simply given her orders to expect her on Thursday, and none of the servants she has left here know more than this. I myself cannot conjecture where she can be; I do not recollect anyone she knows who remains in the country so late in the season.

However that may be, I hope that you will be able to procure me between now and her return certain explanations which may be useful to her, for these odious stories are founded upon the circumstances of M. de Valmont's death, of which you would apparently have been informed if they were true, or at least, it will be easy for you to find out about them, which I beg you to do. This is what people say, or, to express it better, what they are still murmuring; but it cannot be long before it breaks out more openly.

It is said, then, that the quarrel between M. de Valmont and the Chevalier Danceny is the work of Madame de Merteuil, who was unfaithful to both of them; that, as almost always happens, the two rivals began by fighting and did not come to an explanation until afterwards; that this explanation produced a sincere reconciliation; and that, to complete the Chevalier Danceny's knowledge of Madame de Merteuil and also to justify himself completely, M. de Valmont added to his words a whole mass of letters, forming a regular correspondence which he kept up with her, in which she relates about herself the most scandalous anecdotes in the most licentious style.

It is added that Danceny, in his first indignation, showed these letters to anyone who wished to see them and that they are all over Paris now. Two in particular[1] are mentioned; one in which she tells the whole story of her life and principles and which is said to be the height of enormity; the other which entirely justifies M. de Prévan (whose history you will remember) from the proof it gives that he only yielded to the most marked advances on the part of Madame de Merteuil and that the rendez-vous was agreed to by her.

Happily, I have the strongest reasons to believe that these imputations are as false as they are odious. First of all, we both know that M. de Valmont was certainly not devoting himself to Madame de Merteuil and I have every reason to think that Danceny was not doing so either; so it seems proved that she could not have been either the reason or the creator of the quarrel. I do not understand either how it could have been to the interest of Madame de Merteuil (who is supposed to have been in agreement with Prévan) to make a scene which could never have been anything but disagreeable by its publicity and might have become very dangerous to her, since she thus made an irreconcilable enemy of a man who was master of part of her secret and who then had many supporters. Yet it is to be observed that since this adventure not a single voice has been raised in favour of Prévan and that there has been no protest even on his part.

These reflections make me suspect him as the author of the rumours to-day and I regard these base accusations as the work of the hatred and vengeance of a man who, seeing himself ruined, hopes in this way at least to spread doubts and perhaps to cause a useful diversion. But wherever these spiteful rumours come from, the most urgent thing is to destroy them. They would fall of themselves, if, as is probable, M. de Valmont and Danceny did not speak to each other after their unfortunate affair and if no papers were handed over.

In my impatience to verify these facts I sent this morning to M. Danceny; he is no longer in Paris. His servants told my footman that he left last night after a warning he received yesterday and that the place where he is staying is a secret. Apparently he is afraid of the results of this affair. It is therefore only through you, my dear and excellent friend, that I can obtain these details which interest me so much and which may become so necessary to Madame de Merteuil. I renew my request that you will let me have them as soon as possible.

P.S. My daughter's indisposition had no bad result; she sends you her respect.
Paris, 11th of December, 17—.

LETTER CLXIX

The Chevalier Danceny to Madame de Rosemonde

Perhaps you will think the step I am taking to-day a very strange one; but, I beg you, hear me out before you judge me and do not see boldness and temerity where there is only respect and confidence. I do not conceal from myself the wrongs I have done you, and I should not forgive myself for them all my life if I could for a moment think that it would have been possible to avoid them. Be sure, Madame, that though I think myself free from blame I am not free from regrets; and I can say with sincerity that the regrets I have caused you are a considerable part of those I feel myself. To believe these sentiments which I dare to express to you, it will suffice you to do justice to yourself and to know that, without having the honour to be known by you, I have that of knowing you.

Yet, while I groan at a fatality which has caused at once your grief and my misery, I am made to fear that you are given up entirely to thoughts of vengeance, that you are seeking the means of satisfying it even through the severity of the law.

Allow me to point out to you in this respect, that here you are carried away by your grief, since in this point my interest is essentially one with that of M. de Valmont who would himself be involved in the condemnation you provoked against me. I should then think, Madame, that I might rather count upon aid than obstacles on your part in the care I am obliged to take so that this unhappy incident may remain buried in silence.

But this resource of complicity which suits equally the guilty and the innocent is insufficient for my delicacy; while I desire to remove you as an opponent, I claim you as my judge. The esteem of persons one respects is too precious for me to allow yours to be torn from me without defence, and I think I have the means to do so.

Indeed if you agree that vengeance is allowed, let me rather say is a duty, when one has been betrayed in one's love, in one's friendship, in one's confidence above all; if you agree to that, my injuries to you will disappear in your eyes. Do not rely on what I say but read, if you have the courage, the correspondence I hereby place in your hands.[1] The number of original letters among them appear to render authentic those which only exist in copy. Moreover, I received these papers just as I have the honour to hand them to you, from M. de Valmont himself. I have added nothing to them and I have only taken from them two letters which I have taken the liberty to publish.

One was necessary to the common vengeance of M. de Valmont and myself, to which we both had a right and which he expressly charged me with. Moreover, I thought it was rendering a service to society to unmask a woman so really dangerous as Madame de Merteuil and who, as you can see, is the only, the real cause of all that passed between M. de Valmont and me.

A sentiment of justice induced me also to publish the second for the justification of M. de Prévan, whom I hardly know, but who by no means deserved the rigorous treatment he has endured nor the still more formidable severity of public condemnation under which he has suffered since that time, without having anything for his defence.

You will therefore only find copies of these two letters whose originals I must retain. As to the remainder, I think I cannot place in surer hands a collection which it is important for me not to have destroyed, but which I should blush to abuse. By confiding these papers to you, Madame, I think I am serving the persons they concern as well as if I handed them over to themselves; and I save them from the embarrassment of receiving them from me and of knowing that I am informed of their adventures, of which doubtless they wish everybody to be kept in ignorance.

I think I ought to inform you that the enclosed correspondence is only part of a much more voluminous collection from which M. de Valmont took it in my presence; you will find it, when the seals are broken, under the title, which I saw myself, of "Account opened between the Marquise de Merteuil and the Vicomte de Valmont." You will take in this matter the course which your prudence suggests.

I am with respect, Madame, etc.

P.S. Certain warnings I have received and the advice of my friends have determined me to leave Paris for some time; but the place of my retirement, which is kept secret from everyone else, will not be so kept from you. If you honour me with a reply, I beg you to address it to the Commandery of P . . ., under cover to M. le Commandeur de . . . It is from his residence that I have the honour to write to you.

Paris, 12th of December, 17—.

LETTER CLXX

Madame de Volanges to Madame de Rosemonde

I go, my dear friend, from surprise to surprise, from grief to grief. Only a mother can have an idea of what I suffered all yesterday morning; and, though my most cruel anxieties have since been calmed, I still retain a keen affliction, the end of which I cannot foresee.

Yesterday, about ten o'clock in the morning, astonished at not having seen my daughter, I sent my waiting-woman to find what was the reason for this lateness. She returned a moment later in great fear, and frightened me still more by announcing that my daughter was not in her room and that her waiting-woman had not found her there this morning. Imagine my position! I called together all my servants, especially the porter; all swore they knew nothing and could tell me nothing about this event. I then went at once to my daughter's room. The disorder it was in showed me that apparently she had only left that morning; but I found no explanation. I examined her wardrobes and her writing-table; I found everything in its place and all her possessions, except the dress she had gone out in. She had not even taken the little money she had with her.

As it was only yesterday that she heard all that is being said about Madame de Merteuil to whom she is very much attached, to such an extent that she did nothing but weep all evening; as I remembered also that she did not know Madame de Merteuil is in the country; my first idea was that she had wished to see her friend and had been heedless enough to go there alone. But the time which elapsed without her returning brought back all my anxieties. Every moment increased my pain and, while I burned to find out what had happened, I did not dare to make any enquiries in the fear of giving publicity to an event which perhaps I should afterwards wish hidden from everyone. No, I have never suffered so much all my life!

At length, it was not until after two o'clock that I received at once a letter from my daughter and another from the Mother Superior of the Convent of . . . My daughter's letter merely informed me that she had been afraid lest I should thwart the vocation she felt to become a nun and that she had not dared to speak to me about it; the rest was only excuses for having taken this course without my permission, a course, she added, which I should certainly not disapprove if I knew her motives which, however, she begged me not to ask her.

The Mother Superior informed me that, seeing a young person arrive alone, she had at first refused to receive her; but, having questioned her and having learnt who she was, she thought she would be doing me a service by giving shelter to my daughter and thereby saving her from other expeditions, which she seemed determined to make. The Mother Superior, while she naturally offers to return me my daughter, exhorts me, as her profession requires, not to thwart a vocation which she calls "so decided"; she also tells me that she could not give me earlier information of this occurrence owing to the difficulty she had in making my daughter write to me, for her project was to keep everyone in ignorance of her place of retirement. The unreasonableness of children is a cruel thing!

I went to the Convent at once; and after having seen the Mother Superior, I asked to see my daughter; she only could be made to come with difficulty and she trembled very much. I spoke to her in the presence of the nuns and I spoke to her in private; all I could obtain from her in the midst of many tears was that she could only be happy in the Convent; I decided to allow her to remain there, but not yet as one of the Novices, as she desired. I fear that the deaths of Madame de Tourvel and M. de Valmont have too much affected her young mind. However much I respect a religious vocation I could not see my daughter embrace that condition without pain, and even without fear. It seems to me that we already have enough duties to fulfil without creating new ones for ourselves; and, moreover, that at her age we hardly know what is most suitable for us.

My embarrassment is greatly increased by the very near return of M. de Gercourt; ought I to break off so advantageous a marriage? How can we make our children happy if it is insufficient to desire their happiness and to give all our attention to it? You would greatly oblige me if you would tell me what you would do in my place; I cannot decide upon any course; I think nothing is so terrifying as to have to decide the fate of others and in this present situation I am equally afraid of showing the severity of a judge or the weakness of a mother.

I keep reproaching myself for increasing your griefs by speaking to you of mine; but I know your heart; the consolation you could give others would become for you the greatest you could receive.

Good-bye, my dear and excellent friend; I await your two replies in great impatience.

Paris, 13th of December, 17—.

LETTER CLXXI

Madame de Rosemonde to the Chevalier Danceny

After the information you have given me, Monsieur, there is nothing left but to weep and to be silent. I regret to be still alive when I learn such horrors; I blush that I am a woman when I see one capable of such excesses.

In what concerns myself, Monsieur, I will gladly consent to leave in silence and forgetfulness everything connected with these sad events and their consequences. I even hope that they may never cause you any griefs save those inseparable from the unhappy advantage you gained over my nephew. In spite of his faults, which I am compelled to recognise, I feel I shall never be consoled for his loss; but my eternal affliction will be the only vengeance I shall allow myself to take upon you; it is for your heart to perceive the extent of it.

If you will allow me, at my age, to make an observation which is never made at yours, I would say that if we understood our true happiness we should never seek it outside the limits prescribed by the Laws and Religion.

You may be sure that I shall faithfully and willingly guard the collection you have confided to me; but I ask you to authorise me not to hand it over to anyone, not even to you, Monsieur, unless it becomes necessary for your justification. I dare to think that you will not refuse this request and that you now feel that one often regrets having yielded even to the most just vengeance.

I do not pause in my requests, so persuaded am I of your generosity and your delicacy; it would be worthy of both if you also placed in my hands Mademoiselle de Volanges' letters which you have apparently preserved and which doubtless are no longer of interest to you. I know that this young person did you great injuries; but I do not believe you would think of punishing her for them; if only from self-respect you would not degrade the person you had loved so much. It is not necessary therefore for me to add that the consideration the daughter does not deserve is at least due to her mother, to that respectable woman to whom you have many amends to make; for, after all, however you try to delude yourself by a pretended delicacy of sentiment, he who first attempts to seduce a still simple and virtuous heart by that very fact renders himself the first instigator of her corruption and must for ever be held accountable for the excesses and misconduct which follow it.

Do not be surprised, Monsieur, at so much severity on my part; it is the greatest proof I can give you of my complete esteem. You will acquire still further rights to it by consenting, as I desire, to the preservation of a secret whose publicity would do harm to yourself and would carry death to a maternal heart which you have already wounded. And then, Monsieur, I wish to do my friend this service; if I could fear that you would refuse me this consolation, I would ask you first to remember that it is the only one you have left me.

From the Château de . . . , 15th of December, 17—.

LETTER CLXXII

Madame de Rosemonde to Madame de Volanges

If I had been obliged, my dear friend, to send to Paris and wait for the explanations you ask concerning Madame de Merteuil, it would not be possible yet for me to give them to you; and

I should certainly only have received in this way vague and uncertain ones; but I received explanations I did not expect, which I had no reason to expect; and they are but too precise. O my friend! How that woman deceived you!

It is repugnant to me to enter into any of the details of this mass of enormities; but whatever may be said about her, be assured that it is still less than the truth. I hope, my dear friend, that you know me well enough to believe me on my bare word and that you will not require any proofs from me. Let it suffice you to know that masses of proofs exist and that they are now in my hands.

It is not without extreme pain that I also request you not to force me to give you my reasons for the advice you ask me concerning Mademoiselle de Volanges. I urge you not to oppose yourself to the vocation she shows. Certainly, no reason can authorise one to compel another to adopt that state when the subject is not called to it; but sometimes it is very fortunate that she should be; and you see your daughter herself tells you that you would not disapprove if you knew her motives. He who inspires our feelings knows, far better than our vain wisdom can know, what befits each of us; and often that which appears an act of His severity is on the contrary an act of His clemency.

My advice then, which I know will distress you and which for that very reason you will realise is not given you without deep reflection, is that you leave Mademoiselle de Volanges in the Convent, since this determination is her choice; that you encourage rather than oppose the plan she seems to have formed; and that, in expectation of its being carried out, you do not hesitate to break off the marriage you had arranged.

After having fulfilled these painful duties of friendship and in my powerlessness to add any consolation, I have one favour left to ask you; it is that you will not question me any further about these sad events; let us leave them in the oblivion which befits them; and, without seeking for useless and distressing explanations, let us submit ourselves to the decrees of Providence, let us believe in the wisdom of its views, even when it does not permit us to understand them. Good-bye, my dear friend.

From the Château de . . . , 15th of December, 17—

LETTER CLXXIII

Madame de Volanges to Madame de Rosemonde

Oh! My friend! What a fearful veil you wrap about my daughter's fate! And you seem to fear lest I should attempt to raise it! What does it conceal which could more distress a mother's heart than the dreadful suspicions to which you abandon me? The more I know your friendship, your indulgence, the more my tortures increase; twenty times since yesterday I have wanted to leave these cruel uncertainties and to ask you to inform me without circumspection and without evasion; and each time I shuddered with fear, remembering your request that I should not question you. At last, I have decided on a course which still leaves me some hope; and I expect of your friendship that it will not refuse me what I desire—it is to answer me if I have practically understood what you might have to tell me, not to be afraid to tell me everything that maternal indulgence might excuse, that is not impossible to repair. If my misfortunes exceed this measure, I consent to allow you only to explain yourself by your silence. Here then is what I know already and the extent to which my fears go.

My daughter showed some inclination for the Chevalier Danceny and I was informed that she went so far as to receive letters from him and even to answer him; but I thought I had succeeded in preventing any dangerous consequence from this childish error; to-day, when I fear everything, I suppose it was possible that my vigilance may have been deceived and I fear that my daughter has been seduced, has carried her misconduct to its height.

I remember several circumstances which might strengthen this fear. I told you that my daughter was taken unwell at the news of M. de Valmont's misfortune; perhaps the only cause of this sensibility was the thought of the risks M. Danceny had run in this combat. Later, when she wept so much at learning what was said of Madame de Merteuil, perhaps what I thought was the grief of friendship was only the result of jealousy or regret at finding her lover unfaithful. Her last step, I think, can be explained in the same way. We often feel ourselves called to God merely because we feel in revolt against men. Supposing that these are true facts and that you knew them, no doubt you might have thought them sufficient to authorise the severe advice you give me.

Yet, if this is so, while I blame my daughter I think I should still owe it to her to attempt every means to save her from the tortures and dangers of an illusory and temporary vocation. If M. Danceny is not lost to all sense of honour, he will not refuse to repair a wrong of which he is the sole author, and I dare to think that marriage with my daughter is so advantageous that he might be flattered by it, as well as his family.

That, my dear and excellent friend, is the one hope left me; hasten to confirm it, if it is possible for you to do so. You will understand how much I desire that you should reply to me and what a dreadful blow to me your silence would be.[1]

I was about to close this letter when a man of my acquaintance came to see me and told me of the cruel scene endured by Madame de Merteuil the day before yesterday. As I have seen nobody for the last few days I had heard nothing of this occurrence; here is an account of it, as I received it from an eye-witness.

Madame de Merteuil on returning from the country the day before yesterday, Thursday, was set down at the Comédie

Italienne, where she has a box; she was alone and, what must have seemed very extraordinary to her, no man entered it during the whole of the play. When it was over she went, as she was accustomed, to the small drawing-room which was already filled with people; there was at once a murmur, but apparently she did not think she was the cause of it. She noticed an empty place on one of the benches and went to sit down in it; but immediately all the women who were already there rose as if in concert and left her absolutely alone. This marked movement of general indignation was applauded by all the men and increased the murmurs, which, they say, became hooting.

For there to be nothing lacking to her humiliation, her ill-luck would have it that M. de Prévan, who had showed himself nowhere since his adventure, entered the small drawing-room at that moment. As soon as he was observed everyone, men and women, surrounded and applauded him; and he was carried, as it were, before Madame de Merteuil by the public who made a circle around them. I am assured that she preserved an air of seeing and hearing nothing and that she did not even change countenance! But I think that fact exaggerated. However that may be, this situation, so ignominious for her, lasted until her carriage was announced; and at her departure the scandalous hootings were redoubled. It is dreadful to be a relative of this woman. That same evening, M. de Prévan was welcomed by all the officers of his Corps who were present and no one doubts that he will soon regain his post and his rank.

The same person who gave me these details told me that the following night Madame de Merteuil was seized with a strong fever, which was at first thought to be the result of the violent situation in which she had been placed; but since yesterday evening it is known that a confluent small-pox of a very bad kind has shown itself. Really, I think it would be fortunate for her if she died of it. It is also said that this adventure will perhaps do her great harm in her law-suit which will soon be decided and for which they say she needed a great deal of favour.

Good-bye, my dear and excellent friend. In all this I see the wicked are punished, but I find no consolation in it for the unhappy victims.

Paris, 18th of December, 17—.

LETTER CLXXIV

The Chevalier Danceny to Madame de Rosemonde

You are right, Madame, and I shall certainly not refuse you anything which depends upon me and to which you appear to attach some value. The packet I have the honour to send you contains all Mademoiselle de Volanges's letters. If you read them, you will see perhaps not without astonishment how so much ingenuousness and so much perfidy can be united. At least, that is what most struck me at the last reading I have just given them.

But, especially, can one resist the keenest indignation against Madame de Merteuil when one recollects with what dreadful pleasure she gave all her attention to perverting so much innocence and candour?

No, I am no longer in love. I retain nothing of a sentiment so unworthily betrayed; it is not love which makes me seek to justify Mademoiselle de Volanges. But yet, would not that simple heart, that gentle, facile character, have been conducted even more easily towards good than they were led towards evil? What other girl, coming thus from a Convent without experience and almost without ideas, bringing into society (as almost always happens) only an equal ignorance of good and evil—what other girl, I say, would have resisted any better such criminal cunning? Ah! To be indulgent it suffices to reflect on how many circumstances, independent of ourselves, depends the dreadful alternative of delicacy or the degradation of our feelings. You did me justice then, Madame, when you thought that the injuries done me by Mademoiselle de Volanges, though I felt them keenly, do not inspire me with any idea of vengeance. It is sufficient to be obliged to give up loving her! It would cost me too much to hate her.

I needed no reflection to make me desire that everything which concerns her and might harm her should for ever remain unknown to the world. If I have seemed to delay for some time my compliance with your wishes in this respect, I think I need not conceal my motive from you; I wished first of all to be sure that I should not be molested about the consequences of my unhappy affair. At a time when I was asking for your indulgence, when I even dared to think I had some right to it, I should have been unwilling to seem in any way to be buying it by this compliance on my part; I was certain of the purity of my motives and, I confess it, I was proud enough to desire that you should have no doubt of it. I hope you will pardon this delicacy, perhaps too susceptible to the veneration you inspire in me, to the value I set on your esteem.

The same feeling makes me ask you as a last favour to be good enough to tell me if you think I have fulfilled all the duties imposed upon me by the unfortunate circumstances in which I was placed. Once assured of this, my course is plain; I shall leave for Malta; I shall go there to make gladly and to keep religiously the vows which will separate me from the world, of which, while yet so young, I have already so much reason to complain; I shall go and try to forget under a foreign sky the idea of so many accumulated enormities, whose memory can never but sadden and wither my soul.

I am with respect, Madame, your most humble, etc.

Paris, 26th of December, 17—.

LETTER CLXXV

Madame de Volanges to Madame de Rosemonde

Madame de Merteuil's destiny seems at last accomplished, my dear and excellent friend; and it is such that her worst enemies are divided between the indignation she merits and the pity she inspires. I was indeed right to say that it would perhaps be fortunate for her if she died of her small-pox. She has recovered, it is true, but horribly disfigured; and particularly by the loss of one eye.[1] You may easily imagine that I have not seen her again; but I am told she is positively hideous.

The Marquis de . . ., who never misses the opportunity of saying a spiteful thing, speaking of her yesterday, said that her disease had turned her round and that now her soul is in her face. Unhappily, everyone thought the expression a very true one.

Another event has increased her disgrace and her errors. Her law-suit was decided yesterday and she lost it by a unanimous vote. Costs, damages, and interest, restitution of profits—everything was given to the minors; so that the small part of her fortune which was not concerned in this suit is more than absorbed by its expenses.

As soon as she heard this news, although she was still unwell, she made her arrangements and left, alone and by night. Her servants say to-day that not one of them would follow her. It is thought that she went in the direction of Holland.

This departure has caused even more comment than all the rest because she took with her the very valuable diamonds which ought to have been comprised in her husband's estate, her silver, her jewels, in fact everything she could, while she leaves behind nearly 50,000 *livres* of debts. It is a positive bankruptcy.

The family is to meet to-day to discuss what arrangements shall be made with the creditors. Although I am a very distant relative, I have offered to assist; but I shall not be at this gathering, for I have to be present at a still sadder ceremony. Tomorrow my daughter takes the veil as a Novice. I hope you will not forget, my dear friend, that my only motive for thinking myself forced to make this great sacrifice is the silence you have kept towards me.

M. Danceny left Paris nearly a fortnight ago. It is said that he is going to Malta and intends to remain there. There might be yet time to detain him? . . . My friend! . . . My daughter is very culpable then! . . . You will surely pardon a mother for only yielding with difficulty to this dreadful certainty.

What fatality is that which has spread about me recently and has struck me through my dearest possessions! My daughter and my friend!

Who would not shudder at thinking of the miseries which may be caused by one dangerous acquaintance! And what griefs we should avoid if we reflected more upon this! What woman would not flee from the first words of a seducer! What mother could see, without trembling, anyone but herself talking to her daughter! But these tardy reflections never occur until after the event; and one of the most important truths, which is also perhaps one of the most generally recognised, remains stifled and unregarded in the whirl of our inconsequent morals.

Farewell, my dear and excellent friend; I feel at this moment that our reason, so incapable of foreseeing our misfortunes, is still less capable of consoling us for them.[2]

APPENDIX

(The two following letters are contained in the MS. of *Les Liaisons Dangereuses*; they have been printed at the end of several modern editions in France and are given here for completeness' sake. The first—Valmont to Madame de Volanges—was cancelled by Laclos and replaced by the note to *Letter CLIV*. The second—Madame de Tourvel to Valmont—is placed at the end of the MS. as a lost letter just recovered. It was suppressed by the publishers, who substituted the note which now ends the book. Such, at least, is the explanation supplied by M.A. van Bever).

The Vicomte de Valmont to Madame de Volanges

I know you do not like me, Madame; I know also that you have always been against me with Madame de Tourvel and I have no doubt that you are more than ever of the same opinion, I even confess that you have reason to think it well founded; but I address myself to you to ask you to hand Madame de Tourvel the letter I enclose for her and also to ask you to persuade her to read it; to bring her to do it by assuring her of my repentance, of my regrets, above all of my love. I know that this step will seem a strange one to you. It surprises me; but despair seizes any means and does not calculate them. And then an interest so great, so dear, and common to us both, should put every other consideration aside. Madame de Tourvel is dying, Madame de Tourvel is unhappy, she must be given back life, health, and happiness. That is the object to be attained; any way is good which can assure or hasten its success. If you reject the help I offer, you will remain responsible for the result; her death, your regrets, my eternal despair, all will be your work.

I know I have basely outraged a woman worthy of all my adoration; I know that my dreadful actions have alone caused all the ills she suffers; I do not try to dissimulate my faults or to excuse them; but, Madame, do not become their accomplice by preventing me from atoning for them. I plunged the dagger in your friend's heart, but I alone can draw the steel from the wound; I alone know how to cure it. What matters it that I am guilty if I can be useful! Save your friend! Save her! She needs your help, not your vengeance.

Madame de Tourvel to the Vicomte de Valmont

O my friend! what is this disturbance I feel the moment you leave me—a little tranquillity is so necessary to me! How does it happen that I am given over to an agitation which becomes a pain and causes me real terror? Would you believe it? I feel that I need to collect strength and recall my reasoning faculty even to write to you. Yet I tell myself, I repeat to myself, that you are happy; but, this idea, which is so dear to my heart and which you called so aptly the gentle calmative of love, has now become its ferment and makes me succumb beneath a too great felicity; while, if I try to tear myself from this delicious meditation, I immediately fall into those cruel agonies which I have so often promised you to avoid and from which I ought indeed so carefully to preserve myself, since they impair your happiness. My dear, you have easily taught me to live only for you; teach me now how to live absent from you. . . . No, that is not what I meant, rather I would say that absent from you I would not live at all, or at least forget my existence. When I am abandoned to myself, I can endure neither my happiness nor my pain; I feel the need of rest and all rest is impossible to me; vainly have I called upon sleep, sleep flies from me; I can neither occupy myself nor remain idle, by turns a burning fire devours me and a mortal shuddering prostrates me; every movement fatigues me and I cannot remain still. How shall I express it? I should suffer less in the burning of the most violent fever and, without being able to explain it or understand it, I yet feel that this state of suffering only comes from my inability to contain or to direct a crowd of sentiments to whose charm, however, I should think myself happy if I could abandon my whole soul.

At the very moment you had left, I was less tormented; some agitation was mingled with my regrets, but I attributed this to the impatience caused me by the presence of my woman who came in at that moment and whose duties, always too long for my choice, seemed to be prolonged a thousand times more than usual. I wanted to be alone above everything; I did not doubt that, surrounded by such sweet memories, I should find in solitude the only happiness of which I am capable in your absence. How could I foresee that, while strong enough near you to endure the shock of so many different sentiments, so rapidly experienced, alone I could not even endure their memory? I was rapidly undeceived . . . Here, my tender friend, I hesitate to tell you everything . . . yet, am I not yours, entirely yours, and ought I to hide one of my thoughts from you? Ah! That would be quite impossible; only, I claim your indulgence for involuntary faults which my heart does not share; in accordance with my custom, I had sent my woman away before going to bed . . . (*Letter imperfect*).

Notes

INTRODUCTION

1 The best short account of Laclos is the preface written by M. A. Van Bever for his excellent edition of the *Liaisons Dangereuses* (1920).
2 A story told of the Duc de Richelieu with far more probability.
3 "Qui me demanderoit la première partie en l'amour, je repondrais que c'est sçavoir prendre le temps; la seconde la mesme; etencores la tierce; c'est un poinct qui peult tout." *Sur des Vers de Virgile.*
4 M. Van Bever has catalogued nearly thirty editions of *Les Liaisons Dangereuses*, only a few of which are possessed by the British Museum. There is a unique copy of an edition of 1788 in the possession of Mr. Arthur Symons. The "fifty pirated editions" are mentioned on the authority of biographers only; they produce no definite evidence.
5 By this comparison I do not intend to rank Laclos with these eminent men, but merely to point out that with practically equal opportunities in the Revolution, he failed to achieve anything important. His commanding position among the early Jacobins certainly appeared to promise a great future. Perhaps he was a little too old.
6 "For, between ourselves, what you may say will not be worth what Cléon has done."
7 "In town, at court, in the army, the men of wit never have the best shares; the fools have everything, even fame."

8 The reader will already have noticed that the word "Rosine" which ought to rhyme with "délire, sourire" etc. spoils the correct rhyme-scheme. We must suppose that it was written about a lady with a name like "Elvire", which was changed for purposes of publication. The following is a rough translation:

> She is mine! that charming Rosine, her tender heart has crowned my flame: she is mine . . . could I be happier? O voluptuousness! I feel your ecstacy, and my happiness transcends my hopes.
> Palpitating breasts, voluptuous smile, lips of rose and languid glances, Rosine has them all: that charming Rosine, she is mine!
> In artless verses in my amorous songs thus I proclaimed Love's power. Alas! Imagine my extreme pain. Lysis comes, tunes his lyre, and then sings too: "That charming Rosine, she is mine," Compare Bertin, Élégie iv. of *Les Amours*.

9 I need scarcely say that this does not refer to Mr. Saintsbury, whose antipathy to Stendhal and Laclos is well-known and is perhaps only one more of his delightful "préventions." I am thinking of the numerous manuals of literature published for the misguidance of the young.

10 This is not the place to discuss the problem of authorship, but I will only add that Apollinaire's evidence for attributing these dialogues to Crébillon seems to me no evidence at all; they are different in style from Crébillon's other works and they lack the "moral purpose" evident in his other work.

11 Who, of course, eventually turns out to be a missing heiress.

12 In his letters.

13 Notice the similarity of the names "Valville" and "Valmont"; like Valmont, Valville was an "infidèle" but infinitely more tender. The Don Juan of Crébillon's *Egarements* is named "Varsac." Laclos uses the name "Vressac."

14 *Memoires et Correspondence de Mme. d' Epinay*. Quoted by the Goncourts in *La Femme au* XVIII *Siècle*.

15 So Stendhal says.

16 Louis XVI was incapable of leading anything.

17 Witness that charming and benevolent Duc de Penthièvre whose innumerable charities and amiable character made him so universally beloved.

18 See the *Memoires et Correspondences de Mme. d' Epinay*. And the Goncourts' book already referred to.

19 The most serious argument against the political power of women is

seldom alleged; it is that they are naturally despotic, are rarely tolerant and mistake petty reforms for love of liberty. The usual arguments of intellectual incapacity, &c, are absurd.

20 *Le Sofa*, pp.295, 6, 7. *Oeuvres de M. Crébillon le fils*, Londres, 1772. I am confident that the unprejudiced reader will feel that Crébillon is attacking heartlessness and excess, not defending them or inculcating them.

21 *Mémoires du Comte de Grammont*. 1760. Tome I, p. 87.

22 Vauvenargues. *Caractères*. 1747.

23 Goncourts. *Op. Cit. Vol* I, p. 204.

24 *Vie privée du maréchal de Richelieu, contenant ses amours et ses intrigues*. Paris. Buisson. 1791 Vol. III. Quoted by Goncourts.

25 *Le Hasard du Coin du Feu*. Crébillon fils. Oeuvres. Londres 1772. Tome III. pp. 435.

26 See *Letter* CXXXIII.

27 "Dissoluteness of the age," almost the very words used by Mme. de Merteuil in one of her letters.

28 *Les Egarements du Coeur et de l'Esprit*. Crébillon fils. *Oeuvres*, Tome I. pp. 157.

29 Goncourts. *Op. cit.* Vol.I.pp.202, 3, 4.

30 "Méchanceté":

31 The abominable Mme. de Brinvilliers was a kind of Merteuil among poisoners, in the 17th century.

32 See also Mémoirs of Tilly, Richelieu, Mme. d'Epinay and others.

33 "Confessions du Comte * * *." *Oeuvres Complètes de Duclos*. Paris, 1806. Tome VIII. pp. 124–5.

34 It has been rather acutely remarked that a clue to Mme. de Tourvel's character may be found in her confusion of "virtue" with benevolence; a common 18th century fallacy. When Valmont gives money to a poor cottager, she cannot believe that he can be as wicked as people say; for if the wicked are benevolent what is left to the virtuous? But a man may smile and smile and be a villain; Laclos is perhaps the first 18th century writer to point out that mere benevolence and charity are not synonymous with virtue; that even in the worst men there may be some benevolence, as there may be malevolence in the best women.

35 The letter is quoted in full by the Goncourts; I do not envy those who can read it unmoved.

36 See M. Paul Morand's brilliant *Lewis et Irène*. (Paris, 1924) Mr. Wyndham Lewis has also published one or two studies of a rather similar type of "hero."

37 See M. Henri Carré's *La Noblesse de France et l'Opinion Publique au XVIIIe Siècle*. (Paris, 1920).

38 Probably the very worst offenders in the 18th century, were the newly rich 'fermiers généraux." The private life of de la Popeliniere for example, is an edifying spectacle.

39 See *Letter* CXXV, particularly the paragraph beginning, "Hitherto, my fair friend."

40 Is there not a last stab of contempt in making Valmont fall in a duel over a woman by the hand of a boy? Of course, it would be a grave error to suppose that even the "*petits maitres*" were cowards; any history of the 18th century will show that they were extremely brave and excellent soldiers. But in times of peace there was nothing for them to do, except seduce women.

PUBLISHER'S NOTE

1 1782. An ironic comment by Laclos himself.

2 "Présidente." A Président was an important magistrate; his wife had the title of "Présidente." Throughout this translation "La Présidente de Tourvel" will appear as "Madame de Tourvel," since there is no English equivalent to the title of "Présidente."

EDITOR'S PREFACE

1 I must also give notice that I have suppressed or changed all the names of persons mentioned in these letters; and that if among those which I have substituted there should be a name belonging to somebody this will be no more than an error on my part from which no conclusion should be drawn. (C. de L.)

PART I

LETTER I

1 Pupil in the same convent. (C. de L.).

2 In her best clothes.

3 This refers to the elaborate head-dress of the 18th century lady.

4 Attendant at the convent turning-box. (C. de L.).

LETTER II

1 The words *roué* and *rouerie*, which are happily falling out of use in good company, were very fashionable at the time these letters were written. (C. de L.).

2 To understand this passage, the reader must know that the Comte de Gercourt had left the Marquise de Merteuil for the *Intendante de*——,

who had sacrificed to him the Vicomte de Valmont and that it was then the Marquise and the Vicomte became attached to one another. Since this adventure is much earlier than the events dealt with in these letters the whole correspondence about it has been suppressed. (C. de L.).

LETTER IV

1 La Fontaine.

LETTER VI

1 The reader will perceive here that bad taste for puns which was beginning then and is now so popular. (C. de L.) "Sauter le fossé" means "to jump a ditch" and "to take a leap," "to cross the Rubicon."

LETTER VII

1 In order not to weary the reader's patience, a large number of letters from this daily correspondence have been suppressed; those only have been given which appeared necessary to a complete understanding of the events in this group. For the same reason all Sophie Carnay's letters have been supressed as well as several others from the actors in these events. (C. de L.).

LETTER IX

1 Madame de Volanges's error allows us to see that Valmont like other scoundrels did not reveal his accomplices. (C. de L.).

LETTER X

1 "*Petite-maison*," a house kept secretly for lovers' rendezvous, equivalent to the modern "*garçonnière*." I am informed that such refuges are not unknown in England, but my informant could not tell me by what name they are distinguished.

LETTER XIII

1 The Chevalier mentioned in Madame de Merteuil's letters. (C. de L.).

LETTER XVI

1 The letter which speaks of this evening cannot be found. It may be surmised that it was the evening mentioned in Madame de Merteuil's note and is also spoken of in the preceding letter from Cécile Volanges. (C. de L.).

LETTER XXI

1 In 1723, the Comte de Charolais, passing through the village of Anet, saw a "*bourgeois,*" in a night-cap, standing at his door; the Comte shot the *bourgeois* as an amusement. The next day he went to the Regent for a formal pardon and received it, with the remark that the pardon was due to his rank, but would more willingly be granted to anyone who did the same to the Comte. See Carré *Noblesse de France*, p. 293.

LETTER XXII

1 Does Madame de Tourvel not dare to say it was by her orders? (C. de L.).

2 The 18th century, the age of privilege and sensibility, considered charity the greatest and most amiable of virtues. The 20th century, the age of democracy and commerce, considers charity an insult or at best the poor man's "claim" to the rich man's money: The emotions of compassion and gratitude are scarcely involved.

LETTER XXXIX

1 The letters from Cécile Volanges and the Chevalier Danceny are omitted since they are not interesting and relate no event. (C. de L.).

LETTER XLI

1 See *LETTER XXXV*. (C. de L.).

LETTER XL

1 After this insult to the *petit-maitres*, it hardly seems possible that anyone can deny the political-moral motives of Laclos; still less possible that he could be supposed to identify himself with Valmont.

LETTER XLIV

1 Piron, *Métromanie*. (C. de L.).

LETTER XLVI

1 Those who have never had occasion to feel sometimes the value of a word, of an expression, consecrated by love, will find no sense in this phrase (C. de L.).

PART II

LETTER LI

1 This letter could not be found. (C. de L.).
2 The reader must long have guessed from Mme. de Merteuil's morals, how little she respected religion. This whole paragraph would have been suppressed, but it was felt that in showing effects one should not fail to make known the causes. (C. de L.).

LETTER LIV

1 "Petite maison."

LETTER LVII

1 This refers to an anecdote about Ninon de Lanclos. A man named La Chârtre was wildly in love with her, and had to go away. He made Ninon give him a written promise of fidelity. She was, of course, unfaithful and, suddenly recollecting her written promise in the arms of another lover, exclaimed; "Ah! What a promise La Chârtre has!"

LETTER LVIII

1 It is thought that this is Rousseau in *Emile*, but the quotation is not exact and Valmont's application is a false one; and then, had Madame de Tourvel read *Emile*? (C. de L.).

LETTER LXIII

1 Gresset, *Le Méchant*. (C. de L.).

LETTER LXV

1 Monsieur Danceny does not state the matter accurately. He had given his confidence to Monsieur de Valmont before this event. See *Letter LVII*. (C. de L.).

LETTER LXVI

1 An expression referring to a passage in a poem by M. de Voltaire. (C. de L.).

LETTER LXXI

1 Racine, *Tragedy of Britannicus*. (C. de L.).

LETTER LXXV

1 Mlle. de Volanges changed her confident shortly after this, as will be seen, in the following letters; this collection will not contain any of

those she continued to write to her convent friend; they would tell the reader nothing. (C. de L.).

LETTER LXXVI

1 This letter could not be found. (C. de L.).

LETTER LXXIX

1 "L'heure du berger" literally "the shepherd's hour," the common phrase for a lover's rendezvous.
2 "Sept et le va," a gambling term.
3 This end to this edifying adventure was occasionally a fact in 18th century life, in spite of the general tolerance. It will be noticed that the women pay the price. Here again we find Laclos's "feminist thesis" and incidently a justification for the perfidy of the Madame de Merteuils.

LETTER LXXXI

1 It is not known whether this verse and the one before it, "Her arms still open though her heart is closed" are quotations from little known works or whether they are part of Mme. de Merteuil's prose. What makes it probable is the multitude of faults of this kind in all the letters of this correspondence. Only those of the Chevalier Danceny are free from them, perhaps because he sometimes occupied himself with poetry and therefore his more practised ear enabled him to avoid this defect more easily. (C. de L.).
2 It will be seen later, in *Letter CLII*, not what M. de Valmont's secret was, but practically of what kind it was; and the reader will understand that further enlightenment on this subject could not be given. (C. de L.).

LETTER LXXXV

1 See *Letter LXX*. (C. de L.).
2 Reference to Voltaire's tragedy *Zaïre*.
3 Some persons may not know perhaps that a "macédoine" is a mixture of several games of chance among which each dealer has the right to chose when it is his turn. This is one of the inventions of the age. (C. de L.).
4 The Commander of the corps in which M. de Prévan was serving. (C. de L.).

PART III

LETTER XCIII

1 Danceny does not know what this way is; he simply repeats Valmont's expression. (C. de L.).

LETTER XCVIII

1 This states in a sentence one of the principal themes of this novel.

LETTER XCIX

1 Voltaire, comedy of *Nanine*. (C. de L.).

LETTER CI

1 A village half-way between Paris and Madame de Rosemonde's *Château*. (C. de L.)

LETTER CVII

1 The same village mentioned before, half-way between Paris and the Château. (C. de L.).

LETTER CX

1 *Nouvelle Héloïse*. (C. de L.).
2 *Nouvelle Héloïse*. (C. de L.).
3 Regnard, *Folies Amoureuses*. (C. de L.).

LETTER CXII

1 This letter could not be found. (C. de L.).

LETTER CXIII

1 "On ne s'avise jamais de tout!" Comedy. (C. de L.).
2 See *Letter CIX*. (C. de L.).

LETTER CXV

1 "Bonnes", a double entendre; it means "good women", and "maid-servants".

PART IV

LETTER CXXV

1 *Letters CXX* and *CXXII*. (C. de L.).

LETTER CXXXI

1 Or: "Let us share the stakes."

LETTER CXXXIII

1 Du Belloi. Tragedy of *The Siège of Calais.* (C. de L.).

LETTER CXXXVIII

1 *Letters XLVI* and *XLVII.* (C. de L.).

LETTER CXLI

1 "Espèces"—eighteenth century polite slang.

LETTER CLIV

1 Monsieur de Valmont's letter has been suppressed because nothing could be found in the remainder of their correspondence to settle these questions. (C. de L.).

LETTER CLXIV

1 A Magistrate.

LETTER CLXV

1 This casket contained all the letters relating to her adventure with M. de Valmont. (C. de L.).

LETTER CLXVIII

1 *Letters LXXXI* and *LXXXV*, of this collection. (C. de L.).

LETTER CLXIX

1 From this correspondence, from that delivered on the death of Madame de Tourvel, from letters confided to Madame de Rosemonde by Madame de Volanges, the present collection has been formed, the originals of which are in the possession of Madame de Rosemonde's heirs. (C. de L.).

LETTER CLXXIII

1 No answer was sent to this letter. (C. de L.).

LETTER CLXXV

1 This is perhaps the weakest thing in the book; it is perhaps intentionally so. See *Introduction*.
2 Private reasons and other consideration which we feel it our duty to respect, force us to stop here.

For the moment we cannot give the reader the continuation of Mademoiselle de Volanges' adventures nor inform him of the sinister event which completed the misfortunes or the punishment of Madame de Merteuil. Perhaps some day we may be allowed to complete this work; but we cannot make any undertaking to do so, and even if we could, we think the public taste ought first to be consulted, since it has not our reasons for taking an interest in reading these matters.

(*Publishers Note*: 1782)

Bibliography

(This abbreviated description of the various editions of *Les Liaisons Dangereuses* is taken from that published by M. A. van Bever in 1920. Very few of these editions are in the British Museum Library.)

Manuscript

The original MS of the *Liaisons Dangereuses* in the handwriting of Choderlos de Laclos was presented to the Bibliothèque Nationale in 1849, by Madame Charles de Laclos. Its Press-mark is Ms. Français, 12845. It is written in a fine hand, almost without corrections, difficult to read. The original title was *Le Danger des Liaisons*, subsequently altered by Laclos to the one by which we now know it. This MS. shows that the novel was originally divided into two, not four parts.

Printed Editions

Les Liaisons Dangereuses ou Lettres Recueillies dans une Société et publiées pour l'instruction de quelques autres. par M. C. . . . de L. . . . A Amsterdam; et se trouve à Paris, chez Durand Neveu, Libraire à la Sagesse, rue Galande. M.D.CCLXXXII, 4 vols, in-12.

Les Liaisons Dangereuses, &c. . . . Genève, 1782, 4 vols, small in-12.
Les Liaisons Dangereuses, &c. . . . A Neufchatel, 1782, 4 vols, in-12.
Les Liaisons Dangereuses, &c. . . . Amsterdam and Paris, 1784, 4 vols. in-12.
Les Liaisons Dangereuses, &c. . . . Genève, 1784, 1786, 1792, 4 vols. in-12.
Les Liaisons Dangereuses, &c. . . . augmentée d'une Correspondance de l'Auteur avec Madame Riccoboni, et de ses Pièces Fugitives. (Without editor's name or place of address). DCC. LXXXVII (*sic*) 4 vols. (One copy only known, at Nantes).
Les Liaisons Dangereuses, &c. as above. M.DCC.LXXXVIII. 2 vols, in-12. (Unique copy in the possession of Mr Arthur Symons).
Les Liaisons Dangereuses, &c. . . . Amsterdam and Paris, 1788, 4 vols.
Les Liaisons Dangereuses, &c. . . . Genève, 1792 and 1793, 4 vols., in-12 or in-18, with 8 plates by Le Barbier, engraved by Dambrun, Delignon, Halbou, Simonet, and Thomas.
Les Liaisons Dangereuses, &c. . . . Londres (Paris) 1796, 2 vols. in-8. 2 Frontispieces and 13 plates by Monnet, Mademoiselle Gérard, and Fragonard fils, engraved by Baquoy, Duplessis-Bertaux, Dupréel, Godefroy, Langlois, Lemire, Lingée, Masquelier, Patas, Pauquet, Simonet, and Trière. (Reprinted 1815 with date 1796).
Les Liaisons Dangereuses, &c. . . . Paris. L. Duprat-Duverger. 1811, 4 vols, small in-12. (Engravings by Canu). Reprinted 1820 in-18.
Les Liaisons Dangereuses, Paris. Bossange, 1820, 2 vols. in-12.
Les Liaisons Dangereuses, &c. . . . Paris. Impr. L. T. Cellot, 1820, 2 vols. in-8. (Plates by Deveria, but not signed).
Ditto. Paris, Parmentier, 1823.
Ditto. Paris, Libr. associés, 1828.
Ditto. Paris, chez les marchands de Nouveautés. 1833–1834.
Les Liaisons Dangereuses, &c. . . . par C. Delaclos. Bruxelles, Librairie universelle de Rosez, 1869, 2 vols., small in-12.
Les Liaisons Dangereuses, &c. . . . Garnier, (No date).
Les Liaisons Dangereuses, (par C. Delaclos) suivies de: *Les Exercices de Dévotion de M. V. Roch*, par l'Abbé de Voisenon. Paris, Dentu, 1894, in-12. (Selections, contains only 99 letters).
Les Liaisons Dangereuses, &c. . . . Paris, Boulenger, 1894, small in-16.
Les Liaisons Dangereuses, &c. . . . Edition collationnée sur le manuscrit original, suivie d'une notice, de variantes et de lettres inédites, ornée d'une reproduction en héliogravure du portrait de Laclos par Boilly. Paris, Mercure de France, 1903. in-18.
Les Liaisons Dangereuses, &c. . . . Librarie Moderne, Bauche. (Paris), 1907. large in-8.

Les Liaisons Dangereuses, &c . . . Coloured lithographs by Lubin de Beauvais. Paris. Ferroud, 1908. (300 copies only).

Les Liaisons Dangereuses. Edition publiée d'après le texte original, avec une étude sur Choderlos de Laclos et une bibliographie . . . par Ad. van Bever, 20 original etchings by Martin Van Maele. Paris. J. Chevrel, 1908. Large in-8. (225 copies only).

L'Œuvre de Choderlos de Laclos. "Les Liaisons Dangereuses, &c." Ouvrage orné de 12 illustrations d'après les gravures de Fragonard fils, Monnet, Mlle Gerard, Paris, Bibliothèque des Curieux, 1913. in-8.

Les Liaisons Dangereuses. Édition publiée . . . par Ad. van Bever. Two portraits. Paris. G. Crès et Cie, 1919. 2 vols., in-18.

Les Liaisons Dangereuses. Édition publiée . . . par Ad. Van Bever. 17 plates, Paris, G. Crès et Cie, 1920, 2 vols., in-18.